# AN ILLUSTRATED MANUAL OF
## PACIFIC COAST TREES

# An Illustrated Manual of
# PACIFIC COAST TREES

*By* HOWARD E. McMINN

PROFESSOR OF BOTANY IN MILLS COLLEGE

*and* EVELYN MAINO

---◆---

*With lists of trees recommended for
various uses on the Pacific Coast*

*By* H. W. SHEPHERD

ASSOCIATE PROFESSOR OF LANDSCAPE DESIGN
IN THE UNIVERSITY OF CALIFORNIA

UNIVERSITY OF CALIFORNIA PRESS
BERKELEY, LOS ANGELES, LONDON

UNIVERSITY OF CALIFORNIA PRESS
BERKELEY, CALIFORNIA

———

UNIVERSITY OF CALIFORNIA PRESS, LTD.
LONDON, ENGLAND

Second Edition
Thirteenth Cloth Printing 1980
First Paperback Printing 1981

ISBN 0-520-00846-4 cloth
0-520-04364-2 paper

Manufactured in the United States of America

3   4   5   6   7   8   9

The paper used in this publication meets the minimum
requirements of American National Standard for Information Sciences—Permanence of Paper for Printed Library
Materials, ANSI Z39.48–1984.   ∞

# CONTENTS

# PREFACE

REES FORM a very important part of most natural and artificial landscapes. On the Pacific Coast trees occur in many different habitats. Some occur, even though sparse and often much dwarfed, in desert washes or along the intermittent and permanent watercourses of the numerous valleys throughout the entire area. The floors of the valleys in the Coast Ranges, and the foothills of the interior mountains, are spotted with groves of oaks in the south and with oaks and conifers in the north. There are extensive forests in favored locations of the coastal and inland mountain ranges. Trees may be found from sea level to 12,000 feet elevation, where they are usually much stunted. Parks, gardens, orchards, streets, and highways are planted with native and introduced trees.

What are the names of these trees? Where have they come from? What trees can be recommended for planting in different habitats and for giving the best effects in landscaping? What book can I obtain that will aid me in identifying the more commonly cultivated and native trees of California, Oregon, Washington, and British Columbia? These and many other questions pertaining to trees are familiar to foresters, rangers, teachers, landscape architects, park guides, Scout and Camp Fire leaders, gardeners, and professional botanists. For those who ask the questions and for those who are asked, this manual is written, with the hope that the identification of the Pacific Coast trees will be made easier.

With the increase of leisure it is becoming more important for the good of the individual and of the community that all should cultivate some interest outside their life work as an intellectual and recreational avocation. The study of trees as a part of the common environment should supply such an interest for those whose leisure takes them into the out-of-doors, be it in the parks, in the mountains, or touring along the highways.

One difficulty in tree study on the Pacific Coast lies in the fact that no one book gives illustrations and descriptive accounts of both the native and the introduced trees. It is hoped that this book may solve this difficulty. It includes the native trees growing within California, Oregon, Washington, and British Columbia and about four hundred introduced species and varieties.

In preparing the manual the authors have made use of many sources of information. The senior author has botanized much of the area in California and has made trips into Oregon, Washington, and southern British Columbia for the purpose of studying trees in parks and gardens, along streets and highways, and in their native habitats. He takes full responsibility for the material in this Preface, the Introduction, the Keys, and for the brief botanical descriptions which have been written from fresh and herbarium material with careful comparison with published descriptions given in the references listed. Although he has thought it desirable to make the descriptions as nontechnical as possible, he has made no conscious effort to eliminate the more commonly used botanical terminology. Even an elementary study of any field of knowledge, be it in art, music, science, or business, requires a certain specific vocabulary, and if a person is not prepared to learn that vocabulary he is not deserving of the enjoyment which comes as a reward for entering a new field of human interest.

The illustrations are all original and are drawn mostly from fresh specimens, although some recourse was had to herbarium material. The greater number were prepared by Evelyn Maino, and the others by Ethel Craig Sumner, Kathryn Drew Jenkins, Alice Handschiegl, Dorothy G. Harris, Lois Chambers, and Emily Patterson. Besides drawing many illustrations and the maps, Miss Maino has checked all the botanical descriptions against living and herbarium specimens and has given valuable aid throughout the preparation of the manuscript. Both authors are responsible for the photographs and all other portions of the book not mentioned above. Free use has been made of the citations of distribution of the native trees given in *Forest Trees of the Pacific Slope*, by George B. Sudworth, in *Trees of California*, by Willis L. Jepson, and in *Illustrated Flora of the Pacific States*, by LeRoy Abrams. However, more recent field studies have extended the range therein given of some species and reduced it in others.

The genera, under which are listed the species, are arranged alphabetically within the family. The arrangement of families is, in the main, that of Engler and Prantl in *Die Natürlichen Pflanzenfamilien*, but since this manual is written for the identification of the genera and species, no descriptions are given of the families. The scientific names used are, for the most part, those given in *Standardized Plant Names*, published by the American Joint Committee on Horticulture Nomenclature. Synonyms are given whenever the names in that volume are not in accord with those used in the different manuals and floras covering the various parts of the area herein treated and when authors of the local manuals use different names. The names of species not included in *Standardized Plant Names* are taken from other sources given in the reference list.

### ACKNOWLEDGMENTS

I take this opportunity of expressing my appreciation to those who have given of their time and thought, and who have shown courtesies in making available living and herbarium specimens necessary for the completion of this book. Dr. Rimo Bacigalupi read the entire manuscript and made many helpful suggestions and criticisms. Dr. H. L. Mason read that part of the manuscript which treats of the origin and the distribution of native trees. Miss Ethel Crum was most helpful in supplying herbarium material from the University of California Herbarium at Berkeley. Mrs. Lora J. Knight, Mr. C. O. G. Miller, Mr. and Mrs. Linden Naylor, Mr. and Mrs. W. W. Carruth, Mr. Duncan McDuffie, Mrs. W. O. Wayman, Dr. W. Barclay Stephens, Mr. John McLaren, Mr. Eric Walther, Dr. S. A. Watson, Professor H. F. Spencer, Dr. P. M. Rea, Mr. Peter Riedel, Miss Flora Belle Ludington, Miss Mary Mangold, Miss Lisle Roddan, Miss Evelyn Graham, Miss Margot Chamberlain, Miss Eleanor Friedman, and others among my friends and students have rendered assistance in various ways.

The sympathetic interest and encouragement of President Aurelia Henry Reinhardt of Mills College have been greatly appreciated in the preparation of this volume.

To the teachers of botany and park superintendents throughout the Pacific Coast region who have coöperated in collecting material I express my thanks.

I am especially indebted to my wife, Helen R. McMinn, for assistance given during the preparation of the manuscript and the reading of the proof.

H. E. McMinn

*Mills College, California*

# INTRODUCTION

## The Naming of Plants

IN ORDER THAT we may speak intelligently about things, if for no other reason, it is convenient to give them names. The many forms of plant and animal life are classified according to an established system. Those individuals which are most alike are usually grouped into a species. Thus all the individuals of Red Oak are of one species, the Red or *rubra* species of oak, and all the individuals of the Valley Oak belong to the Valley or *lobata* species. Some individuals comprise a variety, a subdivision of the species. Those species which are most alike are brought together into a category called the genus. Thus all species of oak belong to the genus *Quercus,* the Latin name for the oak genus, and all species of chestnut belong to the genus *Castanea.* A further step in classification is the grouping of related genera into families. The oaks (*Quercus*), the chestnuts (*Castanea*), the beeches (*Fagus*), the chinquapins (*Castanopsis*), and other related genera make up the Beech Family, Fagaceae. Further groupings are made, but a discussion of them here would be out of place, and would serve no useful purpose.

The scientific name of a plant consists of the name of the genus followed by that of the species. The name or the abbreviation of the name of the person who properly applied the scientific name to the plant is written after the species. Therefore the scientific name of Valley Oak is *Quercus lobata* Née. The genus is always written with a capital letter, and in this handbook the species, in agreement with the practice of many authors, is uncapitalized. The names appearing in italics beneath the scientific name are to be considered as synonyms.

Two accent marks are used to aid in pronunciation, the grave (`) to indicate the long English sound of the vowel, and the acute (´) to indicate the short or otherwise modified sound.

[1]

The hyphen is used in the common name whenever the last part of the name refers to the common name of another genus, as in Cypress-Pine, the common name of *Sciadopitys verticillata,* and in Tanbark-Oak and Tan-Oak, the common names of *Lithocarpus densiflora.* The genus for pine is *Pinus* and the genus for oak is *Quercus.*

## How to Use the Manual

In determining the names of trees by the use of this manual there are two methods of procedure. Often the tree will be known beforehand to belong to a certain genus, as *Cinnamomum,* oak, or pine. The page upon which the genus is given can be ascertained by using the Index. Following the genus, if there is more than one species, is a key to the different species. The illustrations accompanying the descriptions will aid in identifying the unknown tree. If only a single species of a genus is given, it directly follows the genus.

If the genus is not known, then use must be made of the Key to the Genera. This key is based primarily upon leaf characters, because leaves are found upon trees for a greater length of time than are flowers or fruits. However, when the leaves of two or more species or genera are quite similar, the characters of other plant organs are used. When the key is used, ample material should be at hand and a careful examination of it should be made in order to discover the amount of variation which occurs in the characters. Choose a sample typical of the average adult leaves and proceed with the key. The first choice must be made between I and II. Having made that choice, continue to choose between contrasted subheads, indicated by indentation of lines upon the page or by letters, numbers, or other symbols, until one leads to a genus. The number following the genus refers to the page upon which the genus is described. The key following the genus identifies the species. The descriptions, in keeping with a manual, are brief, usually giving only sufficient information to assure correct identification of the tree. As a matter of interest, a brief note pertaining to its nativeness and distribution is added. The books given in the reference list will give further information about the identified species.

## EXPLANATION OF TERMS

For the guidance of readers not familiar with botanical terminology the following account will serve as an introduction to the study of those plant organs the characters of which are used in the keys and descriptions. The information in this account and that given in the Glossary should give sufficient preparation for the successful use of this manual.

### THE STEM

The stem is the part of the plant that grows upward from the root and bears the other organs. At definite places on the stem, called nodes, the leaves arise. The parts of the stem between the nodes are called internodes. In the axils of the leaves arise the branches from which flowers and fruits are produced. In trees the stem is a single woody trunk and does not branch for some distance above the ground. In shrubs the woody trunks branch at or near the ground and the stems and branches grow to form bush-like plants. Many shrubs grow to heights of twenty feet or more and their stems become six inches or more in diameter. These then are arborescent shrubs.

The characteristics of stems used in the classification of plants are mostly internal. In most stems the woody part occurs in annual concentric layers, known as annual rings, surrounding a small central pith and protected by a separable bark on the outside. Such stems are characteristic of Gymnosperms (pines) and Dicotyledonous-Angiosperms (oaks) and are called exogenous (outside-growing). They normally produce leaves with netted venation. In the stems of palms and palm-like plants there is no separable bark and the woody part is much reduced and occurs in threads or bundles which are scattered throughout the abundant pithy material. This type of stem is called endogenous (inside-growing) and is characteristic of Monocotyledonous-Angiosperms. They produce leaves with parallel veins. Terms applying to the external characteristics of stems are included in the Glossary and can be learned when met with in the keys and descriptions.

### THE LEAF

Leaves are usually green expansions growing out laterally from the stems or branches. In the axil formed by the leaf and branch occurs a bud which produces a new branch. In determining the leaf it is very important that the bud be located because confusion often occurs in distinguishing the leaflets of a compound leaf from simple leaves. Buds are not found at the bases of leaflets. The leaves of Gymnosperms and some Angiosperms are very small, often scale-like or needle-like, and the axillary buds are usually apparently absent. Therefore their structural nature is learned only by practice.

*The simple leaf.*—The most common kind of leaf is a simple leaf. It usually consists of an expanded part, the blade, and a slender stalk, the petiole, which attaches the blade to the branch. If the blade is directly attached to the branch then the leaf is sessile. Outgrowths, called stipules, often occur at the base of the petiole. They may be very small and scale-like, thorn-like, or leaf-like. The midrib extends from the summit of the petiole to the apex of the blade. Along the side of or extending from the midrib are veins which divide into veinlets.

*The compound leaf.*—Whenever a leaf is divided to the midrib into separate parts it is a compound leaf. The separate parts are called leaflets. These may be sessile or petiolulate. If the leaflets are arranged laterally along the common stalk or rachis the leaf is pinnately compound. In bipinnately compound leaves, the ultimate divisions are the leaflets. When there is a single leaflet at the end of the rachis the leaf is odd-pinnate. If two leaflets terminate the rachis the leaf is abruptly or even-pinnate. If all the leaflets arise from the apex of the petiole, like the fingers from the palm of the hand, the leaf is palmately compound.

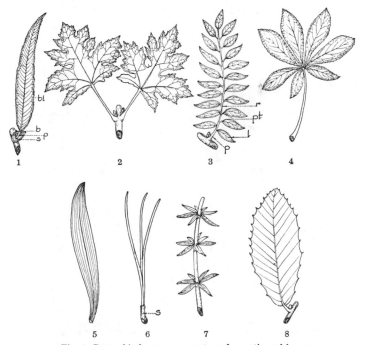

Fig. 1. Parts, kinds, arrangement, and venation of leaves

1. Stem with simple alternate leaf, with netted venation; b, bud; bl, blade; p, petiole; s, stipule. 2. Stem with opposite, palmately veined and lobed leaves. 3. Pinnately compound leaf; l, leaflet; p, petiole; pt, petiolule; r, rachis. 4. Palmately compound leaf. 5. Simple leaf with parallel venation. 6. Fascicled leaves; s, sheath. 7. Stem with whorled leaves. 8 Straight-veined leaf.

*Arrangement of leaves.*—When only one leaf occurs at a node the arrangement is alternate, and if two leaves oppose each other at a node the arrangement is opposite. If three or more leaves are arranged around the stem at a node, like the spokes of a wheel, the arrangement is called whorled or cyclic. Leaves are fascicled when two or more leaves are bundled together and surrounded at the base by a sheath, as in the pines. If no sheath is present the bundled arrangement is called clustered, as are some of the leaves of the Deodar.

*Venation.*—The arrangement of the veins in the leaf-blades is called venation. Leaves are parallel-veined when several

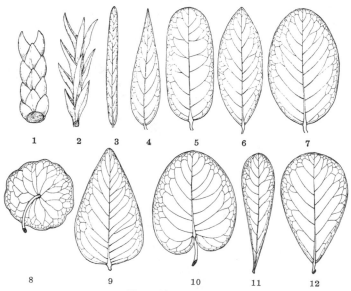

Fig. 2. Shapes of leaves

1. Scale-like. 2. Awl-shaped. 3. Linear. 4. Lanceolate. 5. Oblong. 6. Elliptic. 7. Oval. 8. Orbicular. 9. Ovate. 10. Cordate. 11. Oblanceolate. 12. Obovate.

main veins run parallel to one another. When three or more main veins arise from the summit of the petiole and radiate from one another the venation is palmate. If the secondary veins all branch from the midrib the venation is pinnate and if the secondary veins are very distinct and run straight and parallel to one another the leaves are straight-veined.

*Shapes, margins, tips, bases, surface, texture, and duration of leaves.*—The terms used in describing these leaf characters may be understood by referring to the diagrams which follow and to the Glossary.

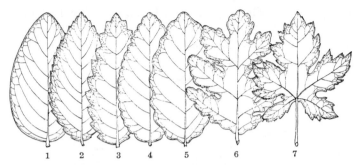

Fig. 3. Margins of leaves
1. Entire. 2. Serrate. 3. Dentate. 4. Crenate. 5. Sinuate. 6. Pinnately lobed. 7. Palmately lobed.

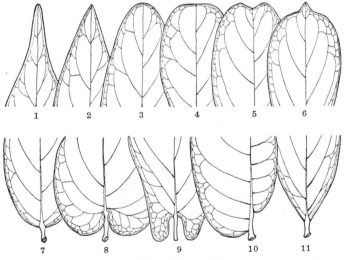

Fig. 4. Tips and bases of leaves
1. Acuminate. 2. Acute. 3. Obtuse. 4. Truncate. 5. Emarginate. 6. Mucronate. 7. Rounded. 8. Cordate or heart-shaped. 9. Auriculate. 10. Oblique or unequal. 11. Cuneate or wedge-shaped.

### THE FLOWER

The flower is that organ of a plant which has to do with the production of seeds. A complete flower is composed of four whorls or cycles of parts borne upon the receptacle. The outermost whorl is the calyx. The individual parts of the calyx are called sepals. The next whorl is the corolla, which is composed of petals. The stamens comprise the next whorl and the pistils form the innermost whorl.

The calyx and corolla taken collectively compose the perianth or floral envelopes. If only one set of floral envelopes is present it is called the calyx and the flower is then apetalous.

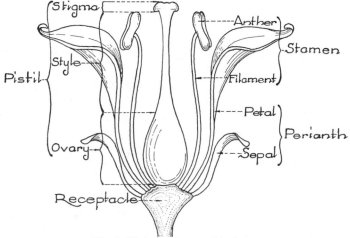

Fig. 5. Diagram of a complete flower

If the petals are joined to one another so as to form a single structure the flower is sympetalous, and if the petals are distinct then the flower is choripetalous or polypetalous.

The stamens and pistils comprise the essential organs. The stamen is made up of a stalk or filament supporting a terminal sac or anther which usually bears a yellow powdery substance called pollen. The pollen contains the male elements, sperms, which unite with the female elements, eggs, within the ovules located in the ovary of the pistil. The transference of the pollen from the anther to the stigma of the pistil is called pollination. This may be accomplished by insects, wind, water, gravity, and other agencies. If the stamens are united into one group they are called monadelphous, and if into two groups, diadelphous. The ovary is the swollen basal part of the pistil which bears one or more ovules. It continues into the style, which terminates in the stigma, the part that receives the pollen. The flower is called perfect if both stamens and pistils are present, staminate if only the stamens are present, and pistillate if only the pistils occur. When the staminate and pistillate flowers are borne upon separate plants, as in the willows, the plants are dioecious, and when they are borne upon the same plants, as in the alders, the plants are monoecious. Pistils may be simple or compound according to the number of carpels involved. The carpel is generally considered a modified leaf bearing ovules. The simple pistil consists of a single carpel; a compound pistil, of two or more united carpels. If the ovary is not adherent to the calyx it is superior; if it is adherent, it is inferior.

### THE ARRANGEMENT OF THE FLOWERS (INFLORESCENCE)

The name inflorescence is given to the arrangement of flowers upon the stems and branches. Flowers are terminal when borne at the ends of the stems or branches and axillary when they arise in the axils of leaves. They may be borne singly or in clusters of varying shapes and sizes. The stalk of a solitary flower or the main stalk of a cluster is called the peduncle. The individual stalklets supporting the flowers of a cluster are called pedicels. Whenever the flowers are without stalklets they are said to be sessile. The simplest type of inflorescence is the raceme, in which the central and indefinite axis bears lateral flowers upon pedicels of nearly equal lengths. If the flowers are sessile upon the elongated axis they form a spike. When the central axis branches and the divisions bear two or more flowers each, the inflorescence is a panicle. If the raceme or panicle is very compact and consists of unisexual flowers, it is called a catkin or ament. When the lower divisions of a panicle or the pedicels of a raceme become elongated, a corymb is formed. When the pedicels or divisions of a raceme or panicle arise from a common base, a simple or compound umbel is formed. Frequently the central axis becomes much shortened and bears sessile flowers in a close cluster known as a head. Small modified leaves, called bracts, often occur in a flower-cluster at the base of the pedicels and

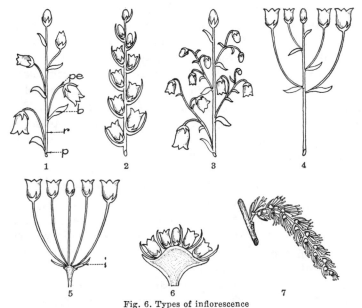

Fig. 6. Types of inflorescence

1. Raceme; p, peduncle; pe, pedicel; b, bract; r, rachis. 2. Spike. 3. Panicle. 4. Corymb. 5. Umbel; i, involucre. 6. Head. 7. Catkin.

below the sepals. If several bracts encircle a flower-cluster as in an umbel or a head they constitute an involucre. The involucral bracts may be scale-like or larger, leaf-like, and often colored. The diagrams (fig. 6) show the common types of inflorescence.

### THE FRUIT AND SEED

The fruit is the matured ovary and any adjacent part that may develop with it. The seed is a matured ovule. The union (fertilization) of the egg within the ovule with the sperm from the germinated pollen grain forms the embryonic plant, stimulates growth of the surrounding tissues, and causes the subsequent ripening of the ovule and ovary.

The kinds of fruits take their names from their texture, origin, and manner of escape of the seeds. The following outline gives the main kinds of fruits used in the keys and descriptions.

A. Kinds of fruits according to texture.
  1. Dry.
      a. *Capsule.* From a compound ovary, splitting along two or more lines.
      b. *Legume.* From a simple ovary, splitting along two lines.
      c. *Follicle.* From a simple ovary, splitting along one line.
      d. *Achene.* A one-seeded fruit from a simple or compound ovary which does not open.
      e. *Nut.* A one-seeded hard fruit from a compound ovary which does not open.
      f. *Acorn.* A nut partly surrounded by a fibrous or woody cup (involucre).
      g. *Samara.* A winged nut, achene, or any other dry indehiscent usually one-seeded fruit, as that of the ash, maple, or elm.
      h. *Cone.* A collection of carpels on a common receptacle, as in pines and firs, also less appropriately applied to the fruits of alders, birches, and similar trees.
  2. Fleshy.
      a. *Drupe.* Usually from a simple ovary, with a fleshy outside (exocarp) surrounding a hard stone or pit (endocarp) which encloses the usually single seed.
      b. *Pome.* From an inferior compound ovary with papery or cartilaginous carpels surrounded by a fleshy receptacle and calyx.
      c. *Berry.* From a compound, superior or inferior ovary, fleshy throughout, with the seeds embedded in the fleshy pulp. The orange, lemon, and other citrus fruits are berries with leathery rinds, called hesperidia (singular, hesperidium).
B. Kinds of fruits according to origin.
  1. *Simple.* From a single, simple or compound ovary.
  2. *Aggregate.* From a collection of ovaries of a single flower ripening into a single mass, as in the blackberry.

3. *Multiple.* From the ovaries of several flowers massed together on or in a common receptacle, as in the mulberry and fig.
4. *Accessory.* From ovaries with adjacent parts such as the calyx and the receptacle developing with them.

C. Kinds of fruit according to manner of escape of seeds.

1. *Dehiscent.* Splitting open.
2. *Indehiscent.* Not splitting open.

## THE ORIGIN AND DISTRIBUTION OF THE PACIFIC COAST TREES

The geographic region included in the title, "Pacific Coast," embodies the states of California, Oregon, Washington, and the province of British Columbia. The trees comprising the forests of this region are, with few exceptions, native species. In several localities of California, forests of eucalyptus have been planted. The trees in the parks and along the streets and highways of California are predominantly introduced species, whereas in Oregon, Washington, and British Columbia many of the species growing in the parks are native, those growing along the streets are mostly introduced, and the highways, with few exceptions, are unplanted. All known native and about four hundred of the introduced species are included in this manual.

### THE NATIVE TREES

A total of 130 species and 16 varieties of trees are native to the Pacific Coast, of which 32, even in their adult form, are often shrub-like. Fifty-nine of these are cone-bearing trees, 84 are broad-leaved trees, and 3 are palm and palm-like trees. Table 1 (p. 13) gives the numerical distribution of the species within the area covered.

It will be observed from the table that a total of 129 species occur in California, 78 in Oregon, 60 in Washington, and 53 in British Columbia. Sixty-three species are limited to California, 3 to British Columbia, 1 to Washington, and none to Oregon. Thirty-eight species occur in all four of the regions.

The present distribution of species depends upon their immediate ancestry, their migrational history, the past and present natural environmental factors, and changed conditions brought about by man. It has been fairly well established that the majority of plants growing naturally in any region are there because their immediate ancestors were there. The presence of the immediate ancestors is probably

TABLE 1

NUMERICAL DISTRIBUTION OF SPECIES WITHIN THE AREA

(C, California. O, Oregon. W, Washington. B, British Columbia)

| Trees | Limited to | | | | | | | | | | Total |
|---|---|---|---|---|---|---|---|---|---|---|---|
| | C | CO | COW | CO WB | O | OW | OWB | W | WB | B | |
| Conifers | 24 | 12 | 1 | 13 | 0 | 1 | 6 | 0 | 0 | 2 | 59 |
| Palm and Palm-like | 3 | 0 | 0 | 0 | 0 | 0 | 0 | 0 | 0 | 0 | 3 |
| Broad-leaved | 36 | 8 | 7 | 25 | 0 | 0 | 5 | 1 | 1 | 1 | 84 |
| Total | 63 | 20 | 8 | 38 | 0 | 1 | 11 | 1 | 1 | 3 | 146 |

ascribable to their migrational history, which was certainly made by the natural environmental factors occurring along the paths of dispersal. Thus the manzanitas (*Arctostaphylos*) and ceanothi, mountain-lilacs (*Ceanothus*), are confined to North America, and the eucalypti (*Eucalyptus*) are confined to Australia and the Malayan region. The absence of a species from a region, however, does not always signify that the ancestors of that species never occurred there. Magnolias, palms, and ginkgoes once thrived in Oregon and Washington, but now none of them is native to those states. They were forced out by a series of geological and climatic changes and by their inability to compete with new species coming into the area.

The migrational history of forest trees suggests important reasons why the species of trees on the Pacific Coast differ from those of the midwestern and eastern parts of the United States. The difference in the species content of these regions is probably to be accounted for by the climatic conditions which governed the migration of the ancestral species into these regions. Only four tree-species, *Betula papyrifera, Populus tremuloides, Salix nigra,* and *Salix amygdaloides,* are common to both regions. These species have very small seeds which are easily dispersed by wind and which are able to germinate and grow into plants adapted to the varying climatic conditions occurring from the northern Atlantic westward to the northern Pacific Coast. All the species of trees on the Pacific Coast belonging to the Beech Family, except the Maul Oak (*Quercus chrysolepis*), are endemic to this region, and irrespective of whether these and the other endemic species and genera are "new-beginners" or "relicts," they originated from species which now occur or once occurred here. The environmental conditions required by these endemics may approach those found elsewhere in the world, but an entirely different group of species is found here because of the absence of natural lanes of dispersal.

That so many trees native to California are not found in Washington and British Columbia is chiefly because of the absence of their immediate ancestors in the northern region. The native acacia, the palm, the yuccas, the mesquite, and other desert trees had their immediate ancestry in the trees of the arid Southwest and northern Mexico.

With respect to the origin of the ancestral and existing species, evidence seems to point to four sources. Many of the species occupying the colder and moister parts of the area are of Boreal origin. The Western Yew (*Taxus brevifolia*), Lodgepole Pine (*Pinus murrayana*), Whitebark Pine (*Pinus albicaulis*), Alpine Fir (*Abies lasiocarpa*), Western Hemlock (*Tsuga heterophylla*), Engelmann Spruce (*Picea engelmanni*), Sitka Spruce (*Picea sitchensis*), Western Larch (*Larix occidentalis*), Nootka-Cypress (*Chamaecyparis nootkatensis*), Quaking Aspen (*Populus tremuloides*), and others have originated in the subarctic and cool-temperate regions of the northern hemisphere. The reason for the presence of these species on many of the higher ridges of the Cascades and the Sierra Nevada and of a few along the coasts of Washington and Oregon is perhaps that the last great glaciation pushed them or their ancestors southward.

A second source of origin is one already referred to, that of the arid Southwest and northern Mexico. With the increasing aridity in the southern part of the Pacific Coast at the close of the glacial period, many species which had originated on the arid plateau of Mexico at the time when species common to the north temperate regions were spreading, migrated northward, and now they or their descendants occupy the desert parts of California. These species are slowly moving northward east of the South Coast Range and the Sierra Nevada and in the San Joaquin Valley.

A third source of origin is the Great Basin and Rocky Mountain region. The Nut Pine (*Pinus edulis*), Limber Pine (*Pinus flexilis*), Hickory Pine (*Pinus aristata*), Colorado Juniper (*Juniperus scopulorum*), Desert Juniper (*Juniperus californica* var. *utahensis*), Fremont Cottonwood (*Populus fremonti*), Western Hackberry (*Celtis douglasi*), Mountain-Mahogany (*Cercocarpus ledifolius*), Arizona Ash (*Fraxinus velutina*), and others have originated in these regions and have migrated westward to the Pacific Coast region.

Sixty-seven species of trees occurring on the Pacific Coast are limited to that area and are therefore endemics. They constitute the fourth source of the native trees. Many of these species are peculiar to the Upper Sonoran Life Zone in California and constitute a part of the flora referred to as the California element by Dr. LeRoy Abrams (*Illustrated Flora of the Pacific States,* Vol. I, p. vii, 1923), who says: "Reasons for the unique character of the flora are to be found in climatic conditions both of the present and the past, and in the isolation brought about by the climatic and physical barriers prohibiting direct communication with the eastern part of the continent." A few of these species have spread beyond the borders of California to the north, east, and south.

One of the most critical environmental factors governing the persistence and geographical distribution of plants is temperature. This factor is dependent upon many other factors, among which are topography, rainfall, the presence of large bodies of water, the direction of prevailing winds, latitude, and altitude. The action and interaction of all these and many other factors are responsible for the existence of environmental regions known as life zones. In mountainous regions, as on much of the Pacific Coast, temperature is often in large measure determined by altitude and consequently the life zones occupy more or less distinct horizontal belts. Dr. C. Hart Merriam (*Life Zones and Crop Zones in the United States,* pp. 1–79, 1898) has worked out these zones for most of North America and his names for them are in general use.

The Lower Sonoran Life Zone is limited on the Pacific Coast to four areas in California—the Colorado Desert, the Mohave Desert, the inner coastal valleys of southern California, and a great part of the San Joaquin-Sacramento Valley. It is characterized by a low rainfall (0 to 5 inches), high summer temperatures (90° to 130°), winter temperatures varying from 15° to 50°, drying winds of great velocity, and altitudes varying between 350 feet below sea level (Death Valley) to 6000 feet elevation on the desert mountains, but usually below 3000 feet elevation. The amounts of precipitation generally become greater with each succeeding life zone and the upper limits of temperature range are lower. Slope exposure, air currents, lingering snow banks, forest fires, rock surfaces, streams with cold water, and the extent of mountain and desert areas are factors which may cause the usually regular arrangement of zones to be altered. Trees are very scarce in this zone and they usually shed their leaves early or have small leaves or possess other means for preserving the small amount of available water. Spanish Dagger (*Yucca mohavensis*), Joshua Tree (*Yucca brevifolia*), California Fan Palm (*Washingtonia filifera*), Catclaw (*Acacia greggi*), Palo Verde (*Cercidium torreyanum*), Desert Ironwood (*Olneya tesota*), Male Palo Verde (*Parkinsonia microphylla*), Honey Mesquite (*Prosopis juliflora* var. *glandulosa*), Screw-bean Mesquite (*Prosopis pubescens*), Smoke Tree (*Parosela spinosa*), Desert-Willow (*Chilopsis linearis*), Fremont Cottonwood (*Populus fremonti*), and Black Willow (*Salix nigra* var. *vallicola*) grow in the Lower Sonoran Life Zone.

The Upper Sonoran Life Zone occupies warm dry areas east of the Cascade Mountains in Washington and Oregon usually below 1500 feet elevation, but in the more southern part of this area and on the more southerly exposures it extends upward to about 2500 feet elevation. In California it comprises the hot dry grassy foothills and the brush-covered lower slopes of the mountains at elevations between 1000 and 3000 feet in the northern and central parts, and in southern California it extends from sea level to nearly 5000 feet and thus occupies most of the coastal area from Santa Barbara to San Diego. The most characteristic trees of this zone in California are Digger Pine (*Pinus sabiniana*), Singleleaf Pine (*Pinus monophylla*), Sargent Cypress (*Cupressus sargenti*), Macnab Cypress (*Cupressus macnabiana*), California Juniper (*Juniperus californica*), Coast Live Oak (*Quercus agrifolia*), Interior Live Oak (*Quercus wislizeni*), Blue Oak (*Quercus douglasi*), Valley Oak (*Quercus lobata*), Mesa Oak (*Quercus engelmanni*), California Buckeye (*Aesculus californica*), Western Redbud (*Cercis occidentalis*), Toyon (*Photinia arbutifolia*), and Hard Tack (*Cercocarpus betuloides*). In Washington and Oregon this zone is mostly occupied by the Sagebrush (*Artemisia tridentata*) and other associated shrubs, but trees of the Western Hackberry (*Celtis douglasi*) and White Alder (*Alnus rhombifolia*) occasionally occur near or along watercourses.

The Arid Transition Life Zone occupies an area with more rainfall than the Upper Sonoran Zone but with less than the Humid Transition Zone and usually occurs at lower elevations than does the last-named. It is a region of hot rainless summers and cold winters. In Washington and Oregon it includes the eastern slope of the Cascade Mountains and areas with similar conditions in the eastern parts of these states. The characteristic trees are Yellow Pine (*Pinus ponderosa*), Oregon Oak (*Quercus garryana*), Madrone (*Arbutus menziesi*), and Mountain-Mahogany (*Cercocarpus ledifolius*). In California it includes part of the main timber belt of the Sierra Nevada (Sierran Transition, by Jepson) and regions having similar conditions in the Coast Ranges and mountains of southern California. The characteristic trees of this area are Yellow Pine (*Pinus ponderosa*), Incense-Cedar (*Libocedrus decurrens*), Sugar Pine (*Pinus lambertiana*), White Fir (*Abies concolor*), Black Oak (*Quercus kelloggi*), Madrone (*Arbutus menziesi*), and the Big Tree (*Sequoia gigantea*). The Big Tree is found in a few localities in the Sierra Nevada.

The Humid Transition Life Zone is characterized on the Pacific Coast by its moist climate, uniform temperature, and luxuriant forests composed of a great number of species. In British Columbia, Washington, and Oregon it occupies the region west of the Cascade Mountains, where the dominant forest tree is Douglas-Fir (*Pseudotsuga douglasi*). Sitka Spruce (*Picea sitchensis*) becomes an important species along the coast, where Beach Pine (*Pinus contorta*) occurs in isolated stands. Other species in this area are Coast Hemlock (*Tsuga heterophylla*), Lowland Fir (*Abies grandis*), Red Alder (*Alnus rubra*), California Wax-Myrtle (*Myrica cali-*

*fornica*), Scouler Willow (*Salix scouleriana*), Bigleaf Maple (*Acer macrophyllum*), Vine Maple (*Acer circinatum*), Oregon Ash (*Fraxinus oregona*), Black Cottonwood (*Populus trichocarpa*), Oregon Oak (*Quercus garryana*), and Oregon Crab Apple (*Malus fusca*). In California and in extreme southwestern Oregon the Humid Transition Life Zone comprises the redwood belt, which occupies a narrow coastal area commonly known as the fog belt, extending from northern San Luis Obispo County, California, to Curry County, Oregon. It ranges in altitude from sea level to nearly 3000 feet elevation. The annual rainfall varies from 20 to 120 inches. The dominant tree in this area is the Redwood (*Sequoia sempervirens*) and associated with it or between areas where it does not form pure stands are Douglas-Fir (*Pseudotsuga douglasi*), Tanbark-Oak (*Lithocarpus densiflora*), Western Hemlock (*Tsuga heterophylla*), Western Yew (*Taxus brevifolia*), Sitka Spruce (*Picea sitchensis*), Lowland Fir (*Abies grandis*), California Wax-Myrtle (*Myrica californica*), and Madrone (*Arbutus menziesi*).

The Canadian Life Zone occupies an ill-defined area above the Transition zones at elevations between 3000 and 5000 feet in the mountains of Washington and Oregon, between 5000 and 7000 feet in the mountains of northern California, and between 6000 and 9000 feet in central and southern California. Along the coast of British Columbia and southward along the coasts of Washington and Oregon to northern California, the Canadian Zone often extends to sea level. Several of the trees occurring in this zone are found also in the zone above or that below. The most characteristic trees in California are Lodgepole Pine (*Pinus murrayana*), Silver Pine (*Pinus monticola*), Jeffrey Pine (*Pinus jeffreyi*), Red Fir (*Abies magnifica*), Lowland Fir (*Abies grandis*), and Quaking Aspen (*Populus tremuloides*). In Washington, Oregon, and British Columbia the Lodgepole Pine, Silver Pine, and Lowland Fir, characteristic of the California Canadian Zone, recur, and the Noble Fir (*Abies nobilis*), Cascade Fir (*Abies amabilis*), Western Larch (*Larix occidentalis*), Engelmann Spruce (*Picea engelmanni*), Western Hemlock (*Tsuga heterophylla*), and Western Yew (*Taxus brevifolia*) are additional species. Sitka Spruce (*Picea sitchensis*), Beach Pine (*Pinus contorta*), and Western Yew (*Taxus brevifolia*) occupy more or less interrupted intervals along the coast from British Columbia to northern California and perhaps should be included in this zone rather than in the Humid Transition Zone.

The Hudsonian Life Zone is the highest zone in which trees are found. It ranges from 5000 to 7500 feet elevation in the northern part of the Pacific Coast, and from 7000 to 11,500 feet in the Sierra Nevada and the mountains of southern California. Mountain Hemlock (*Tsuga mertensiana*), Whitebark Pine (*Pinus albicaulis*), Foxtail Pine (*Pinus balfouriana*), Alpine Larch (*Larix lyalli*), Alpine Fir (*Abies lasiocarpa*), and Nootka-Cypress (*Chamaecyparis nootkatensis*) are characteristic trees of the Hudsonian Life Zone.

The Arctic-Alpine Life Zone occurs above timber line at elevations between 6500 and 14,500 feet. The altitude of the lower limit decreases with increasing latitude.

## THE INTRODUCED TREES

About 1000 species of introduced trees occur on the Pacific Coast. They have been brought in from all those parts of the world in which trees grow naturally. The greater number of these species are rarely seen since they usually occur as specimen trees in collections found in private estates and botanical gardens. The wide range of climatic conditions on the Pacific Coast, varying from the semitropical of southern California to the boreal of northern British Columbia and of the high mountain regions of the entire area, accounts for the large number and distribution of the introduced trees which thrive here.

The three climatic zones of the earth from which the greatest number of trees have been obtained are the north temperate, the south temperate, and the tropical. These zones, occupying definite latitudinal belts, have in common certain factors such as minimum light, heat, and moisture necessary for tree growth. Because of the varying amounts of these factors, the present distribution of oceans and land masses, and the past geological history of the land masses, the zones support different genera and species of trees.

The north temperate zone, with a much more severe and variable climate than the south temperate zone, has been the source of the greatest number of the deciduous broad-leaved trees and conifers. This zone has been divided by plant geographers into two main divisions, Eurasia and north Africa, and the Atlantic and central United States and Canada. Within these divisions are natural botanical provinces which constitute the geographers' regions or areas.

The area known as western and central Europe has been the source of many of the better known genera and species of trees. The English Yew (*Taxus baccata*), European Larch (*Larix europaea*), Silver Fir (*Abies pectinata*), Norway Spruce (*Picea excelsa*), and three species of pines (*Pinus nigra, montana, sylvestris*) are among the conifers introduced from this area. Species of poplar (*Populus*), willow (*Salix*), birch (*Betula*), hornbeam (*Carpinus*), beech (*Fagus*), oak (*Quercus*), hackberry (*Celtis*), elm (*Ulmus*), maple (*Acer*), hawthorn (*Crataegus*), linden (*Tilia*), and ash (*Fraxinus*) are deciduous trees which have come from this area. The Strawberry Tree (*Arbutus unedo*), from the British Isles, and the English Holly (*Ilex europaea*) are broad-leaved evergreen trees frequently planted on the Pacific Coast.

South of western and central Europe the climate is milder and the region supports, in addition to the usual north temperate species, a warm-temperate and subtropical flora. This area is known as the Mediterranean region. From it have been introduced pines (*Pinus pinaster, halepensis, pinea, montana*), the Italian Cypress (*Cupressus sempervirens*), Prickly Juniper (*Juniperus oxycedrus*), the Fig (*Ficus carica*), Grecian Laurel (*Laurus nobilis*), Carob (*Ceratonia siliqua*), Olive (*Olea europaea*), Smallflower Tamarisk (*Tamarix parviflora*), Cork Oak (*Quercus suber*), Hollyleaf Oak (*Quercus ilex*), and Oleander (*Nerium oleander*). The Spanish Fir (*Abies pinsapo*) has come from southern Spain and northern Africa, and the Portugal Laurel (*Prunus lusitanica*) from Spain and Portugal. A single palm (*Chamaerops humilis*), the only palm native to Europe, has been introduced from the warmer parts of the coastal Mediterranean region.

From northern Africa, with a climate similar to that of the opposite coast of southern Europe, have been obtained, in addition to the southern European semitropical species, the Atlas Cedar (*Cedrus atlantica*), Algerian Fir (*Abies numidica*), and the Date Palm (*Phoenix dactylifera*). The Canary and Madeira islands, a part of the Mediterranean region, have been the source of the Canary Island Date Palm (*Phoenix canariensis*), Canary Island Pine (*Pinus canariensis*), the Dragon Tree (*Dracaena draco*), and Madeira-Bay (*Persea indica*).

The eastern Mediterranean area, comparable in latitude and climatic conditions to southern California, besides containing the species common to the Mediterranean region, has been the source of the English Walnut (*Juglans regia*), Pomegranate (*Punica granatum*), Athel (*Tamarix articulata*), Black Mulberry (*Morus nigra*), lindens (*Tilia* sp.), maples (*Acer* sp.), firs (*Abies cephalonica, nordmanniana*), Cedar of Lebanon (*Cedrus libani*), Greek Juniper (*Juniperus excelsa*), and Oriental Spruce (*Picea orientalis*).

One of the most interesting botanical areas of the world is the mountainous region between the eastern Mediterranean area and that of eastern Asia. It is known as the Himalayan region and has been the source of the Himalayan Pine (*Pinus excelsa*), Longleaf Pine (*Pinus longifolia*), Himalayan Spruce (*Picea smithiana*), Bhutan Cypress (*Cupressus torulosa*), Evergreen Dogwood (*Cornus capitata*), and Himalayan Maple (*Acer oblongum*).

To the north and east of the Himalayan region lies the great north temperate region of eastern Asia, comprising the areas of China, Korea, and Japan. Besides having the common European genera of deciduous trees and conifers, this region has been the source of many unusual ornamental warm-temperate trees. Among the most important are the Maidenhair Tree (*Ginkgo biloba*), Chinese Plum-Yew (*Cephalotaxus fortunei*), Chinese-Fir (*Cunninghamia lanceolata*), Mourning Cypress (*Cupressus funebris*), Oriental Arborvitae (*Thuja orientalis*), Chinese Fan Palm (*Livistonia chinensis*), Windmill Palm (*Trachycarpus excelsa*), Weeping Willow (*Salix babylonica*), elms (*Ulmus parvifolia, pumila*), White Mulberry (*Morus alba*), Camphor Tree (*Cinnamomum camphora*), Quince (*Cydonia oblonga*), Loquat (*Eriobotrya japonica*), kumquats (*Fortunella* sp.), Tree of Heaven (*Ailanthus glandulosa*), camellias (*Camellia* sp.), and Royal Paulownia (*Paulownia tomentosa*). Japan, with a heavier rainfall than most of the eastern Asiatic mainland, has been the source of such warm-temperate and subtropical trees as Japanese Plum-Yew (*Cephalotaxus drupacea*), *Podocarpus macrophylla*, Japanese Cryptomeria (*Cryptomeria japonica*), Umbrella-Pine (*Sciadopitys verticillata*), Momi Fir (*Abies firma*), *Chamaecyparis obtusa* and *pisifera*, *Zelkova serrata*, magnolias (*Magnolia hypoleuca, stellata, kobus*), maples (*Acer palmatum, japonica, ginnala*), and Lilac (*Syringa japonica*).

The second division of the north temperate zone, the Atlantic and central United States and Canada, has been the source of many deciduous trees, several conifers, and a few evergreen broad-leaved trees. Among the most important are species of poplar (*Populus*), birch (*Betula*), hawthorn (*Crataegus*), locust (*Robinia*), maple (*Acer*), linden (*Tilia*), hickory (*Hicoria*), oak (*Quercus*), catalpa (*Catalpa*), walnut (*Juglans*), magnolia (*Magnolia*), and pine (*Pinus*). The Tulip Tree (*Liriodendron tulipifera*), American Beech (*Fagus americana*), Osage-Orange (*Maclura pomifera*), Sassafras (*Sassafras variifolium*), Sweet Gum (*Liquidambar styraciflua*), Honey-Locust (*Gleditsia triacanthos*), Yellow-wood (*Cladrastis lutea*), Bull-Bay (*Magnolia grandiflora*), Carolina Cherry-Laurel (*Prunus caroliniana*), Tupelo (*Nyssa sylvatica*), Common Persimmon (*Diospyros virginiana*), Balsam Fir (*Abies balsamea*), Canada Hemlock (*Tsuga canadensis*), American Arborvitae (*Thuja occidentalis*), Red-Cedar (*Juniperus virginiana*), and Bald-Cypress (*Taxodium distichum*) are other characteristic species of this area which have been introduced on the Pacific Coast.

Most of the evergreen broad-leaved trees have come from the south temperate zone. This zone includes the Cape and adjacent regions of South Africa; Australia, except the northern part, which lies in the tropical zone; New Zealand; and southern South America. The species of trees in this zone have evolved under milder climatic conditions than exist in the north temperate zone. These conditions are caused, for the most part, by the latitude in which the south temperate land masses occur and by the ameliorating influences of the oceans which border them. Most of these land masses lie north of 50° S Lat., and those of the north temperate zone extend to 60° N Lat. Three of these principal land masses, Africa, Australia, and South America, are widely separated by oceanic barriers, and thus their extreme isolation has been an important factor in the development of very different genera and species of trees in the three areas.

Among the tree introductions from temperate South America are the Monkey Puzzle (*Araucaria imbricata*), Chilean

Incense-Cedar (*Libocedrus chilensis*), Syrup Palm (*Jubaea spectabilis*), Mayten (*Maytenus boaria*), Soapbark Tree (*Quillaja saponaria*), California Pepper Tree (*Schinus molle*), *Tricuspidaria dependens*, Cherimoya (*Annona cherimola*), and Waxmallow (*Malvaviscus arboreus*).

Temperate South Africa has been the source of *Podocarpus elongata*, Silver Tree (*Leucadendron argenteum*), Cape Pittosporum (*Pittosporum viridiflorum*), *Calodendrum capensis*, *Harpephyllum caffrum*, *Schotia latifolia* and *brachypetala*, and Cape Weddingflower (*Dombeya natalensis*).

From the temperate Australian and New Zealand area have come the well-known genera, *Acacia, Eucalyptus,* and *Pittosporum.* Other genera less widely known but having many fine ornamental species are *Casuarina, Hakea, Macadamia, Hymenosporum, Albizzia, Lagunaria, Sterculia, Callistemon, Eugenia, Leptospermum, Melaleuca, Tristania, Abrophyllum, Castanospermum, Myoporum, Metrosideros* (New Zealand), and *Alectryon* (New Zealand). Of the gymnospermous genera, two species of *Agathis,* one of *Callitris,* and two of *Araucaria* have been introduced into California from this Australian area. The beautiful King Palm (*Loroma amethystina*), seen from Santa Barbara southward to San Diego, California, is native to the rain-forest region of Queensland, Australia.

Although all the area of the Pacific Coast, as delimited in this manual, lies within the latitudinal limits bounding the north temperate zone, there is a considerable area in California in which a warm-temperate climate merges with a semitropical climate. This area includes much of southern California and extends northward to Santa Barbara and gradually diminishes northward to the San Francisco Bay region. Into this area have been introduced many species of the semiarid tropics and a few species of the moister tropics.

The tropical flora of the eastern hemisphere is found in two widely separated areas, Africa and Indo-Malaya, and is composed of distinctly different species. The tropical area of the western hemisphere constitutes a single continental geographical unit, the West Indies and the Galapagos Islands excepted, extending from northern Mexico to southern Brazil, Paraguay, Argentina, and Chile.

Tropical east central and east Africa have contributed to our introduced flora the Abyssinian Banana (*Musa ensete*) and the Cape Palm (*Phoenix reclinata*). The Asiatic tropics, a division of the Indo-Malaya region, has been the source of the Common Banana (*Musa paradisiaca* var. *sapientum*), Purple Bauhinia (*Bauhinia purpurea*), and the Devil Tree (*Alstonia scholaris*). From the Malaya-Polynesian and northern Australian regions have come species of *Eucalyptus, Ficus* (*F. elastica, rubiginosa*), the Silk-Oak (*Grevillea robusta*), Crape-Myrtle (*Lagerstroemia speciosa, indica*), and King Palm (*Loroma amethystina*). One species of *Acacia* (*A. koa*) and one of *Hibiscus* (*H. arnottianus*), from Hawaii, are occasionally planted in California.

The arid semitropics of northern Mexico, Lower California, and southern Arizona have been the source of the Spanish Bayonet (*Yucca aloifolia*), Blue Palm (*Glaucothea armata*), Mexican Washington Palm (*Washingtonia robusta*), Brandegee Palm (*Erythea brandegei*), and Parkinsonia (*Parkinsonia aculeata*). The Avocado (*Persea americana*) and Guava (*Psidium guajava*) are from the moister tropics of Mexico and Central America.

The tropics and semitropics of South America have been the source of the Cocos Palm (*Cocos plumosa*), Pindo Palm (*Cocos australis*), *Cedrela fissilis,* Brazilian Pepper Tree (*Schinus terebinthifolius*), Strawberry Guava (*Psidium cattleianum*), Green-Ebony (*Jacaranda ovalifolia*), and Southern Soapberry (*Sapindus saponaria*). The Spanish Bayonet (*Yucca aloifolia*), Parkinsonia (*Parkinsonia aculeata*), and Soapberry (*Sapindus saponaria*) are also native to the West Indies.

The distribution of the introduced species on the Pacific Coast, as has been indicated, conforms in the main to the varying climatic conditions existing from British Columbia to southern California and from sea level to the high mountain regions of the entire area. Therefore the trees from the tropical, semitropical, and warm-temperate regions are found only in the milder parts of California. The acacias, eucalypti, palms, pittosporums, and others are not found growing successfully in the open north of the California boundary. The deciduous broad-leaved trees and conifers from the cooler parts of the north temperate regions thrive well in British Columbia, Washington, and Oregon. Many of them are successfully grown throughout California because trees of the cool north temperate zone usually adapt themselves with ease to the milder climates of the warm-temperate and semitropical areas. Trees of the south temperate, warm north temperate, and semitropical regions cannot adapt themselves to the colder temperatures prevailing during winter in the northern parts of the Pacific Coast.

# KEY TO THE GENERA

I. Plants palm-like; leaves arising from the summit of the main trunk or its divisions, usually parallel-veined.—MONOCOTYLEDONS.

Leaves not divided..................................................................A, p. 30

Leaves divided.—PALMS.........................................................B, p. 30

II. Plants not palm-like; leaves arising from the branches or branchlets.—DICOTYLEDONS and GYMNOSPERMS.

Leaves apparently absent. (In *Casuarina* and one species of *Tamarix* the leaves are represented by minute teeth-like structures at the joints of the green or grayish, linear, and jointed branchlets.)

A, p. 31

Leaves distinctly present.

Leaves compound.

Leaves 3-foliolate or palmately compound with 3 to 9 leaflets

B, p. 31

Leaves pinnately compound............... ...................................A, p. 32

Leaves simple.

Branches passing through compact cylindrical or spherical clusters of sessile capsules and continuing as foliage shoots.

Stamens united at their bases into 5 groups opposite the petals.................................................MELALEUCA, p. 320

Stamens not united........................................CALLISTEMON, p. 298

Branches not passing through clusters of woody capsules and continuing as foliage shoots.

Leaves typically ⅜-inch or less wide, ovate, narrowly lanceolate, linear, needle-like, awl-shaped, or scale-like. (Most Gymnosperms and a few Angiosperms.)

Leaves not scale-like..........................................................A, p. 34

Leaves scale-like, almost completely covering the branchlets..........................................................................A, p. 37

Leaves typically ⅜-inch or more wide. (Broad-leaved trees—most Angiosperms and *Ginkgo biloba*.)

Leaves opposite or some whorled...............................A, p. 38

Leaves alternate.

Leaf-margins entire or with 1 or 2 coarse teeth or small lobes (or sometimes serrulate in *Arbutus menziesi*)

A, p. 40

Leaf-margins toothed or lobed...................................A, p. 43

[ 29 ]

## A. LEAVES NOT DIVIDED

Leaves 1½ to 3 feet wide, 5 to 20 feet long, pinnately parallel-veined.
$\qquad$ Musa, p. 124

Leaves 1 to 3 inches wide, 6 inches to 6 feet long, parallel-veined.

Mature leaves serrulate, or if entire then filamentous; fruit a cap-sule.............................................................................................Yucca, p. 122

Mature leaves entire, not filamentous; fruit a berry.

Pedicels solitary, with an involucre at base; seeds numerous; fruits blue or whitish...........................................Cordyline, p. 121

Pedicels 2 or 3 together, without an involucre at base; seeds 1 to 3; fruit orange-colored.............................................Dracaena, p. 122

## B. LEAVES DIVIDED.—THE PALMS

### 1. Leaves palmate (Fan Palms)

Petioles spiny, at least toward the base.

Leaves 2 to 3 feet wide; petioles less than 1 inch wide; plants often branching near base.............................................Chamaerops, p. 125

Leaves 3 to 6 feet wide; petioles 1 inch or more wide.

Leaf-blades very blue-glaucous...............................Glaucothea, p. 129

Leaf-blades usually green.

Old leaves hanging against the trunk, forming a compact thatch-like skirt; petioles usually armed for the entire length (sometimes either unarmed or armed only near the base or apex)....................................................Washingtonia, p. 136

Old leaves directed outward and downward, not forming a com-pact skirt about the trunk; petioles usually unarmed (some-times with small blunt spines near base)......Livistona, p. 130

Petioles unarmed or irregularly sharp-serrate for most of their length.

Trunks covered with stiff brown fibers................Trachycarpus, p. 135

Trunks becoming naked, ringed with scars.

Leaves about 3 feet wide, the segments not bending downward 1 foot from apex........................................................Erythea, p. 128

Leaves 3 to 5 feet wide, the segments bending downward about 1 foot from apex........................................................Livistona, p. 130

### 2. Leaves pinnate (Feather Palms)

Trunks usually clothed with old leaf-bases; pinnae folded upward, the lower pinnae reduced to stout spines...........................Phoenix, p. 131

Trunks usually smooth; pinnae not folded upward, none reduced to spines.

Trunk conspicuously swollen at base........Roystonea oleracea, p. 134

Trunk not swollen at base.

Trunks 3 to 6 feet in diameter; leaves erect or spreading
JUBAEA, p. 129

Trunks less than 3 feet in diameter; leaves recurved or drooping.

Inflorescence among the leaves..................................COCOS, p. 126

Inflorescence on the trunk far below the leaves......LOROMA, p. 131

## A. LEAVES APPARENTLY ABSENT

### 1. PLANTS THORNY

Twigs zigzag and thorny at the angles; fruit a flat pod, not constricted between the seeds..............................CERCIDIUM TORREYANUM, p. 248

Twigs not zigzag; fruit-pods not flat; some of the branchlets reduced to spines.

Bark of limbs and twigs ashy gray, coated with a minute white down; spiny branchlets with glands; fruit a small 1-seeded pod
PAROSELA SPINOSA, p. 257

Bark of limbs greenish; fruit a cylindrical pod, constricted between the seeds..................................................PARKINSONIA, p. 255

### 2. PLANTS NOT THORNY

Branchlets jointed, green or grayish green.

Five to 14 teeth (leaves) at a joint...............................CASUARINA, p. 140

One minute tooth (leaf) at a joint..............TAMARIX ARTICULATA, p. 295

Branchlets not jointed; one minute tooth (leaf) at a node; internodes ⅛ inch or less long..................................................TAMARIX, p. 294

## B. LEAVES 3-FOLIOLATE OR PALMATELY COMPOUND WITH 3 TO 9 LEAFLETS

Plants thorny.......................................................ERYTHRINA, p. 252

Plants not thorny.

Leaves opposite.

Leaflets 5 to 9; fruit a capsule with 1 or 2 large polished seeds
AESCULUS, p. 281

Leaflets 3, the basal ones sometimes divided; fruit a double-winged samara............................................ACER NEGUNDO, p. 270

Leaves alternate.

Leaflets finely toothed, aromatic..................................PTELEA, p. 263

Leaflets entire, not aromatic; leaves apparently clustered on very short branchlets....................................................LABURNUM, p. 254

## A. LEAVES PINNATELY COMPOUND

### 1. Leaves opposite

Leaves twice-divided.

Ultimate leaflets attached by very broad bases
LYONOTHAMNUS, p. 225

Ultimate leaflets attached by very small petiolules or sessile
JACARANDA, p. 348

Leaves once-pinnately compound.

a. Leaflets irregularly and coarsely toothed or lobed, the basal leaflets sometimes divided..............................ACER NEGUNDO, p. 270

b. Leaflets regularly and coarsely toothed or lobed, the lobes appearing as secondary leaflets and attached by broad bases
LYONOTHAMNUS, p. 225

c. Leaflets finely serrate or sometimes entire.

Leaflets usually 7 (5 to 9), some oblique at base; fruit small and berry-like................................................................SAMBUCUS, p. 348

Leaflets 5 to 13, not oblique at base; fruit a 1-seeded winged samara................................................................FRAXINUS, p. 334

### 2. Leaves alternate

a. *Leaves twice-pinnately compound* (*or nearly so in* Grevillea)

Plants with thorns or spines.

Leaflets serrate, ¾-inch or more wide..............................ARALIA, p. 327

Leaflets usually entire, less than ¾-inch wide.

Thorns branched, 2 to 4 inches long..........................GLEDITSIA, p. 253

Thorns not branched.

Thorns 2 (sometimes single or none) at the bases of the 2-forked or sometimes 4-forked petioles; pods indehiscent
PROSOPIS, p.257

Thorns single.

Petioles 2-forked, sometimes forking close to the stem and appearing as 2 once-pinnately compound leaves; flowers in branched clusters.

Leaflets 5 to 30 pairs on each fork; pods cylindrical, constricted between the seeds..................PARKINSONIA, p. 255

Leaflets 2 to 4 pairs on each fork; pods flat, not constricted between the seeds....................................CERCIDIUM, p. 248

Petioles not 2-forked; flowers small, in cylindrical clusters
ACACIA GREGGI, p. 236

Plants not thorny.

Leaves more than 8 inches long; leaflets serrate or irregularly toothed, usually over ½-inch long, odd-pinnate.

Leaflets irregularly cut, the divisions entire and pointed; leaves feathery in appearance..................................................GREVILLEA, p. 198

Leaflets regularly toothed, sometimes lobed, the lobes toothed; leaves not feathery.......................................................MELIA, p. 266

Leaves less than 8 inches long; leaflets entire, less than ½-inch long, even-pinnate.

Leaves less than 4 inches wide, with prominent glands on upper side of the leaf-rachis; stamens distinct..................ACACIA, p. 234

Leaves 4 inches or more wide, without prominent glands on upper side of the leaf-rachis; stamens united...............ALBIZZIA, p. 245

b. *Leaves once-pinnately compound*

(1) LEAFLETS EVEN-PINNATE, THE MARGINS ENTIRE

Leaflets typically 1 inch or less long.

Leaflets ½-inch to 1½ inches long, densely tomentose beneath; plants not thorny......................................................................CASSIA, p. 247

Leaflets ½-inch or less long, gray on both surfaces; plants thorny

Leaflets more than 1 inch long. OLNEYA, p. 255

Leaflets 4 to 12.

Rachis winged.........................................................SAPINDUS, p. 284

Rachis not winged .......................................................CERATONIA, p. 247

Leaflets 12 to 22.................................................................CEDRELA, p. 265

(2) LEAFLETS ODD-PINNATE

(a) *Larger leaflets more than 2 inches long*

Leaflets typically entire (or with 1 to 4 small blunt points or lobes near the base in *Ailanthus*).

Leaflets 5 to 11.

Leaflets alternate, petiolulate................................CLADRASTIS, p. 251

Leaflets opposite, sessile or nearly so.

Terminal leaflet more than 2 inches wide....JUGLANS REGIA, p. 157

Terminal leaflet less than 1½ inches wide

SCHINUS TEREBINTHIFOLIUS, p. 267

Leaflets more than 11, with a disagreeable odor when crushed and with 1 to 4 small blunt gland-bearing points or lobes near base; leaves 1 to 3 feet long .....................................................AILANTHUS, p. 264

Leaflets definitely toothed or lobed.

Leaflets coarsely and irregularly toothed, often lobed.

Leaflets divided almost to the midvein into acutely pointed lanceolate lobes........................................................GREVILLEA, p. 198

Leaflets coarsely toothed or lobed..................KOELREUTERIA, p. 283

Leaflets regularly and finely toothed.

Leaflets less than ½-inch wide; fruit a globular red drupe, in clusters........................................................................SCHINUS, p. 266

Leaflets more than ½-inch wide; fruit a nut or winged nutlet.

Leaflets rarely more than 11; husks breaking from the nut into 4 sections................................................................HICORIA, p. 154

Leaflets usually 13 to 27.

Rachis not winged................................................JUGLANS, p. 156

Rachis winged................................................PTEROCARYA, p. 159

(b) *Leaflets typically less than 2 inches long*

Leaflets entire.

Leaflets stalked.

Leaflets less than ¼-inch long..............................PARKINSONIA, p. 255

Leaflets more than ¼-inch long.

Leaflets broadly elliptical, often tipped at apex with an abrupt short point; plants usually thorny....................ROBINIA, p. 259

Leaflets ovate or ovate-lanceolate, tapering at apex; plants not thorny................................................................SOPHORA, p. 261

Leaflets sessile.

Leaflets linear, prickly-pointed............HAKEA SUAVEOLENS, p. 200

Leaflets elliptic or linear-lanceolate, not prickly-pointed

SCHINUS, p. 266

Leaflets toothed, almost sessile.

Leaflets less than ½-inch wide, alike on both surfaces

SCHINUS MOLLE, p. 267

Leaflets ½-inch or more wide, paler beneath..............SORBUS, p. 233

## A. LEAVES NOT SCALE-LIKE

1. BRANCHLETS REDUCED TO SLENDER SPINES 1 TO 3 INCHES LONG......................PAROSELA, p. 256

2. BRANCHLETS NOT REDUCED TO SPINES

a. *Leaves in clusters, fascicles or whorled or some single and spirally arranged on the young leading shoots*

Leaves whorled, short-linear, ½-inch or less long

ACACIA VERTICILLATA, p. 237

Leaves in clusters or fascicles, more than ½-inch long.

Leaves needle-like, usually 2 to 8 inches long (1 to 2 inches long in a few species of pines).

Leaves 2 to 5 (rarely 1) in a cluster surrounded at base by a sheath..............................................................................Pinus, p. 82

Leaves 15 to 35 in terminal clusters not surrounded at base by a sheath, or in whorls in the axils of the branchlets

Sciadopitys, p. 58

Leaves narrow-linear or short and needle-like, usually 1 to 2 inches long, many in clusters on short lateral branchlets, or some single on leading shoots.

Leaves angled, evergreen; cones more than 2 inches long; cone-scales deciduous.........................................................Cedrus, p. 74

Leaves flat, deciduous; cones less than 2 inches long; cone-scales persistent.........................................................................Larix, p. 76

b. *Leaves all single, alternate or opposite or some whorled*

(1) Flowers with perianth segments; fruit a capsule or pod, never a cone or drupe (Angiosperms)

*Leaves less than ¼-inch long, awl-shaped or almost scale-like*

Tamarix, p. 294

*Leaves ¼-inch or more long.*

Leaves ¼-inch to 1 inch long, very short-linear, ovate, elliptic, or short-lanceolate.

Leaves ¼-inch or more wide.

Main vein eccentric; leaves somewhat sickle-shaped; flowers yellow.........................................................................Acacia, p. 234

Main vein central; leaves oblanceolate; flowers white

Leptospermum, p. 319

Leaves less than ¼-inch wide; fruit a 2-valved pod....Acacia, p. 234

Leaves 1 inch or more long.

Seeds embedded in a sticky substance; branches drooping

Pittosporum phillyraeoides, p. 211

Seeds not embedded in a sticky substance.

Leaves with a recurved or hooked apex, entire, 1 to 4 inches long; flowers yellow, in heads........Acacia calamifolia, p. 239

Leaves not hooked at apex.

Leaves serrate, 1 to 1½ inches long.................Maytenus, p. 269

Leaves mostly entire, 2 to 5 inches long.

Fruit a small woody capsule opening at summit

Eucalyptus amygdalina, p. 308

Fruit a long cylindrical pod.............................Chilopsis, p. 347

(2) FLOWERS WITHOUT PERIANTH SEGMENTS; FRUIT A WOODY CONE,
OR BERRY- OR DRUPE-LIKE (GYMNOSPERMS)

(a) *Leaves usually more than ¼-inch wide*

Leaves distinctly rigid and prickly-pointed.
Leaves with 2 light-colored bands beneath, the margins minutely
serrulate ...............................................................CUNNINGHAMIA, p. 58
Leaves without light-colored bands beneath, the margins entire
ARAUCARIA, p. 55
Leaves usually not rigid, commonly leathery, lanceolate, alternate,
with a distinct midrib.................................................PODOCARPUS, p. 51

(b) *Leaves usually less than ¼-inch wide*

*Leaves awl-shaped, ¼- (or rarely ½-) inch or less long,
thickly covering the branchlets.

Leaves with free part less than ¼-inch long, their bases adherent to
the stem.........................................................SEQUOIA GIGANTEA, p. 61
Leaves spreading, free from the stems for at least ¼-inch.
Adult branchlets rope-like, that is, cylindric; leaves decidedly curving upward...............................................ARAUCARIA EXCELSA, p. 57
Adult branchlets not rope-like; leaves nearly straight, extending
from the branches at about a 45° angle.
Leaves 3 or 2 at a node, sharp to the touch, usually with conspicuous white bands above; fruit a fleshy berry-like cone
JUNIPERUS, p. 112
Leaves 1 at a node, spirally arranged, without conspicuous white
bands above; fruit a globose woody cone with toothed cone-
scales................................................................CRYPTOMERIA, p. 57

**Leaves linear or narrowly lanceolate to oblong-linear,
½- (or rarely ¼-) inch to 2 inches long.

*Leaves* when removed leaving circular scars on the branches.

Scars smooth, not raised above the bark; cones erect, the scales de-
ciduous...............................................................................ABIES, p. 63
Scars somewhat raised from the bark, giving the branchlets a slightly
roughened character; cones pendulous, the scales persistent;
bracts protruding...................................................PSEUDOTSUGA, p. 99

*Leaves* when removed not leaving circular scars.

Leaves stiff and distinctly prickly-pointed, with 2 whitish bands or
lines on one surface.
Leaves flat.
Leaves lanceolate, ³⁄₁₆-inch wide, gradually tapering to apex, minutely serrulate on the margins...............CUNNINGHAMIA, p. 58

Leaves linear-oblong, ⅛-inch wide, entire on the margins
TORREYA, p. 54

Leaves usually 4-angled, leaving distinct brown pegs on the branches when falling......................................................................................PICEA, p. 78

Leaves more flexible, often pointed at apex, but not distinctly prickly to the touch.

    Leaves very narrow-linear, ½- (or ⅜-) inch to 1¼ inches long, sessile, not forming flat sprays; foliage sometimes soft to the touch; bracts not protruding........................................CRYPTOMERIA, p. 57

    Leaves linear, twisted at base, forming somewhat flattened sprays.

    *Leaves acutely pointed at apex, attached to the branchlets usually by long bases.*

        Leaves with 2 whitish or light-colored bands beneath or at least not alike on both surfaces.

        Branchlets whorled or opposite; leaves usually more than 1 inch long, abruptly rounded at base, attached to the decurrent part of base by a very short petiole..........CEPHALOTAXUS, p. 50

        Branchlets alternate, seldom apparently whorled; leaves usually 1 inch or less long.

        Leaves with distinct short petioles continuous with the decurrent bases, falling singly......................................TAXUS, p. 52

        Leaves sessile, falling with the branchlets
SEQUOIA SEMPERVIRENS, p. 61

        Leaves alike on both surfaces.................................PODOCARPUS, p. 51

    *Leaves abruptly pointed or rounded at apex.*

        Leaves with distinct short petioles.

        Petioles attached by decurrent bases..........................TAXUS, p. 52

        Petioles attached by very short persistent stalk-like processes raised from the stem, the denuded branchlets roughened.

        Leaves ¾-inch to 1¾ inches long; bracts longer than the scales.......................................................PSEUDOTSUGA, p. 99

        Leaves ¼- to ¾- (rarely 1) inch long; bracts shorter than the scales....................................................................TSUGA, p. 100

        Leaves sessile, ¾-inch or less long, soft to the touch, falling with the branchlets......................................................TAXODIUM, p. 62

## A. LEAVES SCALE-LIKE, ALMOST COMPLETELY COVERING THE BRANCHLETS

### 1. BRANCHLETS FORMING DECIDEDLY FLAT SPRAYS

Leaves apparently 4 at a node; internodes longer than broad
LIBOCEDRUS, p. 117

Leaves 2 at a node, seldom apparently 4 at a node; internodes of
shorter branchlets about as long as broad.

Cones oblong or ovate; cone-scales overlapping; seeds 2 under each
fertile scale; branchlets $1/12$- to $1/8$-inch broad............ THUJA, p. 118

Cones globular; cone-scales toadstool-shaped; branchlets less than
$1/12$-inch broad.

Cones developing within 1 year, usually less than $1/2$-inch in diameter; seeds 2 or rarely as many as 5 to a scale.

CHAMAECYPARIS, p. 103

Cones maturing the second year; seeds many at the base of each
scale.....................................................................CUPRESSUS, p. 106

2. BRANCHLETS ROPE- OR CORD-LIKE, NOT FORMING FLAT SPRAYS

Leaves more awl-shaped than scale-like, always spirally arranged,
$1/4$- to $1/2$-inch long.

Leaves with only their tips free, their bases adherent to the stems
SEQUOIA GIGANTEA, p. 61

Leaves spreading and almost entirely free from the stems throughout their entire length ...........................ARAUCARIA EXCELSA, p. 57

Leaves distinctly scale-like, $1/8$-inch or less long, completely covering
the branchlets.

a. Leaves 3 at a node.

Leaves light green, usually with glands on their backs; cone-scales
fleshy, forming small berry-like fruits; plants usually dioe-
cious........................................................................JUNIPERUS, p. 112

Leaves dark green, without glands on their backs; cones woody,
persistent on the branches for many years; plants monoe-
cious........................................................................CALLITRIS, p. 103

b. Leaves 2 at a node.

Plants monoecious; cone-scales woody...................CUPRESSUS, p. 106
Plants usually dioecious; cone-scales fleshy.........JUNIPERUS, p. 112

c. Leaves 1 at a node..........................................................TAMARIX, p. 294

## A. LEAVES OPPOSITE OR SOME WHORLED

### 1. LEAVES MORE THAN $2\frac{1}{2}$ INCHES WIDE

Leaves palmately veined and lobed, the lobes toothed; fruit a 2-
winged samara........................................................................ACER, p. 269

Leaves pinnately veined, 5 to 12 inches long, the margins entire; fruit
a capsule.

Leaves broadly heart-shaped, finely pubescent above and beneath;
flowers pale blue or violet; fruit an ovoid capsule 1 to 2 inches
long........................................................................PAULOWNIA, p. 343

Leaves broadly ovate, rarely heart-shaped, sometimes finely hairy beneath; flowers white, pink, or yellowish; fruit a slender cylindrical capsule, 6 to 18 inches long..............................CATALPA, p. 344

### 2. LEAVES LESS THAN 2½ INCHES WIDE

#### a. *Leaves coarsely toothed or lobed or some almost entire*

Leaves ¾-inch or less wide.......................................LYONOTHAMNUS, p. 225
Leaves more than ¾-inch wide.

Leaves palmately veined; fruit a 2-winged samara.............ACER, p. 269
Leaves pinnately veined; fruit a leathery drupe......MALADAMIA, p. 201

#### b. *Leaves entire, not lobed*

### (1) LEAVES GRAY, SOMETIMES PUBESCENT OR TOMENTOSE BENEATH

*Leaves 1½ inches or more wide.*

Leaves thin, with lateral veins curving upward from the midrib toward the apex; petioles ½-inch to 1 inch long......CORNUS, p. 328

Leaves thick, with lateral veins nearly parallel; petioles about ¼-inch long.......................................................PSIDIUM GUAJAVA, p. 325

*Leaves less than 1½ inches wide.*

Leaves elliptic or oblong to lanceolate, 1 to 3 inches long; flowers in clusters.

Leaves ⅜- to ⅝-inch wide; fruit a drupe, ½-inch to 1 inch long
OLEA, p. 341

Leaves ½-inch to 1½ inches wide; fruit a capsule
METROSIDEROS TOMENTOSA, p. 325

Leaves oval-oblong to elliptic, prominently veined beneath; flowers solitary; fruit an ellipsoidal or spherical berry, 1 to 3 inches long, crowned by the persistent calyx-lobes.......................FEIJOA, p. 318

### (2) LEAVES NOT GRAY, NOT PUBESCENT OR TOMENTOSE BENEATH

Leaves 3-veined from the base, 1 to 1½ inches long, elliptic-lanceolate..................................................:..METROSIDEROS ROBUSTA, p. 325
Leaves 1-veined from the base.

Leaves 3 to 8 inches long.

Leaves narrowly oblong-lanceolate, less than 1 inch wide, commonly in whorls of 3....................................................NERIUM, p. 342

Leaves ovate-lanceolate to broadly elliptical, more than 1 inch wide.

Leaves thick and leathery; petioles ½-inch or less long
LIGUSTRUM, p. 340

Leaves thin; petioles ½- to ¾-inch long.................SYRINGA, p. 341

Leaves 1 to 3 inches long.
Leaves thick and leathery, evergreen.
*Branchlets angled or slightly winged.*
Branchlets and young foliage tinged with red; fruit a berry
EUGENIA, p. 317
Branchlets and young foliage not tinged with red; fruit a capsule..............................................................LAGERSTROEMIA, p. 296
*Branchlets not angled or winged*............PSIDIUM CATTLEIANUM, p. 326
Leaves thinner, deciduous.............................................PUNICA, p. 297

## A. LEAF-MARGINS ENTIRE OR WITH 1 OR 2 COARSE TEETH OR SMALL LOBES

1. LEAVES WITH REVOLUTE MARGINS, 1 INCH OR LESS LONG; FRUIT A PLUMOSE-TAILED ACHENE............CERCOCARPUS LEDIFOLIUS, p. 219

2. LEAVES WITHOUT REVOLUTE MARGINS

a. *Leaf-blades 3 or more times longer than wide*

Leaves with translucent dots..........................................MYOPORUM, p. 348
Leaves without translucent dots.
Leaves densely covered with silvery silky hairs on both surfaces, sessile.................................................................LEUCADENDRON, p. 200
Leaves without silvery silky hairs on both surfaces.
Fruit a woody capsule opening at apex.
Bark not cinnamon-brown; leaves commonly lanceolate
EUCALYPTUS, p. 300
Bark cinnamon-brown; leaves broadly elliptical..TRISTANIA, p. 326
Fruit not a woody capsule opening at apex.
Foliage when crushed emitting a distinct odor of bay rum
UMBELLULARIA, p. 209
Foliage without odor of bay rum.
Leaves rusty or olive-gray beneath.
Leaves ½-inch to 1½ inches wide.
Fruit a prickly bur................................CASTANOPSIS, p. 169
Fruit a small capsule............................................SALIX, p. 147
Leaves 2 to 3½ inches wide; fruit a cone of follicles
MAGNOLIA GRANDIFLORA, p. 205
Leaves green beneath.
Leaves with one distinct midrib and never with glands on the margins near base.
Leaves 2 inches or less wide.
Leaves more than 4 inches long......................SALIX, p. 147

Leaves usually 4 inches or less long.

Fruit a capsule; the seeds surrounded by a sticky substance..................................PITTOSPORUM, p. 210

Fruit a small berry.....................................LAURUS, p. 207

Leaves more than 2 inches wide; fruit a collection of follicles on an axis.................................MAGNOLIA, p. 202

Leaves with 2 or more main veins from base, or if one then with a distinct gland on the margin near base.

Leaves with a large gland on the margin near base; flowers yellow, in small heads or spikes................ACACIA, p. 234

Leaves without a gland on the margin; axillary buds large, ¼-inch to 1 inch long; flowers crimson, in large globular heads......................................HAKEA LAURINA, p. 199

◯b. *Leaf-blades usually less than 3 times longer than wide*

(1) PLANTS DISTINCTLY THORNY OR SPINY

Plants nearly leafless, ashy gray; branchlets reduced to slender spines.................................................................PAROSELA, p. 256

Plants leafy; foliage dark green; spines in the axils of the leaves
MACLURA, p. 196

◖(2) PLANTS NOT DISTINCTLY THORNY OR SPINY

(a) *Leaves distinctly pubescent, rusty, gray-tomentose, or scurfy beneath*

Leaves more than 1½ inches wide.

Leaves thick, leathery, some with rusty tomentum beneath, 4 to 10 inches long; flowers large, white..MAGNOLIA GRANDIFLORA, p. 205

Leaves thinner, pubescent beneath.

Leaves 2 to 4 inches long.................................................CYDONIA, p. 224

Leaves 3 to 7 inches long.................................DIOSPYROS KAKI, p. 334

Leaves 1½ inches or less wide.

Fruit an acorn; some leaves variously toothed
QUERCUS CHRYSOLEPIS, p. 177

Fruit a leathery capsule; seeds embedded in a sticky substance
PITTOSPORUM, p. 210

Fruit a rusty or gray-tomentose capsule about 1 inch long; stamens numerous, united into a tube................................LAGUNARIA, p. 290

Fruit a legume.....................................ACACIA PODALYRIAEFOLIA, p. 238

◖b) *Leaves glabrous on both sides (or seldom slightly pubescent beneath in* Nyssa)

☉*Leaves round, heart-shaped at base; fruit a legume....CERCIS, p. 249

**Leaves not round.

†Leaves when crushed emitting odor of camphor..CINNAMOMUM, p. 206

††Leaves without odor of camphor.

Leaves usually 1½ inches or less long.

Leaves subsessile.

Leaves ½-inch or less wide..............................Leptospermum, p. 319

Leaves ¾-inch or more wide...............Lagerstroemia indica, p. 297

Leaves distinctly petiolate.

Leaves varying from toothed to entire on the same plant; fruit an acorn........................................................Quercus wislizeni, p. 174

Leaves entire; fruit a leathery capsule
Pittosporum tenuifolium, p. 212

Leaves all more than 1½ inches long.

Leaves distinctly glaucous or pale beneath.

Bark smooth and cinnamon-brown; fruit a red berry
Arbutus menziesi, p. 332

Bark not cinnamon-brown.

Leaves thick and leathery, evergreen; fruit a woody capsule, opening at summit........................................Eucalyptus, p. 300

Leaves thin but firm, deciduous; fruit a juicy berry
Diospyros virginiana, p. 334

Leaves not glaucous beneath.

*Petioles ¾-inch or more long* (some only about ½-inch long in *Persea, Magnolia,* and *Tristania*).

Petioles with milky juice; leaves thick and leathery, very lustrous above...................................................................Ficus, p.193

Petioles without milky juice.

Leaves usually less than 3 inches long, often with 1 or 2 coarse teeth or lobes.........................Sterculia diversifolia, p. 293

Leaves mostly more than 3 inches long.

Fruit a woody capsule.

Bark cinnamon-brown; leaves often crowded at the ends of the branches.......................................Tristania, p. 326

Bark not cinnamon-brown; leaves scattered along the branches...........................................Eucalyptus, p. 300

Fruit not a woody capsule.

Leaves deciduous; fruit a cluster of follicles; flowers large and showy...............................................Magnolia, p. 202

Leaves evergreen; fruit large and fleshy; flowers small
Persea, p. 207

*Petioles mostly less than ¾-inch long.*

Leaves unequal at base, often rough to the touch....CELTIS, p. 186

Leaves equal at base.

Petioles winged or margined, or jointed near summit.

Fruit 1½ inches or more long..............................CITRUS, p. 262

Fruit less than 1½ inches long...................FORTUNELLA, p. 263

Petioles not winged or jointed.

Leaves entire and toothed on the same plant; fruit a drupe.................................................................PRUNUS, p. 228

Leaves entire.

Flowers solitary, more than 2 inches broad; fruit a collection of follicles on an elongated axis..MAGNOLIA, p. 202

Flowers in clusters.

a. Fruit a capsule.

Capsules 1 inch or more long
LAGERSTROEMIA SPECIOSA, p. 297

Capsules less than 1 inch long.

Seeds not winged.

Seeds more than ⅟₁₆-inch long, embedded in a sticky substance................PITTOSPORUM, p. 210

Seeds very minute, not embedded in a sticky substance.............................EUCALYPTUS, p. 300

Seeds winged, not embedded in a sticky substance................................HYMENOSPORUM, p. 215

b. Fruit a drupe, ¼- to ½-inch long.

Leaves deciduous..........................................NYSSA, p. 330

Leaves evergreen......................................PRUNUS, p. 228

c. Fruit a berry.

Berry ¾-inch to 3 inches long; leaves 3 to 7 inches long....................................................DIOSPYROS, p. 333

Berry ¼- to ½-inch long; leaves 2 to 4½ inches long......................................................LAURUS, p. 207

d. Fruit a triangular nut enclosed in a bur..FAGUS, p. 170

## A. LEAF-MARGINS TOOTHED OR LOBED

### 1. LEAVES DISTINCTLY LOBED

a. *Leaves more than 1 foot wide*..TETRAPANAX, p. 328

b. *Leaves less than 1 foot wide*

#### (1) LEAVES NOTCHED OR TRUNCATE AT APEX

Leaves parallel-veined, fan-shaped, usually clustered on short lateral spurs.................................................................................. GINKGO, p. 49

Leaves pinnately veined and lobed, not fan-shaped.
　Leaf-lobes 2 or 4, acute; petioles 2 to 4 inches long
　　　　　　　　　　　　　　　　　LIRIODENDRON, p. 201
　Leaf-lobes 2, rounded; petioles about 1 inch long........BAUHINIA, p. 246

(2) LEAVES NOT NOTCHED OR TRUNCATE AT APEX

(a) *Leaves palmately 3- to 7-veined and lobed*

Leaves rough to the touch.
　Leaves 3 inches or less long, covered beneath with a dense gray, white, or rusty felt; flowers large, yellow..........FREMONTIA, p. 291
　Leaves 3 inches or more long, not covered beneath with a dense felt.
　　Leaves 8 to 12 inches long, the blades deeply 3- or 5-lobed
　　　　　　　　　　　　　　　　　FICUS CARICA, p. 194
　　Leaves 3 to 6 inches long, irregularly lobed or some only deeply serrate........................................................................MORUS, p. 196
Leaves not rough to the touch.
　Leaves white-downy beneath................................POPULUS ALBA, p. 143
　Leaves not white-downy beneath.
　　Leaf-lobes entire.
　　　Leaf-lobes 2 or 3, rounded at apex........................SASSAFRAS, p. 208
　　　Leaf-lobes 3 or 5, acute or acuminate at apex
　　　　　　　　　　　　　　　STERCULIA PLATANIFOLIA, p. 293
　　Leaf-lobes serrate to irregularly and coarsely toothed or rarely entire.
　　　Leaf-lobes regularly glandular-serrate; bark rough and furrowed..............................................................LIQUIDAMBAR, p. 215
　　　Leaf-lobes coarsely and irregularly toothed or nearly entire; bark smooth, grayish or whitish mottled..PLATANUS, p. 216

(b) *Leaves pinnately veined and lobed*

Plants thorny....................................................................CRATAEGUS, p. 220
Plants not thorny.
　Petioles less than 1½ inches long.
　　Leaf-lobes linear, terete; fruit a capsule.HAKEA SUAVEOLENS, p. 200
　　Leaf-lobes not linear or terete; fruit an acorn............QUERCUS, p. 172
　Petioles 1½ inches or more long; fruit a follicle
　　　　　　　　　　　　　　　STERCULIA ACERIFOLIA, p. 293

2. LEAVES VARIOUSLY TOOTHED, BUT NOT DISTINCTLY LOBED

a. *Leaves evergreen*

Fruit an acorn.
　Acorn-cup with slender spreading scales; leaves straight-veined and grayish or rusty-tomentose beneath................LITHOCARPUS, p. 171
　Acorn-cup with closely appressed scales........................QUERCUS, p. 172

Fruit not an acorn.

Leaves pale and usually tomentose beneath, 6 to 10 inches long, remotely toothed along upper half of margin......ERIOBOTRYA, p. 224

Leaves not tomentose beneath.

Leaves spiny or prickly-toothed or some almost entire, undulate.

Spines on margins very coarse, fewer than 20...,.........ILEX, p. 268

Spines on margins slender, more than 20

PRUNUS ILICIFOLIA, p. 229

Leaves not spiny or undulate.

Leaves less than 1 inch wide.

Leaves with translucent dots...........................MYOPORUM, p. 348

Leaves without translucent dots.

Leaves slightly unequal at base......ULMUS PARVIFOLIA, p. 188

Leaves equal at base.

Leaves less than 2 inches long...................MAYTENUS, p. 269

Leaves 2 inches or more long.

All leaves serrate; fruit a waxy berry-like nutlet

MYRICA, p. 153

Some leaves entire; fruit a drupe

PRUNUS CAROLINIANA, p. 230

Leaves 1 inch or more wide.

Leaves with translucent dots.........................MYOPORUM, p. 348

Leaves without translucent dots.

Petioles ½-inch or more long.

Leaves rhomboid, coarsely and irregularly few-toothed

PITTOSPORUM RHOMBIFOLIUM, p. 212

Leaves elliptic or oblong-ovate, regularly, sharply serrate.

Leaves 1 to 2¼ inches wide; fruit berry-like

PHOTINIA, p. 227

Leaves 1½ to 2¼ inches wide; fruit a drupe

PRUNUS LUSITANICA, p. 231

Petioles less than ½-inch long.

Leaves 2 to 3½ inches long.

Leaves broadly elliptic to round-ovate, finely glandular-serrate...............................................CAMELLIA, p. 290

Leaves elliptic to oblong-ovate, sharply serrate

ARBUTUS UNEDO, p. 332

Leaves 3 to 6 inches long; fruit a drupe

PRUNUS LAUROCERASUS, p. 230

b. *Leaves deciduous (or sometimes evergreen in* Cercocarpus)

Plants dioecious; fruit a 2- to 4-valved capsule; seeds with tufts of hairs.

Leaves 2 or more times longer than wide..............................Salix, p. 147

Leaves less than 2 times longer than wide......................Populus, p. 141

Plants with perfect or monoecious flowers.

Fruit an acorn........................................................................Quercus, p. 172

Fruit not an acorn.

Leaves unequal at base.

Petioles 1 inch or more long; fruit in clusters attached to an oblong leaf-like bract................................................Tilia, p. 286

Petioles less than 1 inch long.

Lateral veins straight and parallel.

Leaves green, smooth or rough to the touch; fruit a nutlet surrounded by a membranous wing..............Ulmus, p. 187

Leaves green or bronze-colored, never rough to the touch. Fruit a triangular nut, 1 or 2 enclosed in a prickly bur

Fagus, p. 170

Fruit a large polished nut, not triangular, 1 to 3 in a prickly bur..................................................Castanea, p. 168

Lateral veins curving; leaves often with 3 prominent veins from base.

Leaves rough to the touch.

Leaves conspicuously reticulate-veiny beneath

Celtis, p. 186

Leaves not reticulate-veiny beneath............Zelkova, p. 193

Leaves not rough to the touch.........................Zizyphus, p. 286

Leaves equal at base.

Leaf-margins coarsely toothed with ascending hooked bristles; fruit a nut, 1 to 3 in a prickly bur......................Castanea, p. 168

Leaf-margins without hooked bristles.

Leaves 3-veined from base.

Leaves more than 2 inches wide, often lobed; fruit a cluster of drupelets, blackberry-like......................Morus, p. 196

Leaves less than 2 inches wide, never lobed; fruit a drupe.

Leaves rough to the touch..............................Zelkova, p. 193

Leaves not rough to the touch....................Zizyphus, p. 286

Leaves 1-veined from base.

Branches with spines or thorns....................CRATAEGUS, p. 220
Branches without spines or thorns.

Lateral veins not straight and parallel; fruit fleshy.

Fruit a drupe.................................................PRUNUS, p. 228
Fruit a pome.

Pome apple-shaped....................................MALUS, p. 226
Pome pear-shaped....................................PYRUS, p. 233

Lateral veins straight and parallel.

Leaves pubescent on both surfaces, slightly lobed and
finely double-serrate, cordate at base; fruit a nut
enclosed within a leafy involucre......CORYLUS, p. 167
Leaves glabrous above, not cordate at base (subcordate
in some species of *Betula* and *Rhamnus*).

Flowers perfect, not in catkins.

Leaves less than 2 inches long, cuneate at base
CERCOCARPUS, p. 218
Leaves more than 2 inches long, obtuse to sub-
cordate at base............................RHAMNUS, p. 285

Flowers monoecious, one or both kinds in catkins.

Fruit a single nutlet, subtended by a 3-lobed leafy
bract.............................................CARPINUS, p. 166
Fruit of several nutlets, borne in cone-like catkins.

Cones woody, borne in racemes, the scales per-
sistent.............................................ALNUS, p. 159
Cones not woody, solitary, the scales decidu-
ous.................................................BETULA, p. 162
Fruit a triangular nut enclosed in a prickly bur
FAGUS, p. 170

# DESCRIPTIVE ACCOUNT OF THE TREES

## GYMNOSPERMAE

Plants with the seeds borne upon open scales instead of within closed ovaries

### I. CONE-BEARING TREES AND THEIR ALLIES

#### Ginkgoaceae. Ginkgo Family
##### 1. Ginkgo L. GINKGO

(The Chinese name.)

A genus with a single species.

1. **Ginkgo bíloba** L. GINKGO TREE. MAIDENHAIR TREE. Fig. 7.

A sparsely branched tree, 20 to 100 feet high. Leaves simple, alternate or in clusters of 3 to 5 on short lateral spur-like branchlets, deciduous; the blades fan-shaped, 2 to 3½ inches wide, more or less divided at the broad apex into 2 or more lobes, thick and leathery, almost parallel-veined, the veins dividing by twos; petioles 1 to 2 inches long. Flowers small, dioecious; the staminate borne in loose catkins about 1 inch long; the ovulate in pairs on long stalks, one of which usually fails to develop a seed. Fruit yellowish, drupe-like, with a fleshy outer coat and hard inner coat, ovoid, about 1 inch long.

Native in China. The Ginkgo Tree is one of the most distinctive and interesting of gymnospermous trees. Its deciduous leaves are so suggestive of the Maidenhair Fern (*Adiantum*) that the common name Maidenhair Tree is frequently and appropriately given to this tree. The wedge-shaped leaves and fleshy drupe-like fruit make it seem

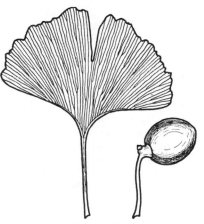

Fig. 7. Ginkgo biloba L.
Bilobed leaf, fleshy seed constituting a drupe-like gymnospermous fruit, × ½.

more like an Angiosperm than a Gymnosperm; however, the fruit is an uncovered seed, which is characteristic of the latter class. The Ginkgo Tree is the sole living representative of a gymnospermous group that flourished in Mesozoic time. It is now almost unknown in the wild state, but is extensively and generally cultivated in China, Manchuria, Korea, Japan, Europe, and the United States. It is a very desirable tree for solitary planting and as a street tree because of its picturesque effects and freedom from insects.

## Taxaceae. Yew Family

### 1. Cephalotáxus Sieb. & Zucc. PLUM-YEW

(From *kephale*, the Greek word for head, and *Taxus;* the plant resembles *Taxus,* with the staminate flowers in heads or clusters.)

Evergreen trees or shrubs with opposite or whorled branchlets. Leaves simple, spirally arranged and spreading to form 2 rows, linear, with a prominent midrib above and with 2 broad glaucous bands beneath, acutely pointed or acuminate, abruptly rounded at the base, short-petioled and decurrent on the branchlets. Flowers small, usually dioecious; the staminate in globose axillary heads of 1 to 8 flowers; the ovulate composed of 2 naked ovules borne on small bracts forming a small cone at the base of the branchlets. Fruit drupe-like, composed of a hard seed completely surrounded by a fleshy envelope.

Five or 6 species, closely allied to *Torreya,* native to eastern Asia and Japan. The plum-yews resemble the torreyas and yews in general appearance, but they differ from the torreyas in having the leaves less spiny-pointed and with the 2 whitish glaucous lines beneath broader than the 3 green lines, and from the yews in having larger leaves not yellowish green beneath. Like the yews and torreyas, the plants are male and female, and in order for fruit to set, male or pollen-bearing plants must grow near the female or fruit-bearing plants. On the Pacific Coast the plum-yews are not nearly so extensively cultivated as the yews. They are usually planted in Japanese gardens or as lawn trees.

KEY TO THE SPECIES

Leaves ¾-inch to 1½ inches long, abruptly pointed, in 2 ranks directed upward................................................................................1. *C. drupacea.*
Leaves 1½ to 3 inches long, gradually tapering to a fine point, in 2 ranks spreading outward................................................................2. *C. fortunei.*

**1. Cephalotaxus drupàcea** Sieb. & Zucc. JAPANESE PLUM-YEW. Fig. 8. A small bush-like tree, 6 to 20 feet high. Leaves ¾-inch to 1½ inches long, abruptly pointed, dark green above, paler beneath and with 2 yellowish or glaucous bands, the 2 ranks directed upward. Fruit purplish green, about 1 inch long. Native to Japan. Occasionally cultivated.

1*a.* Var. fastigiàta Pilger. SPIRAL PLUM-YEW. *C. pedunculata* var. *fastigiata* Carr.

Columnar in habit, with ascending branches and spirally arranged dark green leaves about 2 inches long.

2. **Cephalotaxus fortùnei** Hook. CHINESE PLUM-YEW. Fig. 9.

A small tree or densely branched shrub, 4 to 8 feet high. Leaves 1½ to 3 inches long, dark green, spreading horizontally. Fruit purplish, about 1 inch long. Native to China. Occasionally planted in gardens and parks.

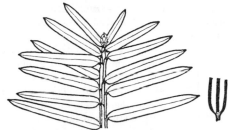

Fig. 8. Cephalotaxus drupacea Sieb. & Zucc. Branchlet, × 1. Part of leaf, lower surface, showing alternating white and green bands, × 1½.

### 2. **Podocàrpus** L'Her.

(From the Greek *podos,* foot, and *karpos,* fruit, in reference to the fleshy fruit-stalks of most species.)

Evergreen trees or shrubs. Leaves simple, alternate or rarely opposite, linear to ovate, entire, flat, sessile, or short-petioled. Flowers small, usually dioecious; the staminate yellow and catkin-like; the ovulate usually solitary, consisting of 1 or 2 scales enclosing the ovule, with several bracts at the base. Fruit drupe- or nut-like, consisting of a single seed borne at the top of a fleshy base composed of the matured fleshy bracts.

About 60 species, native in tropical and subtropical mountains of the West Indies, South America, Africa, Asia, Japan, and Australia. Several species are occasionally cultivated in parks and conservatories. **Podocarpus macrophýlla** D. Don, fig. 10, native to Japan, and **Podocarpus elongàta** L'Her., native to South Africa, are the two species most frequently seen in cultivation.

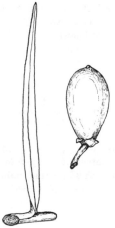

Fig. 9. Cephalotaxus fortunei Hook. Leaf, fruit, × ½.

### 3. Táxus L. Yew

(Ancient Latin name of the yew.)

Evergreen trees or shrubs. Leaves simple, spirally arranged but usually appearing 2-ranked, linear, flat, with distinct but short petioles attached to the branches by decurrent bases, acutely or abruptly pointed at the apex. Flowers small, dioecious; the staminate forming stalked globose heads, each flower composed of 4 to 8 stamens; the ovulate consisting of a single ovule surrounded at the base by several bracts. Fruit a hard seed, incompletely enveloped by a red fleshy outgrowth from its base.

About 6 closely related species, confined to the northern hemisphere, one native on the Pacific Coast. The yews are well known throughout the world because of their extensive use in landscaping and their association with early religious and social life. The dark green foliage and berry-like red fruits of the female plants make the yews very desirable evergreens for ornamental planting. In order to ensure an abundance of fruit it is necessary that a male tree grow in the vicinity of the female trees. *Taxus baccata*, the English Yew, is the best-known species in cultivation, being hardy from British Columbia to southern California.

Fig. 10.
Podocarpus
macrophylla
D. Don.
Leaf, × ½.

#### KEY TO THE SPECIES

Leaves ½- (¼-) to ¾-inch long, about ¹⁄₁₂-inch wide, acute, distinctly 2-ranked, forming flat sprays, with the midrib slightly elevated above..................................................................1. *T. brevifolia.*
Leaves ¾- (½-) inch to 1¼ inches long.

Leaves incompletely 2-ranked, not forming flat sprays, with 2 broad yellowish or pale bands beneath, abruptly pointed..2. *T. cuspidata.*
Leaves distinctly 2-ranked, forming flat sprays, with 2 indistinct pale green bands beneath, abruptly acuminate................3. *T. baccata.*

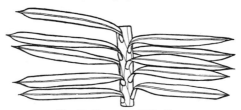

Fig. 11. Taxus brevifolia Nutt.
Branchlet, × 2.

1. **Taxus brevifòlia** Nutt. PACIFIC YEW. WESTERN YEW. Fig. 11.

A small tree, 10 to 40 (or 70) feet high, with yellowish green foliage and spreading or drooping branches. Leaves ½- (or ¼-) to ¾-inch long, thinnish, acute, sharply pointed, yellow-green and shining above, paler beneath, distinctly 2-ranked, forming flat sprays.

Native in cool canyons of the Coast Ranges from Mendocino County, California, northward to southern Alaska, south on the west slope of the Cascades and in the Sierra Nevada to Tulare County, California, east to Idaho and western Montana. Not a common tree and rarely cultivated.

2. **Taxus cuspidàta** Sieb. & Zucc. JAPANESE YEW.

A small tree, 5 to 15 (or 25) feet high, usually of shrubby habit in cultivation, with dark dull green foliage and spreading or ascending branches. Leaves ¾- (or ½-) inch to 1 inch long, thickish, abruptly pointed, dull dark green above, with a prominent midrib, paler and with 2 yellowish or grayish bands beneath, incompletely 2-ranked, the ranks directed upward and with a V-shaped space between.

Native in Japan, Man-

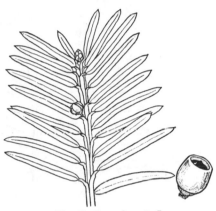

Fig. 12. Taxus baccata L.
Branchlet, seed with fleshy cup-like outgrowth, × 1.

churia, and Korea. Occasionally cultivated in parks and gardens.

3. **Taxus baccàta** L. ENGLISH YEW. Fig. 12.

A small densely branched tree, 10 to 60 feet high, with dark green foliage forming a compact crown. Leaves ¾-inch to 1¼ inches long, abruptly acuminate, dark green and usually shining above, pale beneath, with rather indistinct yellowish or pale bands, 2-ranked, forming flat sprays (except in var. *fastigiata*).

Native in Europe, western Asia, and northern Africa. Extensively planted as an ornamental tree. Many garden forms are in cultivation and the following are commonly seen in gardens and parks:

3*a*. Var. **elegantíssima** Beiss. VARIEGATED ENGLISH YEW.

An erect or spreading and compact form, with brilliantly variegated golden-yellow leaves.

3*b*. Var. **erécta** Loud. Erect English Yew. Broom Yew.

An erect and compact small tree, forming a columnar bush-like plant resembling the Irish Yew, but with smaller leaves.

3*c*. Var. **fastigiàta** Loud. Irish Yew.

A form with crowded erect branches and glossy dark green foliage. Leaves spreading from all around the branchlets. Extensively used

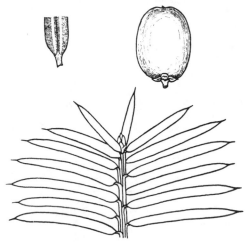

Fig. 13. Torreya californica Torr.
Branchlet, × 1. Upper left, base of leaf, lower surface, with two narrow glaucous bands, × 2. Upper right, fleshy seed constituting the fruit, × ½.

in formal planting for its columnar habit. A rare form of this variety with variegated foliage has been called **T. baccata** var. **fastigiata variegata** Carr.

### 4. **Tórreya** Arn. Torreya

(Named after John Torrey, an American botanist.)

Five closely related species, native to North America and eastern Asia, one on the Pacific Coast.

1. **Torreya califórnica** Torr. California-Nutmeg. Fig. 13.
*Tumion californicum* Greene.

An evergreen tree, 15 to 70 feet high, with whorled branches and glossy dark green foliage. Leaves simple, spirally arranged but distinctly 2-ranked by a twisting of the petioles, linear, 1 to 2½ inches long, flat, rigid, dark green above, with 2 narrow deeply impressed glaucous bands beneath, abruptly tapering to the decurrent petiole-like base, sharp-pointed at the apex. Flowers small, dioecious; the staminate in axillary clusters consisting of 6 to 8 whorls of 4 stamens each; the ovulate solitary and terminal on short lateral branchlets. Fruit drupe-like, green or purplish, ellipsoidal, 1 to 1¾ inches long, consisting of a single seed with a woody outer coat completely surrounded by a thin fleshy envelope.

A native of California, in cool canyons of the Coast Ranges from the Santa Cruz Mountains north to southern Mendocino and Lake counties, and on the western slope of the Sierra Nevada from Tulare County north to Tehama County. California-Nutmeg resembles the yew in general appearance, but the leaves are more rigid and distinctly spiny-pointed. It is only occasionally cultivated as a park tree for its glossy dark green foliage and interesting plum-like seeds. It thrives best in moist shady locations.

## Araucariaceae. Araucaria Family

### 1. Araucària Juss. ARAUCARIA

(Chilean name, from *Araucanos,* the name of the people in whose district one species grows.)

Evergreen trees with whorled branches. Leaves simple, spirally arranged and thickly clothing the branches, scale-like to short ovate-lanceolate, entire. Flowers usually dioecious, in catkins; the staminate catkins of numerous stamens in large cylindrical clusters; the ovulate consisting of numerous scales bearing single ovules. Fruit an ovoid cone, falling apart when mature.

About 12 species, native to South America, Australia, and the Pacific Islands. The following 3 species of araucarias are extensively planted in parks and gardens as specimen trees in the coastal and valley regions of middle and southern California. Only *A. imbricata* thrives well in Oregon and Washington.

#### KEY TO THE SPECIES

Leaves flat, ¾-inch to 2½ inches long, very sharp-pointed.
    Leaves contracted and twisted at base, forming flat sprays
                                          1. *A. bidwilli.*
    Leaves broad at base, well imbricated on the cylindrical branches
                                    2. *A. imbricata.*
Leaves not flat, ½-inch or less long, forming rope-like branches; juvenile
    foliage soft, spreading from the branches.................3. *A. excelsa.*

### 1. Araucaria bidwilli Hook. BUNYA-BUNYA. Fig. 14.

A large handsome tree, 40 to 80 feet high, with very dark green and glossy foliage. Leaves lance-ovate, ½-inch to 3 inches long, firm and sharp-pointed, glabrous and glossy on both surfaces, spreading in 2 rows by a twisting of the contracted bases. Cones pineapple-like, 7 to 10 inches long.

A native of Australia. Cultivated in California parks and gardens.

2. **Araucaria imbricàta** Pav. MONKEY PUZZLE. Fig. 15.

*A. araucana* K. Koch.

A peculiar and striking tree, 15 to 30 (or 100) feet high, with a

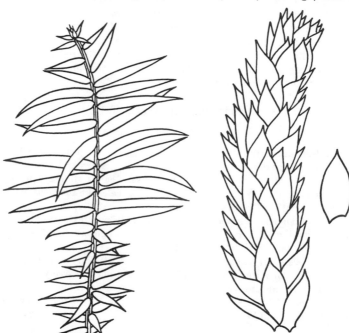

Fig. 14. Araucaria bidwilli Hook.
Branchlet, × ½.

Fig. 15. Araucaria imbricata
Pav.
Branchlet, single leaf, × ½.

straight trunk and stout spreading branches in regular whorls of 5. Leaves ovate-lanceolate, ¾-inch to 2 inches long, very stiff, sharp-pointed, overlapping like shingles and persisting on the branches and trunk. Cones ovoid or globose, 5 to 8 inches long, rarely maturing on the Pacific Coast.

Native in Chile. A very hardy species, cultivated in lawns and parks from San Diego, California, northward to Victoria and Vancouver, British Columbia.

**3. Araucaria excélsa** R. Br. NORFOLK ISLAND-PINE. Fig. 16.

A medium-sized or tall tree, 20 to 70 feet high, with horizontal or slightly upturned branches in regular whorls of 4 to 7. Leaves awl-shaped, ¼- to ½-inch long, curved, sharp-pointed, densely overlapping and clothing the horizontal or drooping branchlets. Cones subglobose, 4 to 6 inches broad, rarely maturing on the Pacific Coast. Native to Norfolk Island. The most commonly cultivated species in California lawns, parks, and gardens. A favorite pot plant.

## Taxodiaceae. Redwood Family

### 1. Cryptomèria D. Don. CRYPTOMERIA

(From the Greek *kryptos,* hidden, and *meros,* part; of obscure meaning, perhaps in reference to its hidden relationship with the cedar.)

One species and several cultivated varieties.

**1. Cryptomeria japónica** D. Don. JAPANESE CRYPTOMERIA. COMMON CRYPTOMERIA. Fig. 17.

A medium-sized or large tree, 30 to 100 feet high, with horizontally spreading branches and evergreen foliage. Leaves simple, spirally arranged, linear-subulate, ½- (or ¼-) inch to 1 inch long, the tips curved inward, compressed laterally, 3- or 4-angled, keeled above and below, decurrent at the base. Flowers small,

Fig. 16. Araucaria excelsa R. Br. Branchlet, × 1.

monoecious, in catkins; the staminate catkins in clusters terminating the branchlets, composed of numerous imbricated stamens; the ovulate solitary at the ends of the branchlets, composed of 20 to 30 scales, each scale bearing 2 to 5 ovules. Fruit a globular red-brown cone, ¾-inch to 1 inch in diameter, with wedge-shaped scales often longitudinally ridged and having 3- to 5-pointed processes at the apex, thus giving the cone a bristly appearance, maturing the first year.

Fig. 17. Cryptomeria japonica D. Don. Left, staminate cones; right, fruiting cone; × 1.

Native to Japan and doubtfully to China. Cultivated in parks and gardens. Japanese Cryptomeria, sometimes called Japanese Cedar, is the national tree of Japan, where it forms extensive forests and is abundantly planted as a street tree. It belongs to the same family of conifers as the genus *Sequoia,* and like that genus its descent can be traced to a very remote geological epoch.

1*a.* Var. élegans Mast. SPREADING CRYPTOMERIA. PLUME CRYPTOMERIA.

A small tree, 8 to 15 feet high, with soft feathery foliage forming a dense compact crown. Leaves linear, ½-inch to 1 inch long, green in summer, changing to copper-red in fall and winter.

Cultivated as a lawn and park tree.

## 2. Cunninghámia R. Br. CHINESE-FIR

(Named after J. Cunningham, who discovered one of the species.)
Two species, one native in southern and western China, and one in Formosa.

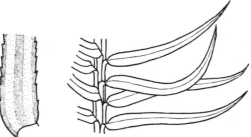

Fig. 18. Cunninghamia lanceolata Hook.
Base of leaf, enlarged to show serrate margin and the broad bands on the under surface, × 2. Branchlet with spirally arranged 2-ranked leaves showing decurrent bases, × 1.

### 1. Cunninghamia lanceolàta Hook. CHINESE-FIR. Fig. 18.

A small evergreen tree, 20 to 30 feet high, with spreading branches drooping at the ends. Leaves simple, densely and spirally arranged and 2-ranked, linear-lanceolate, 1¼ to 2½ inches long, rigid, light green and shining above, with 2 broad white bands beneath, finely serrulate, sharply pointed, with broad decurrent bases. Flowers monoecious, in catkins; the staminate catkins cylindric-oblong, in terminal clusters; the ovulate globose, in small clusters of 1 to 3 terminating the ends of the branches. Fruit a subglobose coriaceous cone, 1 to 2 inches long, with thick pointed scales.

Native in China. Occasionally cultivated in parks and conservatories. It resembles *Araucaria bidwilli* in general appearance and foliage character, but may be distinguished by its narrower minutely serrulate leaves with 2 light-colored bands beneath.

## 3. Sciadópitys Sieb. & Zucc.

(From the Greek *skias, skiados,* umbrella, and *pitys,* pine, in reference to the leaf arrangement.)

A genus of a single species.

**1. Sciadopitys verticillàta** Sieb. & Zucc. UMBRELLA-PINE. Fig. 19.
A very distinctive and handsome evergreen tree, 20 to 60 feet high, with short horizontally spreading branches forming a narrow subcylindric crown. Leaves of 2 kinds—small and scale-like leaves scattered along the shoots but crowded at the ends, and linear flat leaves borne in whorls of 8 to 25 in the axils of the scale-like leaves. The linear leaves composed actually of 2 united leaves 3 to 6 inches long, deeply furrowed at their point of union on both sides, dark green and

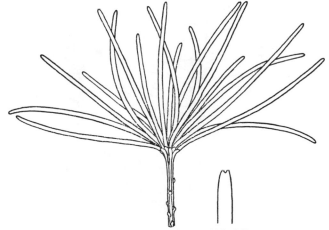

Fig. 19. Sciadopitys verticillata Sieb. & Zucc.
Branchlet, × ½. Tip of leaf, enlarged to show
terminal notch, × 2.

glossy above, with a white band beneath, shallowly notched at the apex, borne on undeveloped spurs like the fascicled leaves of pines. Flowers monoecious, in catkins; the staminate catkins in dense clusters at the ends of the branchlets; the ovulate solitary and terminal, consisting of numerous spirally arranged scales, each scale bearing 7 to 9 ovules. Fruit an oblong-ovoid cone, 2 to 4½ inches long, maturing the second year.

Native in central Japan. Occasionally cultivated in lawns and parks. Umbrella-Pine is a slow-growing tree. The specimens seen in cultivation are usually less than 25 feet high.

#### 4. Sequòia Endl. REDWOOD. SEQUOIA

(Named after Sequoyah, a Cherokee chief of Georgia who invented the Cherokee alphabet.)

Tall evergreen trees with very thick red fibrous deeply grooved bark. Leaves simple, spirally arranged, linear to short-lanceolate or awl-shaped, decurrent on the branchlets. Flowers monoecious, in catkins; the staminate catkins consisting of numerous spirally arranged stamens; the ovulate composed of many spirally arranged scales, each scale with 5 to 7 ovules. Fruit an ovoid woody cone, with divergent scales widening from a narrow wedge-shaped base to a rhom-

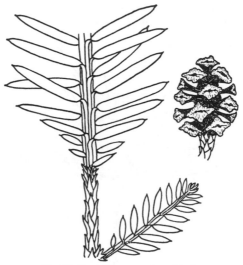

Fig. 20. Sequoia sempervirens Endl.
Portion of branchlet with leaves, cone, × 1.

boidal wrinkled disk-like apex with a depressed center. Seeds flattened, narrowly winged.

Two species, both native to California, one reaching southern Oregon. The two species of living sequoias are easily the most magnificent trees in the world, being the oldest and tallest trees known. Although now limited in their distribution, they once inhabited a great part of the northern hemisphere. Fossil remains of the two living and several extinct species have been found in Alaska, Canada, Greenland, Europe, and Asia. The Bald-Cypress of the southeastern United States is a close relative, but differs in shedding its leaves in the autumn of each year.

KEY TO THE SPECIES

Leaves linear, about ½-inch long, 2-ranked, forming flat sprays
1. *S. sempervirens.*
Leaves awl-shaped, ¼- (or ½-) inch or less long...............2. *S. gigantea.*

**1. Sequoia sémpervìrens** Endl. REDWOOD. Fig. 20.

A tall rapid-growing tree, 100 to 340 feet high, with a pyramidal or narrow crown and horizontally spreading or drooping branches. Trunk 2 to 12 feet in diameter, with reddish brown bark 4 to 12 inches thick. Leaves of 2 kinds: those of the lower branches linear, ½- to ¾-inch long, dark green above, with 2 gray bands beneath, short-petioled, in 2 ranks, forming flat sprays; those of some upper shoots short-linear to awl-shaped, ¼- to ½-inch long, spirally arranged and thickly clothing the branchlets, these then resembling those of *S. gigantea*. Cones ovoid, ½-inch to 1⅛ inches long, borne in clusters terminating the branchlets, ripening the first year.

Native to the fog belt in the Coast Ranges from northern San Luis Obispo County, California, north to southwestern Oregon. Cultivated in parks and gardens, thriving best in cool and moist districts. Redwood is a very handsome tree, growing to greater heights than any other tree in the world. It crown-sprouts and often forms circles of younger trees around the old stump.

**2. Sequoia gigantèa** Decne. BIG TREE. GIANT SEQUOIA. Fig. 21.

Fig. 21. Sequoia gigantea Decne. Branchlet, × ¾. Cone, × ½.

A tall tree, 150 to 300 feet high, with a trunk 10 to 35 feet in diameter and with a pyramidal crown, becoming rounded and broken at summit in age. Bark often 1 to 2 feet thick, deeply furrowed. Branches horizontally spreading or pendulous. Leaves short-lanceolate or awl-shaped, ⅛- to ½-inch long, thickly clothing the stems, their tips free. Cones ovoid, 2 to 3 inches long, ripening the second year.

Native on the western slope of the Sierra Nevada, occurring in 32 groves of varying extent from Placer County southward to Tulare County. Giant Sequoia is the most massive of all trees and often attains ages from 1000 to almost 3000 years. It does not crown-sprout as does *S. sempervirens*. Planted in parks and gardens.

**2a. Var. péndula** Lav. WEEPING BIG TREE.

A peculiar form with long drooping branches closely covering the trunk, producing an effect similar to that of the Italian Cypress. Not commonly cultivated. Seen in Victoria, British Columbia, and in Berkeley, Oakland, and on the grounds of the California Nursery Company at Niles, California.

## 5. Taxòdium Rich. BALD-CYPRESS

(The foliage similar to that of *Taxus*.)

Evergreen or deciduous trees with 2 kinds of branchlets, those on the lower part of the branches deciduous and without axillary buds, those near the apex of the shoots persistent and with axillary buds. Leaves simple, alternate, linear, flat, those of the lateral deciduous branchlets in 2 ranks, those of the persistent branchlets spreading radially. Flowers small, monoecious, in catkins; the staminate catkins forming terminal drooping panicles; the ovulate composed of several imbricated scales, each scale bearing 2 ovules near the base. Fruit a globose cone, maturing the first year, consisting of several spirally arranged thick leathery scales, each scale enlarged at the apex into a 4-sided disk, tapering toward the base into a short stalk. Seeds 2 to each scale, triangular and narrowly winged.

Two species, one in the southeastern United States and one in Mexico. The name Bald-Cypress was applied to these trees because of the yearly shedding of the leaves in the species *Taxodium distichum*. They are closely related to the Redwood, *Sequoia sempervirens*, which they resemble in their alternate linear leaves arranged in 2 ranks, deciduous together with the short lateral branchlets. The branchlets in bald cypresses fall at the close of the first or second year, but in the Redwood they persist for several years. The leaves of the true cypresses are opposite, scale-like, persisting for several years.

KEY TO THE SPECIES

Foliage deciduous; cones about 1 inch in diameter...........1. *T. distichum*.
Foliage evergreen; cones 1¼ to 1½ inches in diameter
                                            2. *T. mucronatum*.

### 1. Taxodium dístichum Rich. COMMON BALD-CYPRESS. Fig. 22.

A tall deciduous tree, 60 to 125 feet high, with light cinnamon-brown shallowly fissured bark, and erect or spreading branches forming a pyramidal crown while young, this becoming rounded and with descending branches at maturity. Leaves narrowly linear, ½- to ¾-inch long, soft, light yellowish green, turning brown each year before falling with the branchlets. Cones about 1 inch in diameter, the scales falling apart at maturity.

Native in swampy regions of the southeastern part of the United States. A handsome ornamental tree with light green feathery foliage, cultivated in parks and gardens.

2. **Taxodium mucronàtum** Ten. MONTEZUMA BALD-CYPRESS.

*T. distichum* var. *mucronatum* Henry.

An evergreen tree similar to *T. distichum,* except that the leaves are shorter and fall with the branchlets the second year. Cones 1¼ to 1½ inches in diameter.

A native of Mexico. Occasionally cultivated in parks and large estates. One specimen of this Bald-Cypress growing at Santa María del Tule in the State of Oaxaca, Mexico, is considered by some au-

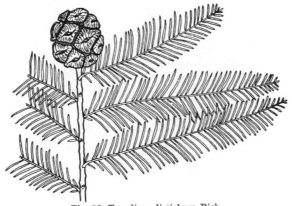

Fig. 22. Taxodium distichum Rich.
Branchlet with cone, × ½.

thorities to be the oldest living tree. It has a trunk diameter of more than 41 feet above the swollen base and is estimated to be more than 4000 years old.

## Pinaceae. Pine Family
### 1. Abies Link. FIR

(The ancient Latin name of the Silver Fir.)

Tall evergreen trees, with spreading whorled branches. Leaves simple, spirally arranged, appearing 2-ranked by a twist or bend at the bases or sometimes radiating in all directions or curving upwards, linear, flat or 4-angled, with 2 white bands beneath, sessile (short-petioled in *A. concolor*), leaving smooth circular scars when falling. Flowers monoecious, in axillary catkins; the staminate catkins pendent on the lower side of the branches, composed of numerous spirally arranged yellow or purple stamens; the ovulate erect on the upper side of the higher spreading branches, composed of many spirally arranged scales, each scale with 2 ovules at the base and subtended by a distinct bract. Fruit an erect woody cone, maturing the first year, the scales, bracts, and seeds falling from the stout straight axis which persists on the branchlet. Seeds winged.

About 35 species, mostly in the mountains of the northern hemisphere. Seven species and one variety are native to the Pacific Coast. The firs are related to the spruces (*Picea*), but are easily distinguished by their erect cones, deciduous cone-scales, and sessile leaves which

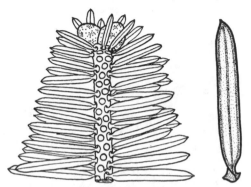

Fig. 23. Abies cephalonica Loud.
Tip of branchlet with sugary buds, upper leaves removed to show distribution of circular scars, × 1½.
Leaf, lower surface, showing white bands, × 2¼.

leave smooth circular scars when falling. Native and exotic species of firs are extensively planted as ornamentals throughout the Pacific Coast, but thrive best in the cooler districts. They are beautiful while young, and are rather short-lived.

KEY TO THE SPECIES

The descriptive leaf characters in the key and diagnoses refer to the leaves of sterile lateral branches. The leaves of cone-bearing branches and on leading shoots toward the upper part of the tree are usually thicker, shorter, more angled, upturned or ascending, and very acutely and often prickly pointed.

Leaves spreading radially in all directions.

    Leaves dark green above, with white bands beneath, ⅝-inch to 1 inch long, sharp-pointed......................................1. *A. cephalonica.*

    Leaves gray-green and with glaucous or white bands on both surfaces, short and stiff, ⅜- to ¾-inch long, acute or blunt....2. *A. pinsapo.*

Leaves not spreading in all directions, but laterally or laterally and upward, sometimes forming flat sprays.

    Leaves gray-green or glaucous and with whitish bands or lines on both surfaces.

        Winter-buds completely covered by resin; bracts of cones not exserted.

Branchlets glabrous; leaves often twisted at the short petiole-
like base, slightly convex on both surfaces......3. *A. concolor.*
Branchlets pubescent; leaves rarely twisted at base.
  Leaves slightly grooved above, flat beneath; cones 2½ to 4
    inches long......................................................4. *A. lasiocarpa.*
  Leaves ridged above, often 4-angled; cones 5 to 8 inches long
                  5. *A. magnifica.*
Winter-buds not completely covered by resin, with free acuminate
  scales at base.
  Leaves deeply grooved above, ridged below, appressed to the
    branchlets near base; bracts of cones exserted....6. *A. nobilis.*
  Leaves ridged above and below, commonly 4-angled.
    Bracts of cones not exserted................................5. *A. magnifica.*
    Bracts of cones exserted............5a. *A. magnifica* var. *shastensis.*
Leaves dark green and shining above.
  Leaves spiny-pointed.
    Leaves 1¼ to 2½ inches long, not bifid at apex; buds ½-inch to
      1 inch long, not resinous......................................7. *A. venusta.*
    Leaves 1 to 1½ inches long, bifid or notched at apex; buds
      ⅛-inch long, slightly resinous................................8. *A. firma.*
  Leaves not spiny-pointed (often sharply pointed and bifid in *A.
    firma*).
    Leaves in flat sprays, those on the upper side of the twigs shorter;
      winter-buds resinous; bracts not exserted........9. *A. grandis.*
    Leaves not in flat sprays.
      Leaves from the lateral and upper surfaces of the branchlets
        spreading outward and upward, forming a V-shaped space
        along the branchlets between their rows.
      Winter-buds very resinous; bracts of cones not exserted;
        cones 1½ to 2½ inches long.....................10. *A. balsamea.*
      Winter-buds not resinous or only slightly so.
        Branchlets short-pubescent; bracts of cones exserted.
          Leaves often sharply pointed and bifid at apex; branch-
            lets slightly grooved; buds slightly resinous
                  8. *A. firma.*
          Leaves rounded and notched at apex.......11. *A. pectinata.*
        Branchlets glabrous and shining; bracts of cones not ex-
          serted......................................................14. *A. numidica.*
      Leaves from the lateral and upper surfaces of the branchlets
        spreading outward and upward, thickly clothing the up-
        per surfaces of the branches, not forming a V-shaped
        space.
      Winter-buds resinous; bracts not exserted....12. *A. amabilis.*
      Winter-buds not resinous.
        Leaves ¾-inch to 1½ inches long; bracts exserted and
          reflexed............................................13. *A. nordmanniana.*
        Leaves ⅝- to ⅞-inch long; bracts not exserted
                  14. *A. numidica.*

## 1. **Abies cephalónica** Loud. GREEK FIR. Fig. 23.

A tree, 50 to 100 feet high. Young branchlets light brown, glabrous.
Winter-buds resinous. Leaves spreading in all directions around the

branches, ⅝-inch to 1 inch long, those of the upper ranks shorter than those beneath, sharp-pointed, dark green and shining above, with a median groove, sometimes with a few stomata near the apex, with 2 white bands of stomata beneath separated by a green ridge. Cones 4 to 6 inches long, the bracts exserted and reflexed.

Native to Greece. Commonly planted as an ornamental lawn tree.

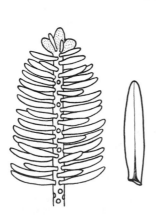

Fig. 24. Abies pinsapo Boiss.
Tip of branchlet showing heavily sugared buds, × 1. Leaf enlarged to show shape, × 2.

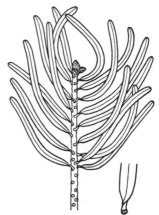

Fig. 25. Abies concolor Lindl. & Gord.
Branchlet showing sugary buds at tip, × ¾. Part of single leaf enlarged to show twist at base, × 1½.

2. **Abies pinsàpo** Boiss. SPANISH FIR. Fig. 24.

A medium-sized tree, 20 to 40 feet high. Young branchlets brown, glabrous. Winter-buds resinous. Leaves spreading in all directions around the branches, short and stiff, ⅜- to ¾-inch long, acute or blunt and not notched at the apex, dark or dull gray-green and stomatiferous and slightly convex above, with 2 white bands of stomata beneath. Cones 4 to 5 inches long, the bracts hidden by the scales.

Native to southern Spain. This species and a glaucous-leaf variety (var. **glauca** Beiss.) are planted as lawn trees. They are easily distinguished from all other firs by the short stiff blunt-pointed leaves spreading uniformly in all directions around the branches.

3. **Abies cóncolor** Lindl. & Gord. W.HITE FIR. Fig. 25.

A forest tree, 60 to 200 feet high. Young branchlets yellowish green, glabrous. Winter-buds completely covered by resin. Leaves irregularly spreading, but mostly spreading outward and curving upward by a bend and twist at the short petiole-like base, ¾-inch to 2½ inches long, short-pointed or rounded or slightly notched at the apex, flattened, dull green or bluish green and glaucous on both surfaces, rarely grooved above, slightly convex and with 2 narrow whitish bands of stomata beneath separated by a greenish band. Cones 2 to 5½ inches long, the bracts hidden by the scales.

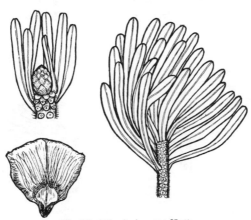

Fig. 26. Abies lasiocarpa Nutt.
Upper left, winter-bud, × 2. Lower left, cone-scale
and bract, × 1. Branchlet, × 1.

Native on mountain slopes in western Oregon, California, Utah, southern Colorado, Arizona, New Mexico, and in the San Pedro Mártir Mountains of Lower California. Sold as a Christmas tree under the erroneous name of Silver Fir or Silvertip Fir. Commonly planted as an ornamental tree in gardens and parks.

4. **Abies lasiocárpa** Nutt. ALPINE FIR. ROCKY MOUNTAIN FIR. Fig. 26.

A tree, 60 to 90 feet high. Young branchlets ashy gray, with short brownish pubescence. Winter-buds small, resinous. Leaves much crowded, irregularly spreading but usually curving upward, those in the median line above closely covering the branches, ¾-inch to 1½ inches long, rounded or acute or slightly notched at the apex, flat, pale bluish green, with whitish stomatal bands on both surfaces, slightly grooved above. Cones 2 to 4 inches long, the bracts hidden by the scales.

Native in the high mountainous regions of western North America from Alaska southward to Oregon and northern California, Idaho, Wyoming, Montana, Utah, Colorado, Arizona, and northern New Mexico. Occasionally cultivated, but usually much stunted and not desirable.

### 5. Abies magnífica Murr. RED FIR. Fig. 27.

A large forest tree, 60 to 200 feet high, with deeply and roughly fissured dark red bark on old trees, and a narrow crown composed of numerous horizontal tiers of short branches. Young branchlets minutely rusty-pubescent. Winter-buds resinous, at least at the tip. Leaves on the lower side of the branches spreading horizontally right and left, those on the upper side curving upward and their bases bent against the branches, not twisted, ¾-inch to 1½ inches long, acute

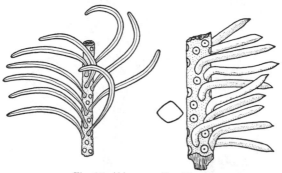

Fig. 27. Abies magnifica Murr.
Left to right: Vegetative branchlet, × 1. Cross section of leaf from fruiting branchlet, × 3½. Fruiting branchlet, × 1.

or rounded at the apex, not notched, somewhat 4-sided, grayish or glaucous-green, with whitish stomatal bands or lines on all surfaces. Cones 5 to 8 inches long, the bracts hidden by the scales.

Native in the Cascade Mountains of southern Oregon, on the western slope of the Sierra Nevada, and in the North Coast Ranges from northern Lake County to Siskiyou County, California, from 5000 to 9000 feet elevation. A very ornamental tree, but not common in cultivation. Easily distinguished by its glaucous foliage, 4-angled leaves, and deeply furrowed dark red bark.

### 5a. Var. shasténsis Lemm. SHASTA FIR.

Winter-buds less resinous. Bracts exserted and reflexed.
Occasionally found throughout the range of the species.

### 6. Abies nóbilis Lindl. NOBLE FIR. Fig. 28.

A tall forest tree, 60 to 225 feet high, with deeply fissured reddish brown bark. Young branchlets minutely rusty-pubescent. Winter-buds resinous at the tip, the outer and basal scales elongated and free. Leaves closely set on the branches, those on the lower side of the branches spreading horizontally right and left, those on the upper surface curving upward, their bases pressed against the branch for a short distance before curving, ¾-inch to 1¼ inches long, rounded at the apex, flattened, more or less glaucous-green or grayish, with stomatal bands or lines on both surfaces, deeply grooved above, ridged beneath. Cones 6 to 10 inches long, the bracts exserted and reflexed.

Native in the Cascade Mountains of Washington and Oregon, where this fir forms extensive forests, and in the Olympic Mountains of Washington. Occasionally cultivated on the Pacific Coast, but extensively planted in the British Isles and the eastern United States.

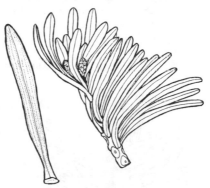

Fig. 28. Abies nobilis Lindl.
Upper surface of leaf showing stomatal bands, × 3. Branchlet, × 1½.

7. **Abies venústa** Koch. SANTA LUCIA FIR. BRISTLECONE FIR. Fig. 29.

A tree, 30 to 100 feet high, with pendulous lower branches almost reaching the ground and short spreading or ascending upper branches

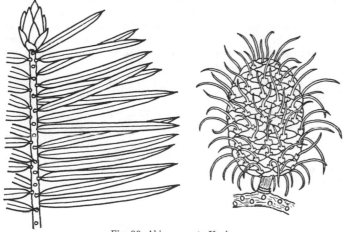

Fig. 29. Abies venusta Koch.
Branchlet, × 1. Cone, × ½.

forming a narrow pyramidal crown extending into a slender spire. Young branchlets greenish, glabrous. Winter-buds ½-inch to 1 inch long, not resinous. Leaves mostly 2-ranked, 1¼ to 2¼ inches long, rigid, spiny-pointed, dark green, glabrous, nearly flat above, not grooved, with 2 broad white stomatal bands and a prominent midrib beneath. Cones 2½ to 4 inches long, the bracts exserted and ending in bristly spines ½-inch to 2 inches long which give the cone a very bristly appearance.

Native in the Santa Lucia Mountains of Monterey County, California. Sometimes cultivated in parks and gardens for its handsome dark green foliage.

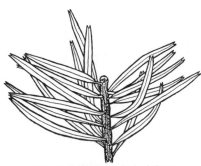
Fig. 30. Abies firma Sieb. & Zucc.
Branchlet, × 1.

8. **Abies firma** Sieb. & Zucc. MOMI FIR. Fig. 30.

A tree, 60 to 150 feet high, with a broad pyramidal crown in its native habitat, usually much reduced in cultivation. Young branches grayish brown, slightly grooved, pubescent in the grooves. Winter-buds small, slightly resinous. Leaves on the under side spreading horizontally to right and left, those on the lateral and upper surfaces spreading outward to form an open V-shaped space above between the rows of leaves, 1 to 1½ inches long, sharp-pointed and bifid at the apex on young plants, obtuse and slightly notched on older plants, broadest about the middle, dark green and shining above, with 2 grayish stomatal bands beneath. Cones 4 to 5 inches long, the bracts exserted but not reflexed.

Native to Japan. Occasionally cultivated as a specimen tree in gardens and parks.

9. **Abies grándis** Lindl. LOWLAND FIR. GIANT FIR. Fig. 31.

A forest tree, 60 to 200 feet high, remaining much smaller in cultivation. Young branches olive-green, pubescent. Winter-buds ovoid, resinous. Leaves usually forming flat sprays, ¾-inch to 2 inches long, those from the upper side usually the shorter, flat, flexible, rounded and notched at the apex, dark green, shining, and grooved above, with 2 white bands of stomata beneath. Cones 2 to 4 inches long, the bracts hidden by the scales.

Native in the coastal mountains from southern British Columbia and Vancouver Island south through Washington and Oregon to

Sonoma County, California, eastward to the Blue Mountains of Oregon and to western Montana. Frequently cultivated for its glossy green foliage.

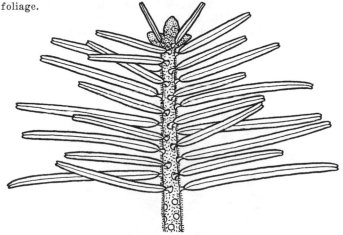

Fig. 31. Abies grandis Lindl.
Tip of branchlet, × 1½.

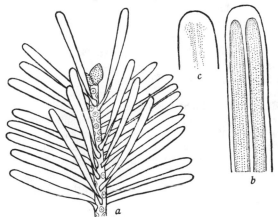

Fig. 32. Abies balsamea Mill.
*a.* Branchlet, × 1½. *b.* Leaf, under surface, × 6. *c.* Apex of
leaf, upper surface, × 6.

10. **Abies balsàmea** Mill. Balsam Fir. Fig. 32.

A tree, 40 to 60 feet high. Young branchlets grayish, short-pubescent. Winter-buds very resinous, reddish. Leaves from the lower side of the branchlets spreading horizontally right and left, those from the lateral and upper surfaces spreading outward and upward forming a V-shaped space between the rows of leaves, ⅝-inch to 1¼ inches long, rounded and slightly notched at the apex, dark green and shining above, often with a few whitish stomata near the apex, with 2 whitish narrow bands of stomata beneath. Cones 1½ to 3 inches long, the bracts usually hidden by the scales.

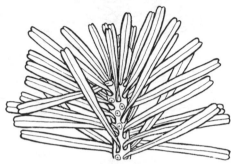

Fig. 33. Abies pectinata DC.
Branchlet, × 2.

Native from Labrador to West Virginia, westward to northern Minnesota and northeastern Iowa. In cultivation it is dwarfed and not very desirable.

11. **Abies pectinàta** DC. Silver Fir. Fig. 33.

*A. alba* Mill.

A medium-sized tree, 30 to 60 feet high. Young branchlets grayish, short-pubescent. Winter-buds not resinous, small. Leaves from the lower side of the branchlets spreading horizontally right and left, those from the lateral and upper surfaces spreading upward and out-ward forming a V-shaped space above between the rows of leaves, ⅝-inch to 1¼ inches long, rounded and notched at the apex, dark green and shining above, with 2 white stomatal bands beneath. Cones 4 to 5 inches long, the bracts exserted and reflexed.

Native in the mountains of central and southern Europe. Occasion-ally cultivated, but not very thrifty.

12. **Abies amábilis** Forbes. Cascade Fir. Lovely Fir. Fig. 34.

A large forest tree, 60 to 225 feet high. Young branchlets gray, stout, densely pubescent. Winter-buds globose, very resinous. Leaves from the lower side of the branchlets spreading horizontally right and left, those from the lateral and upper surfaces spreading out-ward and upward thickly clothing the branchlets, flat, ¾-inch to 1¼ inches long, blunt or notched at the apex, shining dark green and grooved above, with 2 broad whitish stomatal bands beneath. Cones 3½ to 6 inches long, the bracts hidden by the scales.

Native from Alaska south in the Coast Ranges and Cascade Mountains to Oregon. Occasionally cultivated in parks for its dark green and glossy foliage and its narrow pyramidal habit.

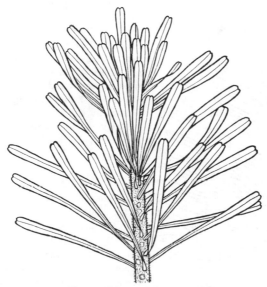

Fig. 34. Abies amabilis Forbes.
Branchlet, × 2.

13. **Abies nordmanniàna** Spach. NORDMANN FIR. Fig. 35.

A symmetrical tree of moderate growth when in cultivation, 20 to 60 feet high. Young branchlets grayish, short-pubescent. Winter-buds not resinous. Leaves spreading horizontally right and left below, those from the lateral and upper surfaces directed outward, upward, and forward, densely covering the branchlets, ¾-inch to 1½ inches long, rounded and notched at the apex, shining dark green and slightly grooved above, with 2 whitish stomatal bands beneath. Cones 5 to 6 inches long, the bracts exserted and reflexed.

Native in Asia Minor, the Caucasus Mountains, and Greece. Frequently cultivated in gardens and parks for its dark green glossy foliage and symmetrical pyramidal form. A very desirable fir for lawn planting.

**14. Abies numídica** De Lannoy. ALGERIAN FIR. Fig. 36.

A small to medium-sized tree, 20 to 60 feet high. Young branchlets glabrous and shining. Winter-buds large, not or only very slightly resinous. Leaves from the lower side of the branchlets spreading to right and left, those from the lateral and upper surfaces spreading outward and upward, often forming a V-shaped space above between the rows of leaves, stout, much crowded, ⅝- to ¾-inch long, rounded or notched at the apex, shining dark green and faintly grooved above, sometimes stomatiferous at the apex, with white stomatal bands beneath. Cones 5 to 7 inches long, the bracts hidden by the scales.

Native to northern Africa. Occasionally cultivated in parks and gardens.

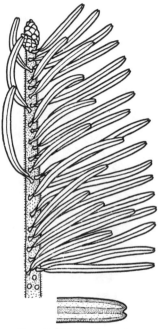

Fig. 35. Abies nordmanniana Spach.
Tip of branchlet, drawn from side to show direction of leaves, × 1. Tip of leaf, lower surface, showing broad white bands, × 3.

## 2. Cédrus Loud. CEDAR

(From the Greek *kedros,* the name of a resinous tree.)

Large evergreen trees with horizontally spreading branches. Leaves simple, spirally arranged on leading shoots but clustered at the ends of the short spur-like branchlets, linear, 4-angled. Flowers monoecious, in catkins; the staminate catkins about 2 inches long, terminal; the ovulate solitary, erect, consisting of numerous spirally arranged 2-ovuled scales. Fruit an ovoid to oblong-ovoid woody cone, 2 to 5 inches long, erect, maturing in 2 or 3 years; scales thin, tightly imbricated, falling away from the stout central axis when mature. Seeds winged.

Three closely related species, in northern Africa, Asia Minor, and southern Asia. *Cedrus* is the genus of true cedars. Many cone-bearing trees belonging to the genera *Juniperus, Chamaecyparis, Thuja, Cupressus,* and *Libocedrus* are often called cedars, but no true cedars are native to the New World. The evergreen needle-like leaves, arranged in clusters of 20 to 40, easily distinguish the genus *Cedrus* from all othcr coniferous genera. The true cedars are highly ornamental trees and are planted on lawns and as avenue trees from San Diego, California, to Seattle, Washington.

### 1. Cedrus deodàra Loud.
DEODAR CEDAR. DEODAR.

A tall graceful tree of pyramidal habit, 50 to 100 feet high, with horizontally spreading branches, pendulous branchlets, and green, yellowish green, or glaucous foliage. Leaves 1 to 2 inches long. Cones 3 to 5 inches long, rounded at the apex. Native in the Himalayas. Abundantly planted in California and occasionally in the Northwest, in parks, lawns, and along drives. The trees in cultivation do

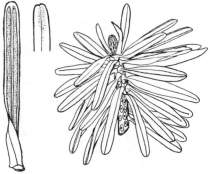

Fig. 36. Abies numidica De Lannoy.
Left to right: Under surface of leaf, × 3.
Upper surface of leaf, × 3. Branchlet, × 1½.

not stand prolonged freezing. Used for living Christmas trees.

### 2. Cedrus atlántica Manetti. ATLAS CEDAR. Fig. 37.

Large trees of open growth, 30 to 100 feet high, with upright leading shoots, horizontally spreading branches drooping in age, and green, bluish green, or glaucous foliage. Leaves ¾-inch to 1 inch long. Cones 2 to 3 inches long, truncate or often concave at the apex.

Native in the Atlas Mountains of northern Africa. A beautiful specimen tree. Planted in parks and gardens throughout the Pacific Coast.

### 2a. Var. gláuca Carr.

Leaves very glaucous. More desirable than the species.

### 3. Cedrus líbani Loud. CEDAR OF LEBANON.
    *C. libanotica* Link.

A large tree, 50 to 120 feet high, with an erect or nodding leading shoot and wide-spreading horizontal branches forming a flat-topped crown in age. Leaves 1 to 1¼ inches long, dark or bright green. Cones 3 to 4 inches long, truncate or often concave at the apex.

Native to Asia Minor and in the Lebanon Mountains of Syria. A picturesque tree of massive growth, cultivated in parks and large estates. Also of interest for its scriptural associations.

Fig. 37. Cedrus atlantica Manetti.
Portion of branchlet with cone, remains of cone after fall of scales, × ½.

### 3. Làrix Adans. LARCH. TAMARACK

(The ancient Latin name.)

Deciduous cone-bearing trees. Leaves simple, in close clusters on the short lateral spur-like branchlets and spirally arranged on the young leading shoots, linear or needle-like. Flowers monoecious, in catkins, appearing before and with the leaves; the staminate catkins globose to oblong, composed of numerous short-stalked anthers; the ovulate composed of several spirally arranged 2-ovuled scales. Fruit an ellipsoidal woody cone, with thin persistent scales subtended by thin bracts, maturing the first year. Seeds winged.

About 10 species, native in the subarctic and colder parts of the north temperate zone. Two species are native on the Pacific Coast. The larches resemble the true cedars (*Cedrus*) in the arrangement of their needle-like leaves, but differ from them in being deciduous. They are occasionally cultivated as specimen trees.

KEY TO THE SPECIES

Bracts shorter than the cone-scales......................................1. *L. europaea.*
Bracts longer than the cone-scales.
    Leaves triangular..............................................2. *L. occidentalis.*
    Leaves quadrangular...............................................................3. *L. lyalli.*

1. **Larix europaèa** DC. EUROPEAN LARCH. Fig. 38.

    *L. decidua* Mill.

A tree, 30 to 70 feet high, with slender horizontal branches and drooping branchlets. Leaves triangular-compressed, ¾-inch to 1¼ inches long, soft, bright green, turning yellow and brown in the autumn. Cones ¾-inch to 1½ inches long, with 40 to 50 scales, the bracts shorter than the scales.

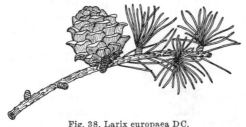

Fig. 38. Larix europaea DC.
Branchlet with cone, × ½.

Native in the mountains of northern and central Europe. Cultivated as an ornamental tree in parks and gardens. Thrives well in dry situations.

2. **Larix occidentàlis** Nutt. WESTERN LARCH. Fig. 39.

A forest tree, 100 to 200 feet high, with dark colored bark, short horizontal or long and drooping branches, and pubescent young branchlets. Leaves flatly triangular, distinctly keeled beneath, 1 to 2 inches long, sharp-pointed, pale green, turning yellow in the autumn, 15 to 30 in a cluster. Cones 1 to 1½ inches long, the bracts longer than the nearly entire scales.

Native on mountain slopes from British Columbia southward to the Cascades of Washington and Oregon, eastward to the Blue Mountains of Oregon, Idaho, and the Rockies of Montana.

3. **Larix lỹalli** Parl. ALPINE LARCH.

A small tree, 30 to 60 feet high, with long crooked branches forming an irregular crown. The 1- and 2-year-old twigs densely woolly. Leaves quadrangular, 1 to 1½ inches long, stiff, pale blue-green, turning yellow and brown in the autumn, 30 to 40 in a cluster. Cones 1 to 2 inches long, the bracts longer than the fringed scales.

Native at high altitudes from British Columbia and Alberta south to Montana, Idaho, and Mount Hood, Oregon.

## 4. Pìcea Dietr. SPRUCE

(The ancient Latin name, from *pix*, pitch.)

Evergreen trees, with thin scaly bark and whorled branches forming a pyramidal crown. Leaves simple, spirally arranged and spreading in all directions, linear, usually 4-angled, acute or sharp-pointed, jointed near the base, the lower part persisting after leaf-fall as brown woody pegs, these giving the branchlets a very rough surface.

Fig. 39. Larix occidentalis Nutt.
Cone, × 2.

Flowers monoecious, in catkins; the staminate catkins composed of numerous stamens; the ovulate composed of several spirally arranged scales, each scale with 2 ovules. Fruit an ellipsoidal pendent woody cone, with numerous spirally arranged thin persistent scales, each scale with 2 seeds at the base and subtended by bracts shorter than the scales.

About 38 species, restricted to the northern hemisphere, of which 5 are native on the Pacific Coast. Two of these, **Picea mariana** B.S.P., BLACK SPRUCE, and **Picea canadensis** B.S.P., WHITE SPRUCE, range from Alaska eastward through Canada and the northern United States. The spruces often are confused with the firs (*Abies*) and false-hemlocks (*Pseudotsuga*), from which they can be distinguished by the persistent leaf-bases, which appear as short spreading woody pegs after leaf-fall. These pegs give the branches a very rough appearance. The bracts subtending the cone-scales are never exserted in spruce cones, but are always exserted in the cones of false-hemlocks and some firs. The cone-scales do not fall apart as they do in the firs.

### KEY TO THE SPECIES

Leaves quadrangular, usually with white stomatic lines on all 4 sides.
    Leaves ¼- to ½-inch long, somewhat appressed to the pubescent branchlets................................................................ 1. *P. orientalis.*
    Leaves ½-inch to 2 inches long, not appressed to the branchlets.
        Some leaves up to 2 inches long; branchlets drooping, glabrous; cone-scales entire................................................2. *P. smithiana.*

None of the leaves more than 1¼ inches long.
    Leaves ½- to ¾- (or rarely 1) inch long; cones 4 to 7 inches long
                                    3. *P. excelsa.*
    Leaves typically 1 inch long; cones 1½ to 4 inches long.
        Leaves soft and flexible; branchlets pubescent
                                  4. *P. engelmanni.*
        Leaves rigid, pungently pointed; branchlets glabrous
                                  5. *P. pungens.*
Leaves somewhat flattened, usually with stomata only on upper surface.
    Leaves rigid, pungently pointed; cone-scales gnawed on the edge;
      branchlets glabrous................................................6. *P. sitchensis.*
    Leaves obtuse; cone-scales entire; branchlets pubescent
                                  7. *P. breweriana.*

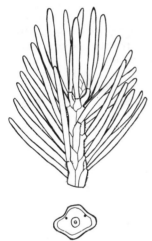

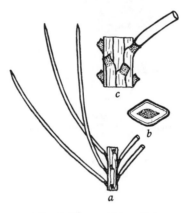

Fig. 41. Picea smithiana Boiss.
*a.* Portion of branchlet, × 1. *b.*

Fig. 40. Picea orientalis Carr.
Branchlet, × 3. Cross section of
leaf, × 10.

Cross section of leaf, × 11. *c.* Portion of branchlet showing peg-like
bases of leaves, × 2 ½.

**1. Picea orientàlis** Carr. ORIENTAL SPRUCE. Fig. 40.

A tall tree, 40 to 100 feet high, with ascending or spreading branches
and spreading or pendulous pubescent branchlets. Leaves ¼- to ½-inch
long, 4-angled, dark green and shining, crowded and somewhat appressed to the branchlets. Cones 2 to 3 inches long; scales entire.

Native to Asia Minor. Planted for its graceful, spreading, and
pendulous branchlets covered with dark glossy leaves.

**2. Picea smithiàna** Boiss. HIMALAYAN SPRUCE. Fig. 41.

    *P. morinda* Link.

A tall tree, 70 to 150 feet high, with widely spreading branches and
numerous drooping glabrous branchlets. Leaves ¾-inch to 2 inches

long, quadrangular, slender, straight or slightly curved. Cones 5 to 7 inches long; scales entire.

Native to the Himalayas. A very handsome tree of broad pyramidal habit and with graceful pendulous branchlets. Cultivated in parks and gardens.

3. **Picea excélsa** Link. NORWAY SPRUCE. Fig. 42.

*P. abies* Karst.

A large tree, 70 to 125 feet high, with spreading branches and usually pendulous glabrous or pubescent branchlets. Leaves ½- to ¾- (or rarely 1) inch long, quadrangular, rigid, dark green, often curved. Cones 4 to 7 inches long; scales denticulate at the apex.

Native to Europe. A great number of garden forms are in cultivation.

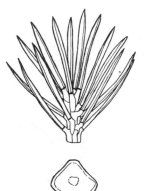

Fig. 42. Picea excelsa Link. Branchlet, × 2. Cross section of leaf, × 10.

4. **Picea éngelmanni** Engelm. ENGELMANN SPRUCE. Fig. 43.

A tall forest tree, 70 to 120 feet high, with spreading or glabrous regularly whorled branches and slender pubescent branchlets forming a narrow compact pyramidal crown. Leaves ½-inch to 1 inch long, slender, soft, flexible, quadrangular, with a strong aromatic odor. Cones 1½ to 3 inches long; scales thin, denticulate at the apex.

Native in the mountains of British Columbia, ranging southward in the Rocky Mountains to Arizona and New Mexico, and in the Cascades of Washington and Oregon to northern California (near Cayton, Shasta County). Cultivated in parks and gardens.

5. **Picea púngens** Engelm. COLORADO SPRUCE. BLUE SPRUCE. Fig. 44.

A tall forest tree, 80 to 100 feet high, with rigid horizontally spreading branches and glabrous branchlets forming a pyramidal crown.

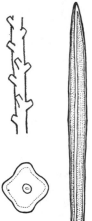

Fig. 43. Picea engelmanni Engelm. Portion of branchlet, × 1. Cross section of leaf, × 10. Leaf, × 3.

Leaves ¾-inch to 1¼ inches long, stiff, curved, spiny-tipped, quadrangular, spreading in all directions. Cones 2½ to 4 inches long; scales thin, denticulate at the apex.

Native in southern Montana, Wyoming, Colorado, Utah, and New Mexico. Very popular as a lawn tree because of the green or silvery white foliage.

6. **Picea sitchénsis** Carr. SITKA SPRUCE. TIDELAND SPRUCE. Fig. 45.

A tall forest tree, 80 to 125 (or 200) feet high, with slender horizontal branches and glabrous branchlets forming a broad pyramidal crown. Leaves ½-inch to 1 inch long, flattened, sharp-pointed, bright green below, whitish on the upper surface. Cones 2½ to 4 inches long; scales thin and erose above the middle.

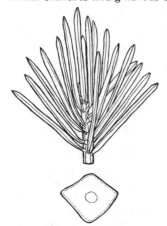

Fig. 44. Picea pungens Engelm. Branchlet, × ½. Cross section of leaf, × 10.

Native to the coastal region from Alaska to northern California. A highly ornamental tree, demanding a moist cool climate for its best development. Much planted in Oregon and Washington.

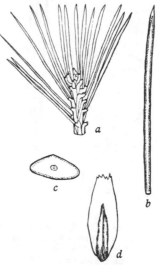

Fig. 45. Picea sitchensis Carr. *a.* Branchlet, × 2. *b.* Upper side of leaf showing stomatal band, × 3. *c.* Cross section of leaf, × 10. *d.* Cone-scale and bract, × 1½.

7. **Picea breweriàna** Wats. WEEP-ING SPRUCE. Fig. 46.

A tree, 30 to 100 feet high, with horizontal or pendulous branches and willow-like pendulous pubescent branchlets often 6 to 8 feet long. Leaves ¾-inch to 1 inch long, almost flat, with whitish bands above. Cones 2 to 4½ inches long; scales thick, entire at the apex.

Native in the Siskiyou Mountains of southern Oregon and northern California, in the Coast Ranges of southwestern Oregon, and on Marble Mountain and the Klamath and Trinity mountains of northern California.

## 5. Pìnus L. PINE

(The Latin name of the pine.)

Evergreen trees, usually with whorled branches and 2 kinds of leaves; the primary leaves thin, scale-like, deciduous; the permanent leaves (called needles) needle-shaped, borne in fascicles of 1 to 5. Flowers monoecious, in catkins; the staminate catkins consisting of numerous stamens; the ovulate composed of spirally arranged scales forming cones, each scale with 2 ovules at the base on the upper side. Fruit a cone composed of woody and usually well imbricated scales, maturing the second year. Seeds with a terminal wing, or wingless.

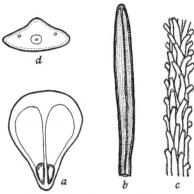

Fig. 46. Picea breweriana Wats.
*a.* Cone-scale with seeds, × 2. *b.* Leaf, upper surface, × 2. *c.* Portion of branchlet showing peg-like bases, × 1. *d,* Cross section of leaf, × 10.

About 80 species, restricted to the northern hemisphere, 20 of which are indigenous to the Pacific Coast. The pines are more widely known by all peoples of the world than any other group of conifers. They surpass all other cone-bearing genera in numbers of species, range of distribution, area occupied, and importance to civilization. The needle-like leaves, in fascicles of 1 to 5, separate the pines from all other trees.

### KEY TO THE SPECIES

A. Leaves 1 (rarely 2) in a fascicle.....................................1. *P. monophylla.*
B. Leaves 2 (rarely 3) in a fascicle.
    Leaves 1 to 2½ (to 3) inches long; cones 1 to 1¾ inches long.
        Cone-scales thin, with slender prickles.
            Bark thick, not flaking off in scales..........................2. *P. contorta.*
            Bark thin, flaking off in scales..............................3. *P. murrayana.*
        Cone-scales thick, the prickles early deciduous..............4. *P. edulis.*
    Leaves 3 (2½) to 9 inches long.
        Cones 2 to 3½ inches long.
            Cone-scales with triangular or conical projections terminated by
                very stout spines or almost spur-like; leaves 4 to 6 inches
                long.......................................................................5. *P. muricata.*
            Cone-scales smooth or with very short prickles.
                Leaves 2½ to 4 (or 6) inches long; cones somewhat deflexed;
                cone-scales without prickles..........................6. *P. halepensis.*
                Leaves 3 to 8 inches long; cones not deflexed; cone-scales with
                minute prickles...............................................7. *P. remorata.*
        Cones 3½ to 8 inches long.

Cones conic-oblong or oblong-ovoid, 4 to 8 inches long, tenaciously persistent on the branches; cone-scales with raised triangular projections; leaves 4 to 9 inches long ........................................8. *P. pinaster.*
Cones broadly ovoid, 3½ to 5½ inches long; cone-scales flat and smooth; trunk free of branches for some height, the branches forming an umbrella-like crown..............................9. *P. pinea.*
C. Leaves 3 (rarely 2) in a fascicle.
   Cones breaking through at base when falling and leaving a few scales on the branches.
      Cones 3 to 8 inches long; cone-scales with thin prickles, never with spurs.
         Prickles pointed outward; cones 3 to 5 inches long, prickly to the touch.............................................................10. *P. ponderosa.*
         Prickles not pointed outward; cones 5 to 8 inches long, not prickly to the touch..............................................11. *P. jeffreyi.*
      Cones 6 to 14 inches long; cone-scales prolonged into conspicuous spurs or hooks.
         Foliage dark green, dense, stiff, straight; branchlets roughened, thick, dark brown or nearly black; cones 9 to 14 inches long; trunk continuing as one main stem....................12. *P. coulteri.*
         Foliage sparse, distinctly grayish, drooping; branchlets thin, gray; cones 6 to 10 inches long; trunk usually branching into 2 or more secondary trunks.....................13. *P. sabiniana.*
   Cones not breaking through at base when falling.
      Leaves 9 to 12 inches long, slender, drooping, light green or gray; cones cylindric-ovoid..................................14. *P. canariensis.*
      Leaves 3 to 6 inches long; cones unsymmetrical.
         Cones broadly ovoid, 2½ to 4½ inches long, sessile..15. *P. radiata.*
         Cones oblong-ovoid, 3 to 6 inches long, short-stalked ..............................................................................16. *P. attenuata.*
D. Leaves typically 4 in a fascicle.....................................17. *P. parryana.*
E. Leaves 5 in a fascicle.
   Leaves 1 to 4 inches long.
     1. Cones 1 to 5 inches long.
       Scales with terminal umbo, without prickles; cones almost sessile.
         Cones subglobose, 1 to 3 inches long; scales very thick at apex ..............................................................................18. *P. albicaulis.*
         Cones subglobose to ovoid-oblong, 3 to 5½ inches long; scales slightly thickened at apex...............................19. *P. flexilis.*
       Scales with dorsal umbo, with prickles; cones stalked.
         Cones 2½ to 5 inches long; scales with minute prickles about ⅛-inch long...............................................20. *P. balfouriana.*
         Cones 2½ to 3½ inches long; scales with prickles about ¼-inch long...................................................................21. *P. aristata.*
     2. Cones 6 to 8 inches long..........................................22. *P. monticola.*
     3. Cones 10 to 18 inches long....................................23. *P. lambertiana.*
   Leaves 5 to 10 inches long.
      Cones broadly ovoid, 4 to 6 inches long; leaves dark green, stiff ..............................................................................24. *P. torreyana.*
      Cones cylindric, 5 to 10 inches long, on stalks 1 to 2 inches long; leaves drooping, grayish green, soft....................25. *P. excelsa.*

1. **Pinus monophýlla** Torr. SINGLELEAF PINE. ONELEAF PIÑON. Fig. 47.
   *P. cembroides* var. *monophylla* VOSS.

A small round-headed tree, 10 to 25 (or 40) feet high. Needles usually solitary, stiff, spiny-pointed, 1 to 2 inches long. Cones subglobose, 2½ to 3½ inches long; scales thickened at the apex. Seeds about ½-inch long, wingless.

Native on arid slopes in the mountains of eastern and southern California, Utah, Nevada, and Arizona. In California this pine occurs on the eastern slope of the Sierra Nevada from Alpine County southward to Kern County, and on the western slope in a few restricted localities, and in the mountains of San Bernardino, Ventura, Los Angeles, Riverside, and San Diego counties. Rarely seen in cultivation. The seeds are sold as "pine nuts."

Fig. 47. Pinus monophylla Torr.
Cone, seed, × ⅔.

2. **Pinus contórta** Dougl. SHORE PINE. BEACH PINE.

A small tree, 15 to 30 feet high, with rather stout branches forming a round-topped crown. Bark thick, dark, becoming rough and fissured. Leaves in twos, stiff, twisted, 1¼ to 2 inches long, dark green. Cones narrowly ovoid to conic-ovoid, somewhat oblique, 1¼ to 2 inches long; scales tipped by slender fragile prickles. Seeds ⅛-inch long, with a wing about ½-inch long.

Native in Mendocino, Humboldt, and Del Norte counties, California, and extending northward along the coast of Oregon and Washington to Alaska. The trees growing on the Mendocino "White Plains" are much dwarfed, having very slender cane-like trunks 2 to 5 feet high and numerous very small slender cones. These trees have been named **P. contorta** var. **bolánderi** Vasey.

3. **Pinus murrayàna** Balf. LODGEPOLE PINE. TAMRAC PINE. Fig. 48.
   *P. contorta* var. *murrayana* Engelm.

A slender pyramidal tree, 30 to 80 feet high, with a dense symmetrical crown. Bark very thin, flaking off in scales. Leaves in twos, 1 to 2½ inches long, stiff, twisted. Cones oblong-ovoid, becoming globose when mature, 1 to 1¾ inches long; scales tipped by short slender prickles. Seeds about ⅛-inch long, with a wing ½-inch long.

Native in the Rocky Mountain and Pacific Coast states. Lodgepole Pine is common around mountain meadows or on moist slopes and often extends to timber line, where it becomes much dwarfed. In California it occurs in the Sierra Nevada from 6500 to 11,000 feet elevation in the south and central part, and 5000 to 7500 feet elevation in the north. In southern California it is found in the San Bernardino, San Jacinto, and Sierra Madre mountains. Rarely cultivated.

4. **Pinus edùlis** Engelm. NUT PINE. Fig. 49.

*P. cembroides* var. *edulis* Voss.

A small tree, 20 to 40 feet high, with stout spreading branches forming a round-top-

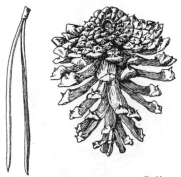

Fig. 48. Pinus murrayana Balf. Fascicle of leaves, cone, × ¾.

ped crown. Leaves in twos, 1 to 2 inches long, stiff, sharp-pointed, dark green. Cones globose, 1 to 2 inches long; scales thick. Seeds ⅜- to ½-inch long, wingless.

Native in the arid mountains of Wyoming, south to Arizona, New Mexico, and Texas. Also in the New York Mountains of eastern San Bernardino County, California. A handsome small tree, occasionally cultivated in parks and lawns in the warmer parts of the Pacific Coast. The seeds are sold in the markets as "pine nuts."

Fig. 49. Pinus edulis Engelm. Branchlet with cone, × ⅔. Seed, × 1⅛.

5. **Pinus muricàta** D. Don. BISHOP PINE. Fig. 50.

A tree, 45 to 75 feet high, with a round-topped or flat crown. Bark rough and brown. Leaves in twos, 4 to 6 inches long, rigid, usually twisted, dark green. Cones conic-ovoid, oblique, 2 to 3½ inches long and almost as thick, globose when mature, borne in circles of 3 to 5; scales on the outer side terminated by conical tips continued into stout prickles or curved spurs. Seeds about ¼-inch long, with a wing ½- to ⅔-inch long. Native in the coastal regions of Humboldt,

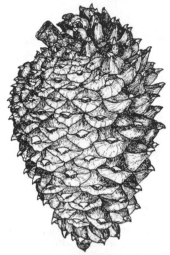

Mendocino, Sonoma, Marin, Monterey, San Luis Obispo, and Santa Barbara (mountains north of Lompoc) counties, California, on Santa Cruz and Cedros islands, and in Lower California from Ensenada to San Quentin. Planted as a windbreak and sometimes cultivated in parks.

**6. Pinus halepénsis** Mill. ALEPPO PINE. Fig. 51.

A tree, 30 to 60 feet high, with short branches forming an open round crown. Leaves in twos (rarely in threes), 2¼ to 5½ inches long, slender, light green. Cones conic-ovoid, nearly symmetrical, 2½ to 4 inches long, short-stalked, spreading or deflexed; scales without prickles. A native of the Mediterranean region. Planted in parks and gardens.

Fig. 50. Pinus muricata D. Don.
Cone, × 1.

**7. Pinus remoràta** H. Mason. SANTA CRUZ ISLAND PINE. Fig. 52.

A slender medium-sized tree, 25 to 60 feet high, with short horizontal branches and dark yellow-green foliage forming an open crown, this becoming flat-topped in age. Bark dark, rough, and furrowed. Leaves in twos, 3 to 8 inches long, rather stiff, twisted, dark yellow-green. Cones ovoid, 2 to 3½ inches long, dark brown, standing straight out from the branches, sessile or very short-stalked; scales slightly thickened at the apex, with minute prickles. Seeds about ¼-inch long, with a thin wing about ⅜-inch long.

Native on Santa Cruz, Santa Rosa, Guadalupe, and Cedros islands.

Fig. 51. Pinus halepensis Mill.
Cone, × ½.

**8. Pinus pináster** Ait. CLUSTER PINE. Fig. 53.

A tree, 40 to 100 feet high, with spreading or pendulous branches forming a pyramidal crown. Leaves in twos, 4 to 9 inches long, stiff, usually twisted, glossy, green. Cones oblong-ovoid, nearly symmetrical, 4 to 8 inches long, clustered, short-stalked.

Native to the Mediterranean region. One of the most commonly cultivated pines in gardens and parks in warmer regions.

9. **Pinus pínea** L. ITALIAN STONE PINE. Fig. 54.

A tree, 30 to 70 feet high, with long ascending or horizontally

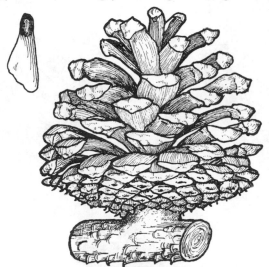

Fig. 52. Pinus remorata H. Mason.
Cone, seed, × 1.

spreading branches forming a parasol-like crown. Leaves in twos, 3½ to 7 inches long, stiff, bright green. Cones broadly ovoid, 3½ to 5 inches long.

Native to the Mediterranean region. Planted in gardens and parks. Hardy only in California.

10. **Pinus ponderòsa** Dougl. WESTERN YELLOW PINE. Fig. 55.

A large forest tree, 60 to 200 (rarely 230) feet high, with spreading or drooping branches forming a narrow cylinder-shaped crown. Bark yellowish brown, commonly fissured into rounded ridges or on old trees into large smooth or scaly plates. Leaves in threes, 5 to 10 inches long, dark yellowish green. Cones reddish brown, oblong-ovoid, 3 to 5 inches long, breaking through the base when falling; scales thickened at the apex, with triangular projections terminated by sharp outward-pointed prickles. Seeds ¼- to ⅓-inch long, with a wing ¾-inch to 1 inch long.

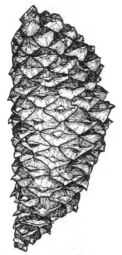

Fig. 53. Pinus pinaster Ait.
Cone, × ½.

Fig. 54. Pinus pinea L.
Cone, seed, × ½.

Fig. 55. Pinus ponderosa Dougl.
Cone, × ½.

Fig. 56. Pinus jeffreyi Murr.
Cone, × ½.

Fig. 57. Pinus coulteri D. Don.
Cone, seed, × ¼.

Native in the western United States and Canada and in northern Mexico. On the Pacific Coast, Western Yellow Pine is the most widely distributed forest tree throughout the mountains at elevations from 5000 to 7000 feet in the south, 2500 to 6000 feet in the central part, and 500 to 6000 feet in the north. An excellent timber tree. Sometimes planted in parks.

### 11. Pinus jéffreyi Murr. JEFFREY PINE. Fig. 56.

*P. ponderosa* var. *jeffreyi* Vasey.

Very similar to the species described just above, but can be distinguished by its cones, which are 5 to 8 inches long, and by the inward-pointed prickles on the scales.

In its typical form, Jeffrey Pine occurs at higher elevations in the mountains than Western Yellow Pine. It is usually in the Canadian Life Zone. Its range extends from the southern Cascades and the Siskiyou Mountains of Oregon southward to the higher North Coast Ranges, and in the Sierra Nevada to the higher mountains of southern California, and to the San Pedro Mártir Mountains of Lower California.

### 12. Pinus còulteri D. Don. COULTER PINE. BIGCONE PINE. Fig. 57.

A medium-sized tree, 40 to 80 feet high, with thick roughened nearly black branchlets and dark green foliage. Bark dark brown to nearly black, with an irregular network of ridges and fissures, often flaking off in thin scales. Leaves in threes, 5 to 12 inches long, stiff, dark bluish green, with sharp points. Cones broadly oblong-ovoid, 9 to 14 inches long, yellowish brown, breaking through the base when falling; scales prolonged into stout spur-like projections at the apex, some much curved and hook-like. Seeds ½- to ⅔-inch long, with a wing about 1 inch long.

Restricted to California, in the South Coast Ranges and the mountains of southern California, where it occurs on dry rocky slopes from 1000 to 6000 feet elevation. The most northern native location for this pine is Nortonville, Contra Costa County. Occasionally cultivated in parks and large estates.

### 13. Pinus sabiniàna Dougl. DIGGER PINE. Fig. 58

A tree, 40 to 80 feet high, with the main trunk usually divided into 2 or more secondary trunks and with sparse grayish drooping foliage. Bark dark gray, roughly and irregularly fissured. Branchlets thin, with gray bark. Leaves in threes, 7 to 13 inches long, slender, usually drooping from the ends of the branchlets. Cones broadly oblong-ovoid, 6 to 10 inches long, light chocolate-brown, breaking through at the base when falling; scales prolonged at the apex into sharp, stout, often hooked, spurs. Seeds about ¾-inch long, with a very hard shell and a short wing ¼- to ½-inch long.

Native on the dry foothills and lower mountain slopes of northern and central California, from 100 to 3000 feet elevation (5000 feet in the south), in the Upper Sonoran Life Zone. Most commonly associated with the Blue Oak (*Quercus douglasi*). Occasionally cultivated throughout the warmer parts of California.

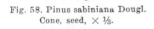

Fig. 58. Pinus sabiniana Dougl. Cone, seed, × ⅓.

### 14. Pinus canariénsis C. Smith. CANARY ISLAND PINE. Fig. 59.

A tall handsome tree, 60 to 80 feet high, with gray-green drooping foliage and slender spreading branches forming a broad round-topped crown. Bark reddish brown, slightly and irregularly fissured. Leaves in threes, 9 to 12 inches long, slender, gray-green, in dense tufts at the ends of the branchlets. Cones oblong-ovoid, 4 to 8 inches long; scales irregularly 4-sided at the apex, with short obtuse

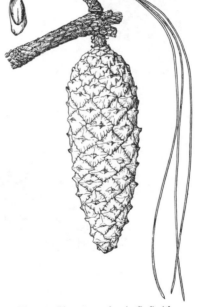

Fig. 59. Pinus canariensis C. Smith. Cone, fascicle of leaves, × ⅓.

projections. Seeds about ⅓-inch long, with a wing about ¾-inch long.

Native to the Canary Islands. Cultivated in parks and gardens, especially in southern California.

Fig. 60. Pinus radiata D. Don.
Cone, × ½.

Fig. 61.
Pinus attenuata Lemm.
Cone, × ½.

Fig. 62. Pinus parryana Engelm.
Cone, seed, × 1.

15. **Pinus radiàta** D. Don. MONTEREY PINE. Fig. 60.

A beautiful tree, 40 to 100 feet high, with dark green foliage and stout spreading branches forming a compact, often irregular, round-topped crown. Bark usually fissured, dark brown to almost black. Leaves in threes (rarely in twos), $3\frac{1}{2}$ to 6 inches long. Cones broadly ovoid, unsymmetrical, 2 to 5 inches long, in cycles of 3 to 5, persisting for many years; scales on the outer side near the base with a rounded apex. Seeds about $\frac{1}{4}$-inch long, with a wing $\frac{3}{4}$-inch long.

Native in San Mateo, Monterey, and northern San Luis Obispo counties, California, and on Santa Rosa, Santa Cruz, and Guadalupe islands. Planted in parks and gardens and, especially about San Francisco Bay, for reforestation.

16. **Pinus attenuàta** Lemm. KNOBCONE PINE. Fig. 61.

*P. tuberculata* Gord.

A slender tree, 10 to 40 (rarely to 80) feet high, with yellow-green foliage and ascending or spreading branches forming an open, rounded crown. Leaves in threes, 3 to 5 inches long, slender. Cones narrowly oblong-ovoid, unsymmetrical, 3 to 6 inches long; scales on the outer side and near the base with a pyramidal apex armed with a prickle; scales near the summit and on the inner side with a flat or depressed apex. Seeds about $\frac{1}{4}$-inch long, with a wing $\frac{3}{4}$-inch to 1 inch long.

Native on rocky slopes and ridges in scattered localities in southern Oregon and, in California, in the Coast Ranges, the Sierra Nevada north of Mariposa County, and the San Bernardino Mountains of southern California. Rarely cultivated.

17. **Pinus parryàna** Engelm. PARRY PIÑON PINE. Fig. 62.

*P. cembroides* var. *parryana* Voss. *P. quadrifolia* Parry.

A low short-trunked tree, 15 to 30 feet high, with bluish green foliage and spreading branches becoming crooked in age. Bark reddish brown, divided into broad connected ridges and furrows, covered by close scales. Leaves usually in fours (1 to 5), $\frac{3}{4}$-inch to $1\frac{1}{2}$ inches long, rather stiff, incurved, whitish on the inner surfaces, bluish green on the backs. Cones subglobose, $1\frac{1}{4}$ to 2 inches long, brown; scales thickened at the apex and with a ridged knob. Seeds about $\frac{5}{8}$-inch long, rather thin-shelled, with a very narrow wing remaining attached to the scale when the seeds fall.

Native in the Santa Rosa Mountains of Riverside County and the desert mountain slopes of San Diego County, California. Ranges south to northern Lower California. The nut-like seeds are used for food by the Indians and are sold as "pine nuts."

**18. Pinus albicáulis** Engelm. WHITEBARK PINE. Fig. 63.

Usually a low spreading tree, 6 to 20 feet high or to 60 feet in favored locations, with tough flexible branchlets. Bark thin, whitish, smooth. Leaves in fives, 1 to 2½ inches long, stiff, dark green. Cones subglobose or ovoid, 1 to 3 inches long, dark purple, very resinous, with the scales thickened at the apex and terminated by a blunt point. Seeds ⅓- to ½-inch long, wingless.

Native to the high mountains of the western United States and Canada. Whitebark Pine is the characteristic timber line tree from

7000 to 12,000 feet elevation, where it often forms table-like masses in exposed situations. In the white granite setting of the Hudsonian and lower Arctic-Alpine life zones, this tree makes a striking and pleasing effect with its numerous purplish red staminate catkins against the dark green foliage.

**19. Pinus fléxilis** James. LIM-BER PINE.

A low much branched tree, 20 to 30 (or 50) feet high, with a short thick trunk and dark yellow-green foliage densely set at the ends of the branches. Bark on the branches thin, smooth, grayish, on old trunks 1 to 2

Fig. 63. Pinus albicaulis Engelm. Cone, seed, × 1.

inches thick, dark brown, breaking into plates covered by thin scales. Leaves in fives, 1 to 2½ inches long, stout, stiff, often curving, dark green. Cones ovoid to subcylindric, 3 to 8 inches long, light brown, short-stalked; scales greatly thickened at the apex and often curved. Seeds ⅜- to ½-inch long, dark reddish brown, speckled with black, with narrow wings persistent on the scales after the fall of the seeds.

Native to desert mountain slopes at high elevations on Mount Pinos, the San Jacinto, San Bernardino, Sierra Madre, San Gabriel, Panamint, and Inyo mountains of southern California, and on the eastern slope of the Sierra Nevada south of Mono Pass in Mono and Inyo counties. Also widely distributed over the Rocky Mountains from British Columbia and Alberta south to New Mexico and Arizona, and in the San Pedro Mártir Mountains of Lower California.

20. **Pinus balfouriàna** Murr. Foxtail Pine. Fig. 64.

A low subalpine tree, 20 to 35 (or 60) feet high, with irregular long

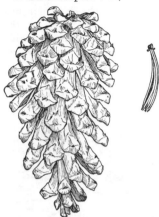

upper branches and spreading stout and short lower branches forming a bushy crown. Bark at first smooth and whitish, becoming reddish brown and checked into flat irregular ridges or plates. Leaves in fives, ¾-inch to 1½ inches long, stout, stiff, curved, bright green on the outer surfaces, whitish on the inner surfaces, thickly clothing and appressed to the ends of the branches, these thus resembling a fox's tail. Cones ovoid, 2½ to 5 inches long, deep purple or reddish brown, short-stalked; scales thickened and somewhat 4-sided at the apex, armed with a minute prickle. Seeds about ⅜-inch long, with a thin wing about ¾-inch long.

Fig. 64.
Pinus balfouriana Murr.
Cone, fascicle of leaves, × ½.

Native to two widely separated regions in the high mountains of California, one on the watersheds of the Kings and Kern rivers in the southern Sierra Nevada at 9000 to 12,000 feet elevation, and the other in the North Coast Ranges on Scott Mountains in Siskiyou County and on Mount Eddy southward to the Yollo Bollys in Tehama County, at elevations from 5000 to 8000 feet.

21. **Pinus aristàta** Engelm. Hickory Pine. Bristlecone Pine. Fig. 65.

A short-trunked tree, 20 to

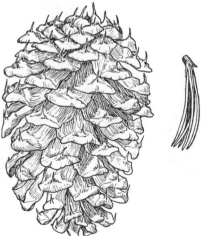

Fig. 65. Pinus aristata Engelm.
Cone, fascicle of leaves, × ¾.

60 feet high, with deep green foliage clustered at the ends of the short stout branches, forming a bushy crown. Bark on old trunks reddish or dark brown, shallowly fissured and scaly, on the branches

and young trunks smooth and whitish. Leaves in fives, 1 to 1½ inches long, dark green and lustrous on the backs, whitish on the inner surfaces. Cones ovoid, 2½ to 3½ inches long, purplish brown, nearly sessile; scales thickened at the apex and armed with a bristle-like, very fragile, prickle. Seeds about ¼-inch long, light brown, with a thin wing ⅜- to ½-inch long.

Native to the Inyo, Panamint, and White mountains of the desert region of southeastern California at elevations between 8000 to 12,000 feet, and on high peaks in Colorado, Utah, southern Nevada, and northern Arizona.

Fig. 66. Pinus monticola D. Don. Cone, × ⅓.

22. **Pinus montícola** D. Don. WESTERN WHITE PINE. SILVER PINE. Fig. 66.

A tall tree, 50 to 100 (or 150) feet high, with bluish green somewhat glaucous foliage and slender horizontal or pendulous branches forming a narrow pyramidal crown. Bark thin, smooth, gray, becoming checked and fissured into nearly square plates. Leaves in fives, 1 to 3½ inches long, very slender. Cones narrowly cylindrical, 6 to 8 inches long, purplish or green when young, stalked, pendulous, borne in clusters of 1 to 6 on the ends of the higher branches; scales thin, scarcely thickened at the apex. Seeds about ⅓-inch long, with a wing 1 inch long.

Native in the higher mountains from southern British Columbia southward through Washington, Oregon, and Idaho to the Sierra Nevada of California. This tree is rarely cultivated.

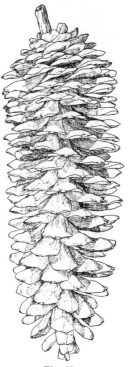

Fig. 67. Pinus lambertiana Dougl. Cone, × ¼.

23. **Pinus lambertiàna** Dougl. SUGAR PINE. Fig. 67.

A forest tree, 70 to 150 (or 200) feet high, with a straight trunk devoid of branches for a great height and horizontally spreading branches arching downward and forming a flat-topped crown. Bark thin, smooth, and grayish at first, becoming thick, deeply and irregularly fissured into long plate-like ridges breaking away into loose purple-brown or reddish scales. Leaves in fives, 2 to 3½ inches long, slender but rigid, dark bluish green. Cones oblong-cylindric, 11 to 20 inches long, pendulous, on stalks from the ends of the higher

Fig. 68. Pinus torreyana Parry.
Cone, × ½.

branches; scales thin, scarcely thickened at the apex. Seeds ⅛- to ½-inch long, with a wing about twice as long.

Native to western Oregon, California, and the San Pedro Mártir Mountains of Lower California. In California, the Sugar Pine has its greatest development on the western slope of the Sierra Nevada from 3000 to 6000 feet elevation at the north and 5000 to 8000 feet at the south. It also occurs on other high mountains of northern California, the Coast Ranges, and in southern California. Rarely cultivated.

24. **Pinus torreyàna** Parry. TORREY PINE. Fig. 68.

Usually a small tree, 20 to 40 (or 60) feet high, with stout spreading or ascending branches. Bark thick, irregularly fissured into broad flat ridges covered by thin compact red-brown scales. Leaves in fives, 7 to 11 inches long, stiff, dark green. Cones chocolate-brown, broadly ovoid, 4 to 6 inches long; scales with a thickened triangular apex, the basal ones terminated by short pyramidal projections. The seeds ½- to ¾-inch long, with a wing ¼-.to ½-inch long.

Native on the bluffs at Del Mar, San Diego County, California, and on Santa Rosa Island. Occasionally cultivated in parks.

25. **Pinus excélsa** Wall. HIMALAYAN PINE. Fig. 69.

A tree, 50 to 125 feet high, with wide-spreading branches and drooping grayish or blue-green foliage. Leaves in fives, 4 to 8 inches long, slender. Cones narrow-cylindrical, 6 to 10 inches long, with thin scales. Seeds about ¼-inch long, the wing ¾-inch long.

Native in the Himalayas. Cultivated in parks and gardens.

OTHER LESS COMMON INTRODUCED SPECIES

26. **Pinus longifòlia** Roxbg. LONGLEAF PINE.

Leaves in threes, 8 to 12 inches long, light green. Native to the Himalayas.

27. **Pinus nìgra** Arnold. AUSTRIAN PINE. *P. laricio* Poir.

Leaves in twos, 3 to 7 inches long, stiff, dark green. Native to Austria.

28. **Pinus montàna** var. **mùghus** Willk. MUGHO PINE.

Usually a small bushy tree, 6 to 10 feet high. Leaves in twos, 1 to 2 inches long. Native in central and southern Europe.

29. **Pinus púngens** Lamb. TABLE MOUNTAIN PINE.

Leaves in twos or threes, 1¼ to 2¾ inches long. Native in the mountains of the eastern United States.

Fig. 69. Pinus excelsa Wall. Cone, × ⅓.

30. **Pinus stròbus** L. WHITE PINE.

Leaves in fives, 3 to 5 inches long, very slender, glaucous on the inner surfaces. Native in eastern North America.

31. **Pinus sylvéstris** L. SCOTCH PINE.

Leaves in twos, 1½ to 2¾ inches long. Native in Europe and eastward to western Asia and northwestern Siberia.

32. **Pinus thúnbergi** Parl. JAPANESE BLACK PINE.

Leaves in twos, 3 to 4½ inches long, stiff, bright green. Native to Japan.

### 6. Pseudotsùga Carr. FALSE-HEMLOCK

(From the Greek *pseudos,* false, and the Japanese word *Tsuga,* the hemlock tree, indicating the relationship of these trees with hemlocks.)

Evergreen trees with irregularly whorled branches. Leaves simple, spirally arranged and spreading in 2 rows, linear, flat, short-petioled. Flowers monoecious, in catkins; the staminate catkins axillary, composed of numerous stamens; the ovulate erect, terminal or axillary, consisting of many spirally arranged scales, each scale with 2 ovules at the base. Fruit a pendulous oblong-ovoid woody cone, with thin scales subtended by exserted 3-lobed bracts, the middle lobe long and narrow. Seeds winged.

Four species, 2 in western North America, 1 in southern Japan, and 1 in southwestern China and Formosa. Two species are native on the Pacific Coast. The genus *Pseudotsuga* is related to the firs (*Abies*), hemlocks (*Tsuga*), and spruces (*Picea*), but differs from the firs in the persistent cone-scales, and from the hemlocks and spruces in having the bracts exserted beyond the cone-scales. After leaf-fall the branches exhibit circular scars somewhat similar to those of firs, but the scars are slightly raised and give the branches a finely roughened character.

KEY TO THE SPECIES

Cones 1½ to 3½ inches long; leaves blunt at apex...............1. *P. douglasi.*
Cones 4 to 7 inches long; leaves acute at apex.................2. *P. macrocarpa.*

### 1. Pseudotsuga doúglasi Carr. DOUGLAS-FIR. Fig. 70.

*P. taxifolia* Britt. *P. mucronata* Sudw.

A large forest tree, 70 to 250 feet high, with clear trunks often to 150 feet, surmounted by a pyramidal crown of horizontal branches and pendulous branchlets. Leaves ¾-inch to 1½ inches long, about ¹⁄₁₆-inch wide, dark or bluish green above, with 2 grayish bands beneath; petioles twisted. Cones ovoid, 1½ to 3½ inches long, the bracts conspicuously exserted.

Native in the mountainous regions of the western United States, British Columbia, and northern Mexico. Douglas-Fir attains its greatest development in western Oregon and Washington. In California it occurs in the Coast Ranges from Del Norte County southward to Monterey County, from sea level to about 4000 feet elevation, and in the Sierra Nevada as far south as the San Joaquin River, from 2500 to 6000 feet elevation. A fine lumber tree, sold under the name of "Oregon Pine." Commonly cultivated in parks and gardens, especially in the Northwest. Several distinct forms with various shades of foliage, habits of growth, and leaf-lengths are in cultivation.

2. **Pseudotsuga macrocárpa** Mayr. BIGCONE-SPRUCE.

A tree, 30 to 60 (or rarely 90) feet high, with elongated, horizontal or pendulous branches forming a broad pyramidal crown. Leaves similar to those of *P. douglasi*, but more acute or acuminate, ¾-inch to 1¾ inches long. Cones 4 to 7 inches long, the bracts only slightly protruding beyond the scales, or their central lobes long-exserted at the base of the cone.

Native on the mountain slopes of southern California, from the Santa Inez Mountains of Santa Barbara County southward to the

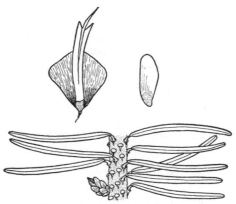

Fig. 70. Pseudotsuga douglasi Carr.
Cone-scale and bract, seed, portion of branchlet, × 1.

Cuyamaca Mountains of San Diego County, at elevations of from 3000 to 5000 feet. Also in the San Pedro Mártir Mountains of Lower California. Rarely cultivated.

### 7. **Tsùga** Carr. HEMLOCK

(The Japanese name for the hemlock.)

Evergreen trees with horizontal or often pendulous branches. Leaves simple, spirally arranged, usually appearing 2-ranked by a twisting of the petioles, linear, flat or angular, abruptly narrowed into short green petioles jointed on decurrent persistent peg-like bases. Flowers monoecious, in catkins; the staminate catkins axillary, globose, composed of numerous subglobose anthers; the ovulate terminal, erect, composed of several nearly circular scales about as long as the membranaceous bracts subtending them, each scale with 2 ovules at the base. Fruit a short-cylindric or ovoid usually pendulous cone, with thin flexible ovate-oblong persistent scales much longer than the bracts, maturing the first year. Seeds winged.

Nine species in temperate North America, Japan, central and western China, Formosa, and the Himalayas. Two species are native to the Pacific Coast. The hemlocks, in their native habitats, vary from small to large forest trees with straight trunks terminated by nodding leading shoots. They are often confused with the firs (*Abies*) and spruces (*Picea*), but the leaves of hemlocks have short green petioles attached by appressed persistent peg-like bases. Hemlocks are graceful ornamental trees in the Pacific Northwest, but are not satisfactory for use in the warmer parts of California.

KEY TO THE SPECIES

Leaves flat, 2-ranked, forming flattened sprays, with 2 white bands beneath, green above; cones 1 inch or less long.
  Cones sessile, ½-inch to 1 inch long; the white bands of the lower leaf-surfaces broader than the green margins; branchlets distinctly hairy................................................1. *T. heterophylla.*
  Cones stalked, ½- to ¾-inch long; the white bands of the lower leaf-surfaces narrower or about as broad as the green margins; branchlets pubescent or glabrous..........................2. *T. canadensis.*
Leaves convex or keeled above, spreading from the branchlets in all directions, glaucous on both surfaces; cones 1½ to 3 inches long................................................3. *T. mertensiana.*

1. **Tsuga heterophýlla** Sarg. WESTERN HEMLOCK. COAST HEMLOCK. Fig. 71.

A large tree, 90 to 200 feet high, with short slender drooping branches and brownish branchlets, usually coated with fine pale hairs, forming a narrow pyramidal crown. Leaves ¼-inch to ¾-inch long, ¹⁄₁₂-inch or less wide, distinctly grooved and dark green above, with 2 white bands beneath. Cones cylindric to ovoid, sessile, ½-inch to 1 inch long, solitary and pendulous at the ends of the branchlets.

Native in northwestern California, Oregon, Washington, British Columbia, Alaska, Idaho, and Montana. In California, Western Hemlock occurs near the coast in scattered localities from Del Norte County southward to the

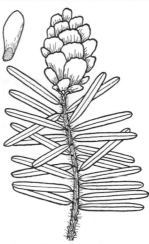

Fig. 71. Tsuga heterophylla Sarg.
Branchlet with cone, × 1.
Seed, × 2.

vicinity of Elk Creek, Mendocino County. In Oregon and Washington it inhabits the Coast Ranges and Olympic Mountains, extending eastward to the Cascades. Rarely cultivated.

2. **Tsuga canadénsis** Carr. CANADA HEMLOCK. Fig. 72.

A tree, 60 to 70 feet high, with long slender horizontal or pendulous branches and slender yellowish brown pubescent young branchlets,

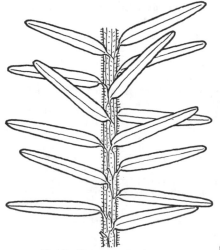

becoming reddish brown and glabrous in age, forming a broad pyramidal crown. Leaves ⅓- to ¾-inch long, about 1/16-inch wide, dark yellow-green and lustrous above, with 2 narrow white bands beneath. Cones ovoid, ½- to ¾-inch long, on slender stalks about ¼-inch long.

Native from Nova Scotia and the mountains of the northeastern United States west to Minnesota

Fig. 72. Tsuga canadensis Carr. Branchlet, × 2.

and south along the Appalachian Mountains to northern Alabama. Occasionally cultivated.

3. **Tsuga mertensiàna** Sarg. MOUNTAIN HEMLOCK. Fig. 73.

A tree, 30 to 90 feet high, with slender pendulous branches and pubescent drooping branchlets forming a loose pyramidal crown terminated by a long drooping leading shoot. Leaves ½- to ¾- (or 1) inch long, about 1/16-inch wide, bluntly pointed, rounded above and

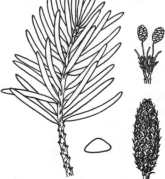

Fig. 73. Tsuga mertensiana Sarg. Branchlet, × 1. Cross section of leaf, × 5. Young ovulate cone, × 1½. Mature staminate cones, × 1.

seldom slightly grooved, glaucous on both surfaces, standing out from all sides of the branchlet. Cones cylindric, 1½ to 3 inches long, narrowed toward the apex and base, sessile.

Native in the mountains of northwestern North America. On the Pacific Coast, Mountain Hemlock occurs in the Sierra Nevada from Bubbs Creek, Tulare County, California, northward along the high slopes to Mount Shasta, thence west to the Klamath, Trinity, and Siskiyou mountains, northward on both slopes of the Cascade and Olympic mountains of Oregon and Washington to British Columbia and Alaska. It also occurs in northern Idaho and Montana. It is usually a subalpine tree inhabiting protected places at the heads of canyons facing north or east, between 6500 and 11,000 feet elevation. Rarely cultivated.

## Cupressaceae. Cypress Family

### 1. Callitris Vent. CYPRESS-PINE

(From the Greek *kallos*, beautiful, and *tris*, thrice, referring to the whole tree and the arrangement of leaves and cones in whorls of three.)

About 15 species, native to Australia, New Caledonia, and Africa.

1. **Callitris robústa** R. Br. CYPRESS-PINE. Fig. 74.

A bushy evergreen tree, 15 to 70 feet high, with the trunk often dividing close to the ground into several main divisions, and with short erect branchlets thickly covered by minute leaves. Leaves simple, scale-like, acute, dark green, in whorls of 3 or rarely 4. Flowers small, monoecious, in small catkins. Fruit a globular woody cone, ½- to ¾-inch in diameter, composed of 6 valvate scales of unequal size, maturing the second year but persisting on the branches for many years after opening. Seeds many to each scale, winged.

Fig. 74.
Callitris robusta R. Br.
Cone, × 1. Leaves on
branchlet, × 4.

Native to Australia. Occasionally cultivated. The dark green, almost black, foliage and woody cones persisting on the stems resemble somewhat those of the Monterey Cypress. Cypress-Pine differs, however, in its valvate cone-scales and its arrangement of leaves in threes.

### 2. Chamaecýparis Spach

(From the Greek *chamai*, on the ground, and *kuparissos*, cypress, referring to the resemblance to a low cypress.)

Evergreen trees, with nodding leading shoots, thin scaly or deeply furrowed bark, spreading branches, and flattened or ultimately terete branchlets. Leaves simple, scale-like, ovate, acuminate or obtuse, opposite (lateral and facial), thickly clothing the branchlets. Flowers small, monoecious, in terminal catkins; the staminate catkins oblong-ovoid, composed of numerous decussate stamens; the ovulate sub-globose, usually composed of 6 decussate peltate scales, each with 2 to 5 erect ovules. Fruit an erect globose woody cone, maturing at the end of the first year. Seeds 1 to 5 under each scale, winged.

Six species, one on the Atlantic and two on the Pacific coast of North America, and three in Japan and Formosa. *Chamaecyparis* is closely related to *Cupressus,* but differs in having smaller cones which mature the first year. Several garden forms, known as retinosporas, retain the linear spreading leaves characteristic of the juvenile state.

### KEY TO THE SPECIES

Leaves conspicuously glandular; staminate catkins red or yellow.
> Leaves with whitish marks beneath, the lateral leaves longer than the facial; branchlets disposed in horizontal planes; staminate catkins red............................................................................1. *C. lawsoniana.*
> Leaves green or bluish on both surfaces, a few linear, the lateral not longer than the facial; branchlets not in strictly horizontal planes; staminate catkins yellow.................................2. *C. thyoides.*

Leaves not or inconspicuously glandular; staminate catkins yellow.
> Leaves with whitish marks beneath.
>> Leaves on leading shoots with spreading tips, the lateral and facial leaves on the main axis about equal in length, acuminate
>> 3. *C. pisifera.*
>> Leaves closely appressed, obtuse, the lateral leaves on the main axis longer than the facial ones...........................................4. *C. obtusa.*
> Leaves dark blue-green on both surfaces...................5. *C. nootkatensis.*

**1. Chamaecyparis lawsoni-àna** Parl. LAWSON-CYPRESS. PORT ORFORD-CEDAR. Fig. 75.

A forest tree, 75 to 200 feet high, with horizontal or pendulous branches ending in broad, flat, drooping, fern-like sprays forming a narrow pyramidal crown. Leaves bright green or glaucous, conspicuously glandular above, soft to the touch, on the main axis the lateral longer than the facial ones. Staminate catkins bright red. Cones globose, about ¼-inch in diameter; scales 8 or 10, each with 2 to 4 seeds.

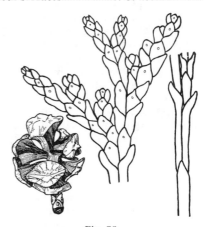

Fig. 75.
Chamaecyparis lawsoniana Parl.
Cone, lateral branchlet, portion of leading shoot, × 3.

Native to southwestern Oregon and northwestern California. Lawson-Cypress is one of the most beautiful of the larger cone-bearing trees. The following garden forms are in cultivation: Var. **álumi** Beiss. Columnar and with dense glaucous foliage. Var. **erécta** Sudw. Columnar and with dense rich green foliage. Var. **péndula** Beiss. With pendulous branchlets and dark green foliage.

2. **Chamaecyparis thyoìdes** Britt. WHITE-CEDAR.

A medium-sized to large tree, 20 to 40 (or 70) feet high. Branchlets very thin and slender. Leaves spreading at the tips, glandular.

Fig. 76.
Chamaecyparis pisifera Sieb. & Zucc.
Branchlet, × 3.

Fig. 77.
Chamaecyparis obtusa Sieb. & Zucc.
Branchlet, × 3.

Native to the eastern United States. Occasionally cultivated. The variety **ericoìdes** Sudw. (*Retinospora ericoides* Hort.) has linear heather-like leaves and is usually a compact shrub or small tree. Forms with glaucous and golden yellow branchlets are occasionally grown in parks and gardens.

3. **Chamaecyparis pisífera** Sieb. & Zucc. SAWARA RETINOSPORA. Fig. 76.

A small tree, 6 to 20 feet high, with horizontally spreading branches forming a narrow pyramidal crown. Lateral and facial leaves on the main stem of about equal length, with whitish marks beneath, acuminate. Cones ¼-inch or less in diameter.

Native to Japan. Several forms varying in color of foliage and habit of growth are in cultivation. The form *"plumosa viridis,"* sold by the California nurseries, has deep green, fern-like foliage.

**4. Chamaecyparis obtùsa** Sieb. & Zucc. HINOKI-CYPRESS. Fig. 77.

A small tree, in cultivation 10 to 30 feet high, with horizontal branches and flattened pendulous branchlets arranged in horizontal planes. Leaves not conspicuously glandular; obtuse, bright green and glossy above, with white lines beneath. Staminate catkins yellow. Cones globose, ¼- to ⅜-inch in diameter; scales 8 or 10, each with 2 (rarely 3 to 5) seeds.

Native to Japan. Several forms differing in habit of growth and color are in cultivation.

**5. Chamaecyparis nootkaténsis** Sudw. NOOTKA-CYPRESS. YELLOW-CEDAR. Fig. 78.

A forest tree, 70 to 100 feet high, with an open narrowly conical

Fig. 78. Chamaecyparis
nootkatensis Sudw.
Cone, leaves, × 2.

crown, drooping branches, scarcely flattened branchlets, and a nodding whip-like leading shoot. Leaves blue-green on both surfaces, usually not glandular, the lateral and facial ones on the main axis of about equal length, those on leading shoots with spreading acute points, forming squarish branchlets harsh and prickly to the touch. Staminate catkins yellow. Cones subglobose, about ½-inch in diameter; scales 4 or 6, each with 2 to 4 seeds.

Native on the coast and islands of southeastern Alaska and British Columbia, extending southward in the coastal mountains and in the Cascades of Washington to northern Oregon.

### 3. Cupréssus L. CYPRESS

(Ancient Latin name, derived from the Greek *kuparissos*.)

Evergreen trees or shrubs, with aromatic foliage. Leaves simple, scale-like, opposite, thickly clothing the branchlets. Flowers monoecious, in catkins; the staminate catkins composed of 6 to 12 decussate stamens; the ovulate globose. Cones globose or nearly so, consisting of 3 to 7 pairs of woody peltate scales, each scale with numerous seeds, maturing the second year.

About 14 species, from Central America to California and Arizona, and from southern Europe to southeastern Asia. Five species and 3 varieties are native on the Pacific Coast. The genus *Cupressus* is related to *Chamaecyparis*, but can be distinguished by its larger cones, which mature the second year; by the cone-scales, which bear 4 or more seeds; and by the branchlets, which are irregularly ramified (with few exceptions) instead of being arranged in distinctly flat sprays. The cypresses are hardy on the Pacific Coast in California only, where they are extensively planted for windbreaks, hedges, and as specimen trees.

**1. Cupressus lusitánica** Mill. PORTUGUESE CYPRESS. Fig. 79.

A tree, 20 to 60 feet high, with spreading or drooping branches and more or less pendulous irregularly ramified branchlets. Leaves acute, not glandular, slightly free at the tips, glaucous. Cones about ½-inch in diameter, glaucous; scales with a prominent-pointed and usually hooked umbo.

Native to Mexico. Introduced into Portugal; hence the name, Portuguese Cypress. The varieties **bénthami** Carr. and **knightiàna** Rehd. are also cultivated. Their drooping branchlets are feathery and fern-like.

2. **Cupressus fùnebris** Endl. MOURNING CYPRESS. FUNERAL CYPRESS. Fig. 80.

*C. pendula* Lamb.

A tree, 30 to 60 feet high, with wide-spreading branches and distinctly flattened drooping branchlets. Leaves light yellowish green,

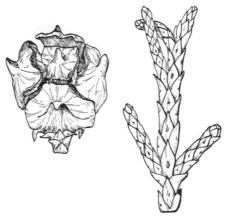

Fig. 79. Cupressus lusitanica Mill.
Cone, × 2. Leaves, × 4.

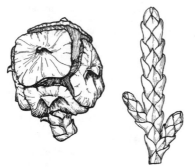

Fig. 80. Cupressus funebris Endl.
Cone, × 2. Leaves, × 3.

Fig. 81. Cupressus torulosa D. Don.
Cones, leaves, × 2.

acute, not glandular. Cones ⅓- to ½-inch in diameter; scales with a short-pointed umbo.

Native to China. Cultivated as a specimen tree for its pendulous flat branchlets.

**3. Cupressus torulòsa** D. Don. BHUTAN CYPRESS. Fig. 81.

A tree, 30 to 100 feet high, with horizontally spreading and drooping branchlets, the ultimate branchlets slightly compressed. Leaves bright or bluish green, obtuse or rarely acute, not glandular or with closed depressed resin pits. Cones ½- to ¾-inch in diameter; scales 8 or 10, the basal ones more overlapping than peltate.

Native in the Himalayas. Occasionally cultivated.

**4. Cupressus guadalupénsis** Wats. GUADALUPE CYPRESS. Fig. 82.

*C. macrocarpa* var. *guadalupensis* Mast. *C. forbesi* Jepson.

A small tree, 15 to 30 feet high, with red-brown smooth inner bark below the exfoliating reddish brown scales, and slender squarish or rounded drooping branchlets. Leaves light or bluish green, obscurely glandular. Cones ¾-inch to 1¼ inches in diameter; scales 6 or 8, very thick.

Native on dry mountain slopes of Orange County, California, south to

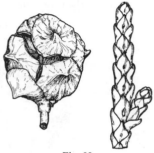

Fig. 82.
Cupressus guadalupensis Wats.
Cone, × 1. Leaves, × 3.

Lower California, and on Guadalupe Island. Cultivated in southern California. It is probable that the trees growing in Orange and San Diego counties are sufficiently distinct from those on Guadalupe Island to be classified as *Cupressus forbesi* Jepson.

**5. Cupressus macnabiàna** Murr. MACNAB CYPRESS. Fig. 83.

A small tree, 15 to 40 feet high, with an open pyramidal crown, or

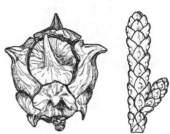

Fig. 83. Cupressus macnabiana Murr.
Cone, ×1½. Leaves, × 3.

sometimes bushy. Leaves with conspicuous dorsal resin pits, commonly glaucous, obtuse. Cones ½- to ¾-inch in diameter, reddish brown, the uppermost pair of scales usually with conical or hooked projections.

Native on the dry hills and flats of the inner North Coast Range of California from Napa County north to Siskiyou County, east to the lava beds of Modoc County, south

in scattered locations to Yuba County, and on the south fork of the Cosumnes River near Aukum, Amador County. Also in Josephine County, Oregon (on Steve Peak, according to Gorman).

*5a.* Var. **nevadénsis** Abrams.

Leaves light green. Cones ¾-inch to 1 inch in diameter, light gray-brown.

Native on the Piute Mountains, Kern County, California.

*5b.* Var. **bàkeri** Jepson. BAKER CYPRESS.

A slender tree. Cones about ½-inch in diameter, glaucous, the scales with short conical projections.

Native on the lava beds of Modoc County, California.

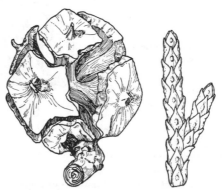

Fig. 84. Cupressus glabra Sudw.
Cone, × 1½. Leaves, × 3.

**6. Cupressus glàbra** Sudw. SMOOTH CYPRESS. ARIZONA CYPRESS. Fig. 84.

*C. arizonica* Hort. *C. arizonica* var. *bonita* Lemm.

A small or medium-sized tree, 20 to 60 feet high, with the bark on young trees separating into large thin deciduous flakes leaving a smooth red-brown surface, on older trees fissured and fibrous. Leaves acute, with conspicuous resinous glands on the back (or inconspicuously glandular), very glaucous when young. Cones ¾-inch to 1⅓ inches in diameter; scales 6 or 8.

Native in central and southern Arizona. A very hardy species and extensively cultivated in southern California. It is related to **C. arizonica** Greene, but that species has smaller cones and the leaves are usually glandless.

**7. Cupressus sémpervìrens** L. ITALIAN CYPRESS.

A tree, 20 to 60 feet high, with erect or horizontal branches and dark green foliage forming a pyramidal crown. Cones about 1 inch in diameter.

Native in southern Europe and western Asia. The variety **stricta** Ait. (var. *fastigiata* Beiss.), with erect branches which form a narrow columnar crown, is extensively used in formal landscaping.

**8. Cupressus goveniàna** Gord. GOWEN CYPRESS. Fig. 85.

*C. pygmaea* Sarg.

A small tree, or sometimes shrub-like, 4 to 20 (or rarely 80) feet high, with squarish branchlets and light to dark green foliage. Leaves not glandular or a few with closed resin pits. Cones ½- to ¾-inch long. Seeds black.

Native to the Monterey Peninsula and the Mendocino "White Plains" east of Fort Bragg and Mendocino

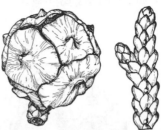

Fig. 85. Cupressus goveniana Gord. Cone, × 2. Leaves, × 4.

City, California, where it often grows no higher than 2 to 10 feet and forms a "pigmy forest." The trees growing on the Mendocino "White Plains" and those grown from seeds collected there have a whip-like leading shoot, whereas the plants at Monterey are more bush-like.

**9. Cupressus sárgenti** Jepson. SARGENT CYPRESS. Fig. 86.

A small tree, 10 to 40 feet high, or sometimes shrub-like, with erect or spreading branches and thickish obscurely squarish branchlets forming an open crown. Leaves not glandular or some with closed resin pits. Cones ¾-inch to 1 inch in diameter. Seeds reddish brown.

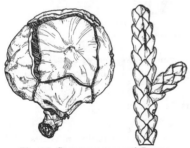

Native on the dry mountain slopes and in the valleys of the Santa Lucia, Santa Cruz, and San Rafael mountains and of Marin, Sonoma, Napa, Lake, and Mendocino counties, California. Rarely cultivated.

Fig. 86. Cupressus sargenti Jepson. Cone, × 1½. Leaves, × 2.

*9a.* Var. **dúttoni** Jepson.

A medium-sized to large tree, 40 to 70 feet high.

Native on Cedar Mountain, Alameda County, California.

**10. Cupressus macrocárpa** Hartw. MONTEREY CYPRESS. Fig. 87.

A tree, 20 to 75 feet high, with horizontal branches, rope-like branchlets, and dark green foliage forming a broad spreading crown. Leaves obtuse, not glandular. Cones subglobose, 1 to 2 inches long. Seeds reddish brown.

A species of restricted distribution, native to the headlands about Carmel Bay, California, where many trees exhibit irregularly shaped and often prostrate crowns because of exposure to the strong ocean winds. Monterey Cypress is the most extensively cultivated species

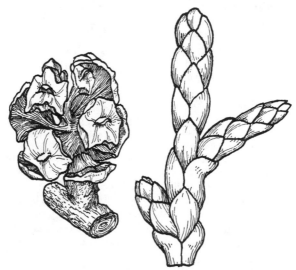

Fig. 87. Cupressus macrocarpa Hartw.
Cone, × 1. Leaves, × 5.

in California. It is used for hedges, windbreaks, and as a park tree. In the vicinity of San Francisco Bay most of the trees are being killed by a wood borer.

### 4. Juníperus L. JUNIPER

(The ancient Latin name.)

Aromatic evergreen trees or shrubs. Leaves simple, in cycles of 2 or 3, of 2 kinds—short-linear to linear-lanceolate and extending radially, or scale-like, closely appressed and covering the cord-like branchlets. Flowers small, usually dioecious; the staminate in small axillary catkins consisting of many stamens; the ovulate consisting of few opposite pairs of fleshy scales, each scale with 1 or 2 ovules. Fruit an ovoid berry-like cone, usually with fleshy coalesced scales, maturing the first, second, or third year.

About 35 species, widely distributed in the northern hemisphere, of which 4 tree-species are native to the Pacific Coast. The genus

*Juniperus* is closely related to *Cupressus* and often the two genera are difficult to distinguish without fruits. The juvenile leaves of both genera are short-linear, but those of *Juniperus* have white bands on their upper surfaces. The junipers usually seen in cultivation are dwarfed or prostrate shrub-like plants.

KEY TO THE SPECIES

The leaf characters used in this key refer to the adult foliage.

Leaves uniformly short-linear, ⅓- to ¾-inch long, radially spreading, spiny-pointed, not glandular..........................................1. *J. communis.*

Leaves all scale-like and imbricated, or both kinds—that is, the very short-linear and the scale-like—on the same plant.

Leaves of 2 kinds, some very short-linear, spreading, and spiny-pointed at their tips, others scale-like.

Scale-like leaves mostly acute or acuminate; linear leaves opposite, or on leading shoots, in threes; fruit blue-black, maturing the first year..........................................2. *J. virginiana.*

Scale-like leaves obtuse; linear leaves commonly in threes; fruit brownish purple, maturing the second year..........3. *J. chinensis.*

Leaves all scale-like and closely appressed.

Fruit maturing the first year; leaves acute, in threes
2. *J. virginiana.*

Fruit maturing the second year.

Leaves glandular on the backs.

Fruit bluish black; cotyledons 2; species of Transition, Canadian, or Hudsonian life zones.

Leaves obscurely glandular or without glands, entire
4. *J. scopulorum.*

Leaves distinctly glandular, finely serrulate under a lens, usually in threes..........................................5. *J. occidentalis.*

Fruit reddish brown beneath the bloom; cotyledons 4 to 6.

Fruit ¼- to ⅜-inch in diameter; introduced species
6. *J. excelsa.*

Fruit ⅜- to ⅝-inch in diameter; native species of the Upper Sonoran Life Zone....................................7. *J. californica.*

Leaves usually glandless on the backs.

Fruit reddish brown beneath the bloom; leaves serrulate
7a. *J. californica* var. *utahensis.*

Fruit bluish black; leaves entire......................4. *J. scopulorum.*

### 1. Juniperus commùnis L. COMMON JUNIPER. Fig. 88.

A small slender tree, 8 to 35 feet high, with a bushy pyramidal crown. Leaves in whorls of three, short-linear or linear-lanceolate, ⅓- to ¾-inch long, sharp-pointed, concave and with 2 broad white bands on the upper surface. Fruit globose, about ¼-inch in diameter, blue-glaucous, maturing the second or third year.

Native to northern and central Europe, extending eastward through northern Asia to Korea and Japan, and in the northeastern United States. A variable species with several geographic and garden forms, commonly used in landscaping.

The variety **hibérnica** Gord. has upright dark green branches forming a narrow columnar crown. It is more desirable than the species as a small ornamental tree.

2. **Juniperus virginiàna** L. RED-CEDAR. Fig. 89.

A small tree, 10 to 30 (or 60) feet high, with ascending or spreading branches. Leaves in threes, of 2 kinds, the juvenile very short-linear, spreading, and spiny-pointed, the adult scale-like, acute or acuminate, and well imbricated. Fruit globular, ¼- to ⅓-inch in diameter, maturing the first year.

Fig. 88. Juniperus communis L.
Branchlet with fruit, × 3.

Native to the eastern and central United States and Canada. In its typical form *J. virginiana* is one of the largest junipers. Many garden forms varying in color of foliage and habit of growth are in cultivation.

3. **Juniperus chinénsis** L. CHINESE JUNIPER. Fig. 90.

Typically a small tree, 10 to 30 feet high, with rather slender ascending branches. Leaves in twos or threes, of 2 kinds, the juvenile very short-linear, pointed, spreading, and with 2 white bands on the upper surface, the adult scale-like, obtuse at the apex, and appressed to the branchlets. Fruit globular, about ¼-inch in diameter, brownish purple, maturing the second year.

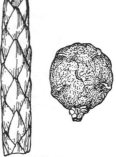

Fig. 89.
Juniperus virginiana L.
Leaves, × 10. Fruit, × 3.

Native to China and Japan. There are numerous garden forms of high ornamental value, varying in habit of growth, form, and color of young and old foliage.

**4. Juniperus scopulòrum** Sarg. COLORADO JUNIPER. ROCKY MOUNTAIN JUNIPER. Fig. 91.

*Sabina scopulorum* Rydb.

A tree, 25 to 50 feet high, often branched at the base into several secondary trunks. Leaves usually in twos, scale-like, obscurely glandular. Fruit globular, about ¼-inch in diameter, blue-glaucous, maturing the second year.

Native in the foothills of the Rocky Mountains from Alberta to Texas, extending

Fig. 90. Juniperus chinensis L.
Leaves, × 6. Fruit, × 3.

westward to British Columbia, Washington, eastern Oregon, Nevada, and Arizona. Rarely cultivated.

**5. Juniperus occidentàlis** Hook. SIERRA JUNIPER. WESTERN JUNIPER. Fig. 92.

*Sabina occidentalis* Heller.

Fig. 91.
Juniperus scopulorum Sarg.
Fruit, leaves, × 3.

A tree, 10 to 60 feet high, with deeply furrowed reddish brown shreddy bark, often with very large spreading branches forming a low broad crown. Leaves in threes and scale-like, glandular. Fruit globular to oblong-ovoid, about ¼-inch long, bluish black, with a whitish bloom, maturing the second year.

Native in the high mountains and plateaus of Washington, Oregon, western Idaho, and California. A very picturesque tree, often with a gnarled and grotesque form; in the Sierra Nevada growing even in granite crevices of the wind-swept mountain ridges from 6500 to 11,000 feet elevation.

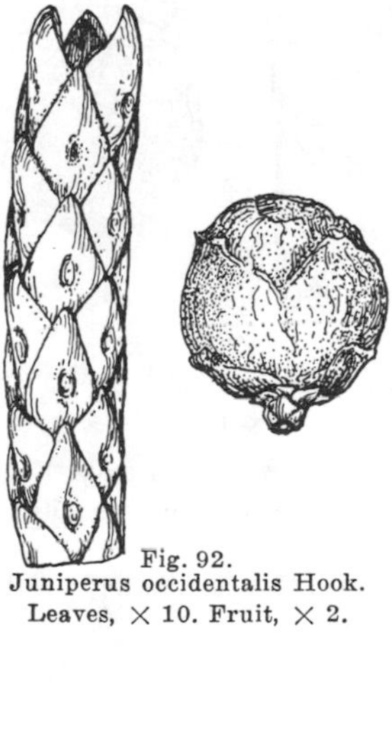

Fig. 92.
Juniperus occidentalis Hook.
Leaves, × 10. Fruit, × 2.

**6. Juniperus excélsa** Bieb. GREEK JUNIPER.

A tree, 20 to 50 feet high, with ascending or spreading branches. Leaves of 2 kinds, the juvenile leaves in threes, spreading, linear,

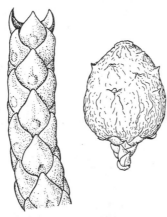

about ¼-inch long, with 2 white bands above, the adult scale-like, acute, glandular, closely appressed. Fruit globose, about ⅓-inch in diameter, purplish brown, covered with a bloom.

Native to Greece and western Asia. The variety **strícta** Rollisson is a columnar form with glaucous juvenile foliage, suitable for planting against dwellings.

**7. Juniperus califórnica** Carr. CALIFORNIA JUNIPER. Fig. 93.

*Sabina californica* Antoine.

A small tree or large shrub, 10 to 30 feet high, with several secondary trunks from near the base. Leaves in threes, scale-like, distinctly glandu-

Fig. 93. Juniperus californica Carr.
Leaves, × 10. Fruit, × 2.

lar-pitted on the back. Fruit globose to oblong, ⅜- to ⅝-inch long, reddish brown beneath the whitish bloom, maturing the second year.

Native on the higher desert mountain and foothill slopes from Lower California northward in California to the Mohave Desert and the Tehachapi Mountains, in the inner Coast Range to Tehama County, and on the western slope of the Sierra Nevada in Tulare and Kern counties. Rarely cultivated.

*7a.* Var. **utahénsis** Engelm. DESERT JUNIPER. UTAH JUNIPER. Fig. 94.

*J. utahensis* Lemm. *Sabina utahensis* Rydb.

A large arborescent shrub or small tree, 3 to 15 feet high, with a bushy

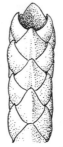

Fig. 94.
Juniperus californica var.
utahensis Engelm.
Leaves, × 10. Fruit, × 2.

habit. Leaves in twos or threes, scale-like, usually without glands. Fruit globular, 3/16- to ⅜-inch in diameter, reddish brown under the whitish bloom, maturing the second year.

Native on the desert ranges of Inyo County, California, and on the arid mountain slopes of the Great Basin region.

### 5. Libocédrus Endl. INCENSE-CEDAR

(From the Greek *libas,* drop or tear, and *kedros,* referring to the resinous character.)

Eight species, the genus distributed in western North and South America, New Zealand, New Guinea, New Caledonia, Formosa, and southwestern China. One species is native on the Pacific Coast. *Libocedrus* is sometimes confused with *Thuja* and *Chamaecyparis* because of the arrangement of the branchlets in flat sprays, but can be distinguished from these genera by the leaves, which are arranged apparently in cycles of four.

### 1. Libocedrus decúrrens Torr. CALIFORNIA INCENSE-CEDAR. Fig. 95.

An aromatic evergreen forest tree, 50 to 150 feet high, with a tapering trunk from a broad base, thick fibrous cinnamon-brown

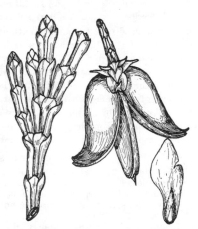

deeply ridged bark, erect or spreading branches, and flattened branchlets arranged in one plane. Leaves simple, scale-like, 1/12- to 1/4-inch long, closely appressed to the branchlets by decurrent bases, only the tips free, cyclic and 4-ranked, the lateral pair keel-shaped and nearly covering the flattened inner pair, the 2 pairs of about equal length. Flowers monoecious, in solitary and terminal catkins; the staminate catkins composed of 6 to 8 pairs of decussate stamens; the ovulate consisting of 2 to 4 pairs of opposite scales, only one pair

Fig. 95. Libocedrus decurrens Torr.
Leaves, × 3. Cone, × 1. Seed, × 1.

bearing ovules at the base. Fruit a small oblong-ovoid reddish brown cone, 3/4-inch to 1 inch long, pendulous, maturing the first year, with one large pair of seed-bearing scales separated by a united pair and usually with a small reflexed pair at the base. Seeds 1/3- to 1/2-inch long, with 2 unequal coalesced wings.

Native in southern Oregon, the Sierra Nevada, the higher Coast Ranges, the mountains of southern California, and Lower California. Usually inhabiting mountain slopes in the main timber belt, from 2500 to 7000 feet elevation. A very handsome tree of columnar or pyramidal habit, frequently cultivated.

The species **L. chilénsis** Endl., fig. 96, is occasionally seen in cultivation as a small bush-like tree, 6 to 15 feet high. The lateral leaves are much longer than the facial leaves. Native to Chile.

Fig. 96.
Libocedrus chilensis Endl.
Branchlet, × 3.

### 6. Thùja L. ARBORVITAE

(From the Greek *Thya*, a resinous tree.)

Evergreen aromatic trees, with thin scaly bark, erect or spreading branches, and flattened branchlets disposed in one plane. Leaves simple, scale-like, decussate, closely appressed and covering the branchlets, the lateral pair compressed and nearly covering the facial ones. Flowers monoecious, in terminal solitary catkins; the staminate catkins composed of 4 to 12 decussate and peltate stamens, the ovulate composed of 6 to 12 erect scales in opposite pairs. Fruit an ovoid to oblong cone, with 6 to 12 acute scales and 2 seeds to each scale, maturing the first year. Seeds winged or wingless.

Six species, native to North America, eastern Asia, and Japan. All but one of the species are in cultivation and scores of named garden varieties are sold by nurserymen as ornamentals. Juvenile forms of bushy habit and with spreading linear leaves turning brownish in winter are sold under the trade name, Retinospora.

KEY TO THE SPECIES

Branchlets in vertical planes; cone-scales thick and leathery; seeds wingless..................................................................................1. *T. orientalis.*
Branchlets in horizontal planes; cone-scales thin; seeds winged.
    Leaves conspicuously glandular, not whitish beneath; cones usually with 4 fertile scales................................................2. *T. occidentalis.*
    Leaves not or inconspicuously glandular, usually whitish beneath; cones usually with 6 fertile scales................................3. *T. plicata.*

**1. Thuja orientàlis** L. ORIENTAL ARBORVITAE. Fig. 97.

A small bushy tree, with the main trunk branching near the base and with spreading and ascending branches forming a compact narrow pyramidal crown. Leaves yellowish green or often bright green above, usually glandular, disposed in flat vertical fan-like sprays. Cones oblong-ovoid, ½-inch to 1 inch long, usually with 6 leathery scales having recurved horn-like tips, bluish before ripening. Seeds ovoid, brown, wingless.

Native to northern and western China and Korea. Many garden forms of this species are extensively cultivated. Some are distinguished by the color of the foliage and others by their form of growth.

2. **Thuja occidentàlis** L. AMERICAN ARBORVITAE. Fig. 98.

A small tree, 10 to 40 feet high, with a short trunk often dividing into 2 or 3 stout secondary stems, short horizontal branches ascending

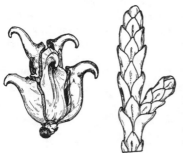

at the tips, and flat yellow-green branchlets disposed in horizontal planes forming a narrow compact pyramidal crown. Leaves bright green above, yellow-green below, those on the leading shoots long-pointed and conspicuously glandular, those on the lateral branchlets short-pointed, without glands or inconspicuously glandular. Cones oblong-ovoid, ⅓- to ½-inch long, with 6 to 10 thin pointless scales.

Fig. 97. Thuja orientalis L.
Cone, × 1. Leaves, × 4.

Seeds with thin wings about as wide as the body.

Native to southeastern Canada and the northeastern United States. The number of named varieties, among which are the following— *pyramidalis, compacta, robusta, fastigiata, filicoides, pendula, argentea, alba, aurea, lutea, lutescens,* and *aureo-variegata* — will give some idea of the diversity in habit of growth and color.

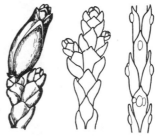

Fig. 98. Thuja occidentalis L.
Left to right: cone, leaves on lateral branchlet, leaves on leading shoot, × 2.

3. **Thuja plicàta** D. Don. GIANT ARBORVITAE. CANOE-CEDAR. Fig. 99.

   *T. gigantea* Nutt.

A forest tree, attaining a height of 200 feet and a diameter of 15 feet, with a trunk tapering from and often strongly buttressed at the base, short horizontal branches usually pendulous at the ends, and slender much compressed branchlets forming a narrow

Fig. 99.
Thuja plicata D. Don.
Cone, × 1. Leaves, × 1½.

pyramidal dense crown. Leaves bright green and glossy above, usually with whitish spots beneath, those on the leading shoots long-pointed and conspicuously glandular, those on the lateral branchlets short-pointed and without glands or obscurely glandular. Cones clustered near the ends of the branches, much reflexed, oblong, about ⅜-inch long, with 8 or 10 thin leathery scales. Seeds winged, about 3/16-inch long, slightly notched at the apex, often 3 on each fertile scale.

Native along the northwestern coast of North America from Alaska to Humboldt County, California, and eastward in British Columbia, northern Washington, and Idaho to Montana. Giant Arborvitae is one of the most popular native trees cultivated in parks and gardens. A form *"aurea,"* with golden green foliage, is an exceptionally fine specimen tree of pyramidal habit. Variety **péndula** Schneid. has slender pendulous branches. Variety **fastigiàta** Schneid. has a columnar habit.

# ANGIOSPERMAE

Plants with the seeds borne within closed ovaries

## II. PALM AND PALM-LIKE TREES (MONOCOTYLEDONS)
## Liliaceae. Lily Family

### 1. Cordyline Comm. DRACENA

(From the Greek *kordyle,* a club, in reference to the thickened roots.)

Tropical-appearing trees with a single trunk branching at the summit when mature. Leaves simple, long-acuminate, stiff or leathery, entire, usually crowded in clusters at the ends of the trunk or branches. Flowers small, numerous, white, on solitary pedicels arranged in large terminal panicles; ovules numerous in each of the 3 cells of the ovary. Fruit a globose white or blue berry. Six to 10 species, native in the tropics of both hemispheres. The plants of this genus are commonly known in cultivation as Dracena-Palms, but they are not members of the genus *Dracaena,* nor are they true palms.

Fig. 100. Cordyline australis Hook., Oakland, California.

KEY TO THE SPECIES

Leaves rather lax, 1 to 2 inches wide, green beneath, with 12 to 20 veins on each side of the midvein....1. *C. australis.*

Leaves stiff, 2 to 5 inches wide, blue-glaucous beneath, with 40 to 50 veins on each side of the midvein.................................2. *C. indivisa.*

### 1. Cordyline austràlis Hook. GREEN DRACENA. Fig. 100.

*Dracaena australis* Forst.

A tall, arborescent, palm-like plant, 15 to 40 feet high, branching at the close of each flowering period. Leaves 1½ to 3 feet long, 1 to 2½ inches wide, with 12 to 20 fine veins on each side of the indistinct midvein, dull green on both surfaces, erect or spreading, drooping against the trunk and turning brown in age. Flowers white, fragrant, in erect or drooping terminal panicles. Berry white or bluish white.

122 *Pacific Coast Trees*

Native to New Zealand. Planted as park and street trees in California.

2. **Cordyline indivisa** Kunth. BLUE DRACENA.

*Dracaena indivisa* Hort.

An arborescent palm-like plant, 10 to 25 feet high. Leaves erect, spreading, and drooping, 2 to 6 feet long, 2 to 5 inches wide, with 40 to 50 rather heavy and conspicuous veins on each side of the prominent reddish midvein, blue-glaucous beneath.

Native to New Zealand. Occasionally cultivated in gardens and parks in California.

2. **Dracaèna** L. DRACENA

(In reference to the female dragon; the dried juice is supposed to resemble the dragon's blood.)

About 50 species of woody plants, native in tropical regions, chiefly of the eastern hemisphere.

1. **Dracaena dràco** L. DRAGON TREE. Fig. 101.

A large arborescent plant, 25 to 60 feet high, with the

Fig. 101. Dracaena draco L., Anaheim, California.

trunk branching at the summit. Leaves sword-shaped, 1 to 2 feet long, 1 to 1¾ inches wide, entire, leathery, glaucous, crowded in clusters at the ends of the branches. Flowers small, numerous, greenish white, on pedicels in clusters of 2 or more, arranged in large terminal panicles; ovules 1 in each of the 3 cells of the ovary. Fruit a globose orange-colored berry.

Native of the Canary Islands. A hardy species, occasionally cultivated in parks and private grounds in California.

### 3. **Yúcca** L.

(A modification of *Yuca*, the name of the manihot, a cassava erroneously applied to this genus.)

Acaulescent or arboreous evergreen plants. Leaves simple, dagger- or bayonet-like, long-pointed, serrulate or filamentous on the margins, arranged in spiral rosettes at the surface of the ground or at the ends of the trunk or branches. Flowers large, cup-shaped, white or creamy, fragrant, opening at night, usually pendulous, borne in large erect panicles. Fruit a 3-celled capsule, not winged. Seeds numerous, in 2 rows in each cell.

About 30 species, native to North America and the West Indies. In addition to the following more or less tree-like species, **Yucca baccàta** Torr., which is a nearly stemless plant 2 to 3 feet high, is frequently cultivated in the warmer parts of California.

Fig. 102. Yucca brevifolia Engelm., near Victorville, California.

KEY TO THE SPECIES

Leaves 1 foot or less long
　　　　　1. *Y. brevifolia.*
Leaves more than 1 foot long.
　Margins of leaves denticulate or irregularly serrulate....2. *Y. aloifolia.*
　Margins of leaves entire, with curly shreddy filaments
　　　　　3. *Y. mohavensis.*

1. **Yucca brevifòlia** Engelm. JOSHUA TREE. Fig. 102.
　　*Y. arborescens* Trel.

A tree, 15 to 30 feet high, with a columnar much branched trunk forming a weird and picturesque open crown. Leaves 6 to 10 inches long, denticulate, not shreddy-filamentous. Flowers greenish white, about 2 inches long, borne in dense panicles about 1 foot long.

Native on the Mohave Desert of California, northward to the desert regions of Inyo and eastern Kern counties, and eastward to Utah. Rarely cultivated.

2. **Yucca aloifòlia** L. SPANISH BAYONET.

A slender-stemmed plant, 6 to 25 feet high, with a simple or branched trunk. Leaves 1 to 2½ feet long, 1 to 2 inches wide, very stiff and sharp-pointed, denticulate or irregularly serrulate. Flowers white or creamy, about 3 inches broad when open, usually pendulous, in erect panicles 1 to 2 feet long.

Native to the West Indies, the southern United States, and Mexico. Frequently cultivated in desert-like gardens. Variety **marginàta** Bommer has creamy white leaf-margins.

3. **Yucca mohavénsis** Sarg. SPANISH DAGGER. Fig. 103.

An arborescent plant, 4 to 15 feet high, with a simple or branched trunk. Leaves 1 to 3 feet long, 1 to 2 inches wide, yellowish green, concave above, the margins entire and with curled shreddy filaments.

Fig. 103. Yucca mohavensis Sarg., Mohave Desert, California.

Flowers white, in erect panicles 1 to 1½ feet long.

Native to the desert mountain regions of southern California, east to Arizona and Nevada, and south to Lower California. Rarely cultivated.

## Musaceae
### Banana Family
#### 1. **Mùsa** L. BANANA

(Named in honor of Antonio Musa, physician to Octavius Augustus, first Roman emperor.)

Large coarse perennials of tropical appearance with 1 to several stems. Leaves simple, usually very large, pinnately parallel-veined, entire, arranged spirally in a rosette at the summit of a trunk formed by the sheathing leaf-bases. Flowers monoecious, in spike-like clusters, each cluster subtended by a single large colored bract, the staminate flowers in the upper clusters, the pistillate in the lower. Fruit large, cylindric with abruptly tapering ends, usually fleshy.

About 60 species and more than 200 cultivated varieties, native to tropical Asia, Africa, Australia, and the adjacent islands.

The genus *Strelitzia*, Bird of Paradise Flower, which also belongs to the Banana Family, may be confused with the genus *Musa* because of its banana-like leaves. It may be distinguished, however, from *Musa* by the 2-ranked sheathing leaves which form a flattened trunk-like base and by the large bisexual very irregular flowers. **S. nícolai** Regel & Koch is the largest species grown in southern California.

KEY TO THE SPECIES

Leaves 10 to 20 feet long, acute at base, the midrib red; the free part of the petiole 6 inches or less long........................................1. *M. ensete.*
Leaves 4 to 10 feet long, rounded at base, the midrib green; the free part of the petiole 1 to 2 feet long..2. *M. paradisiaca* var. *sapientum.*

1. **Musa ensèté** Gmel. ABYSSINIAN BANANA. Fig. 104.

*Ensete edule* Horan.

Large tropical plants, 20 to 40 feet high, with a stem 10 to 20 feet high, swollen at the base, not stoloniferous. Leaf-blades oblong, 10 to 20 feet long, 2 to 3 feet wide, spreading and arching downward, acute at the apex and base, the midrib red; the free part of the petioles 6 inches or less long. Fruit 2 to 3 inches long, not edible.

Native in the mountains of Abyssinia. Cultivated as an ornamental plant in the frostless parts of California. Not hardy in the north.

2. **Musa paradisìaca** var. **sapiéntum** Kuntze. COMMON BANANA. Fig. 105.

*M. sapientum* L.

Plants 10 to 25 feet high,

Fig. 104. Musa ensete Gmel., Berkeley, California.

stoloniferous, the stems usually in clumps. Leaf-blades oblong, 5 to 9 feet long, 1½ to 2 feet wide, erect or ascending, rounded at the base, the midrib green; the free part of the petioles 1 to 2 feet long. Fruit edible.

Native to India. Cultivated as an ornamental plant in southern California. Most of the bananas of commerce are obtained from this variety and its many horticultural forms.

## Palmaceae. Palm Family
### 1. **Chamaèrops** L. DWARF FAN PALMS

(The Greek for dwarf bush.)

One species of the Mediterranean region:

### 1. Chamaerops hùmilis L. HAIR PALM. Fig. 106.

A low bushy fan palm, 10 to 25 feet high, usually suckering and forming a clump, but sometimes with a single trunk 1 to 3 feet high. Leaf-blades 2 to 3 feet wide, glaucous when young, parted ⅓ to ⅔ the distance to the base into narrow rigid divisions; petioles 2 to 3 feet long, slender, usually armed with stout straight or hooked spines. Flowers, small, yellow, perfect, borne in panicles among the leaves.

The only palm native to Europe. Very hardy, cultivated on lawns and as a tub plant. Single-stemmed plants can be had by removing the suckers.

Fig. 105.
Musa paradisiaca var. sapientum Kuntze,
Whittier, California.

### 2. Còcos L. COCOS PALMS

(From the Portuguese *cocos*, monkey, in reference to the resemblance of the coconut to a monkey's face.)

Small or large monoecious feather palms, with slender or thick and ringed trunks often covered with the bases of the leaves. Leaves pinnately dissected into numerous lanceolate segments; rachis triangular, acute above, convex below; petioles concave above, smooth or spiny on the margins. Flowers numerous, creamy white, in large panicles borne among the leaves. Fruit subglobose, ovoid, or ellipsoidal, terete or somewhat 3-angled, often fibrous-coated as in the Coconut.

About 50 species, native to the tropics and subtropics of South America, one species extending around the world. The genus *Cocos* has been divided into several genera, a segregation leaving the Coconut, *C. nucifera* L., the only species in the original genus. Since two of the ornamental Cocos Palms are widely grown in southern California under the name *Cocos* and because *Standardized Plant Names* includes these in the genus *Cocos,* it seems unwise to segregate them in this manual.

KEY TO THE SPECIES

Tall plants with erect-spreading dark green leaves 9 to 15 feet long
1. *C. plumosa.*

Dwarf plants with strongly arched and recurved sage-green leaves 8 to 12 feet long..................................................................2. *C. australis.*

1. **Cocos plumòsa** Hook. PLUME PALM. COCOS PALM. Fig. 107.
*Arecastrum romanzoffianum* Becc.

A tall feather palm, 15 to 40 feet high, with a smooth slender trunk 8 to 12 inches thick, ringed at intervals of about 12 inches and covered near the summit with a few hanging leaves and remains of the dead petioles. Leaf-blades erect-spreading, 9 to 15 feet long, somewhat recurving; the leaf-divisions 14 to 18 inches long, usually not more than 1 inch wide, soft and flexible, alike on both surfaces; petioles 3 to 6 feet long, sheathing the trunk for most of their length. Flower-clusters 2 to 3 feet long, borne among the leaves. Fruit sub-globose, about 1 inch long, orange-colored.

Native to Brazil. The most extensively cultivated palm along streets, highways, and in parks and gardens in

Fig. 106. Chamaerops humilis L.,
Santa Barbara, California.

southern California. Very desirable because of its erect-spreading crown of long feathery green leaves. Does not endure heavy frosts.

2. **Cocos austràlis** Mart. PINDO PALM. Fig. 108.

A small feather palm, 10 to 20 feet high, with usually a short trunk clothed with the remains of the old petioles, or, if taller, then smooth and ringed below. Leaf-blades 8 to 12 feet long, strongly arched and recurved; the leaf-divisions linear, 1 to 2 feet long, ¾-inch to 1½ inches wide, rather stiff, bluish or gray-green; petioles unarmed. Panicle of flowers borne among the leaves. Fruit ovoid, about ¾-inch long, edible, borne in large clusters.

Native to Paraguay. A hardy species, planted in California as a

Fig. 107. Cocos plumosa Hook.,
Whittier, California.

street tree and in lawns for its graceful arched gray-green leaves.

### 3. Erythèa Wats.

(From Erythea, one of the Hesperides, Daughter of Evening.)

Three species of fan palms, native to Mexico and one to Guadalupe Island.

### 1. Erythea edùlis Wats. GUADALUPE PALM. Fig. 109.

*Brahea edulis* Wendl.

A fan palm, with a stout columnar trunk 10 to 35 feet high and about 15 inches in diameter, covered with a smooth corky cracked bark ringed by leaf-scars. Leaf-blades about 3 feet long, green on both surfaces, to-mentose at first, with 70 to 80 folds divided to ⅓ or ½ the length of the blade, the divisions 1 to 1½ inches wide, torn or irregularly cleft at the apex; petioles stout, 3 to 6 feet long and about 1 inch wide, unarmed or with some very short and blunt spine-teeth. Panicles of perfect flowers borne among the leaves. Fruit globose, about 1 inch in diameter, shining, black, with edible pulp, borne in long drooping clusters.

Native on Guadalupe Island off the coast of Lower California. Cultivated in the coastal and warm valley counties of California for

Fig. 108. Cocos australis Mart.,
Santa Barbara, California.

its handsome growth of dark green leaves.

#### 4. **Glaucothèa** Cook

(*Glaucothea,* Gray Goddess; the name was suggested by the relationship of this genus to *Erythea.*)

A single species, usually included in the genus *Erythea.*

##### 1. **Glaucothea armàta** Cook. BLUE PALM. Fig. 110.

*Erythea armata* Wats. *Brahea armata* Wats.

A fan palm, with a stout robust trunk 18 to 35 feet high and to 3 feet in diameter, naturally covered above by a thatch-like skirt of old leaves. Leaf-blades 2½ to 3½ feet long, silvery blue,

Fig. 109. Erythea edulis Wats.
Reproduced by permission from *Trees and Shrubs of California Gardens* by C. F. Saunders.

with 30 to 50 folds divided nearly to the middle; petioles about 3 feet long, 1 to 2 inches wide, with numerous stout more or less hooked spine-teeth. Flowers perfect, dull purple, in panicles among the leaves. Fruit a globose black drupe, about ¾-inch in diameter.

Native to Lower California. A slow-growing palm cultivated in gardens and parks in southern California.

#### 5. **Jubaèa** H. B. K.

(Named for Juba, king of Numidia.) A single species.

##### 1. **Jubaea spectábilis** H. B. K. SYRUP PALM. MONKEY-COCONUT. Fig. 111.

Fig. 110. Glaucothea armata Cook, Santa Barbara, California.

A massive feather palm, with a columnar trunk to 30 feet high and 3 to 5 feet in diameter, covered with scars of the fallen leaf-bases and bearing a crown of erect-spreading foliage. Leaf-blades 6 to 10 feet long, green, with numerous segments 1½ to 2 feet long and about 1 inch wide; petioles short, not spiny. Flowers monoecious, borne in panicles among the leaves. Fruit a globose-ovoid drupe, 1 to 1½ inches long, fibrous, yellow.

Fig. 111. Jubaea spectabilis H. B. K., Huntington Botanical Garden, San Marino, California.

Native to Chile. Occasionally grown in central and southern California. The Chileans make "palm-honey" from the sap of the large trunks. The resemblance of the fruits to small coconuts has given rise to the name Monkey-Coconut.

6. **Livistòna** R. Br.

(Named in honor of Patrick Murray, of Livistone, near Edinburgh.)

About 18 species of fan palms, native to eastern Asia, Malaya, the Philippines, New Guinea, and Australia.

1. **Livistona chinénsis** R. Br. CHINESE FAN PALM. Fig. 112.

A tall fan palm, with an erect trunk 18 to 25 feet high and 12 to 16 inches in diameter, smooth, ringed, and bearing a crown of erect-spreading leaves which hang downward but stand out from the trunk in age. Leaf-blades 3 to 5 feet long, dark green and glossy, divided to the middle or deeper into numerous segments which bend downward about 1 foot from the apex; petioles 3 to 7 feet long, 1 to 2 inches wide, unarmed in old leaves except for a few short blunt teeth near the base. Flowers perfect, small, in long-peduncled clusters appearing among the leaves. Fruit an ellipsoidal drupe, ½- to ¾-inch long, nearly black.

Native to China. Occasionally cultivated in California gardens and parks. The leaves are somewhat similar to those of the Washington Fan Palm, but they never hang against the trunk, and the petioles are greatly narrower and scarcely armed.

### 7. **Loròma** Cook

(Probably from the Latin *lorum,* in reference to the strap-like forms of the fibers surrounding the seed.)

Two species, native to Australia.

### 1. **Loroma amethystìna** Cook. KING PALM. Fig. 113. *Seaforthia elegans* Hort.

A large feather palm, with a columnar trunk 20 to 30 feet high and 14 to 24 inches in diameter, smooth except for the ringed leaf-scars, the upper part composed of long smooth green sheathing petioles. Leaf-blades 10 to 20 feet long, spreading and recurved, green, with numerous flat segments 1 to 1½ feet long and 2 to 3 inches wide; petioles mostly sheathing the trunk, unarmed.

Fig. 112. Livistona chinensis R. Br., Oakland, California.

Flowers monoecious, purplish pink, in panicles arising from the trunk far below the leaves. Fruit a subglobose drupe, about ½-inch long, red.

Native to Australia. Planted in southern California, where, according to O. F. Cook (*Journal of the Washington Academy of Sciences,* Vol. V, No. 1, p. 116), this palm is sold as *Archontophoenix alexandrae* Wendl. & Drude, *A. cunninghamiana* Wendl. & Drude, and *Seaforthia elegans* Hook. Horticulturists usually use the last of these names.

### 8. **Phoènix** L. DATE PALM

(From the Greek name of the tree, applied by Theophrastus, perhaps in reference to Phoenicia.)

Dioecious feather palms, either trunkless, or with short or long and erect or reclining trunks. Leaves spreading and recurved from the summit of the trunk, divided into many lanceolate segments which fold lengthwise and upward; rachis flattened laterally, convex on the back; petioles armed with shortened stout spiny leaf-segments. Flowers dioecious, small, yellowish, in large panicles borne among the leaves. Fruit a fleshy drupe or berry, with grooved stones.

About 12 species, native to tropical or subtropical Asia and Africa. The identification of species of the genus *Phoenix* is often difficult because the plants are dioecious and the technical characters often used refer to both the fruits and the staminate flowers. The genus *Phoenix* can be distinguished from all other genera of feather-leaved palms by the upward and lengthwise folding of the leaf-segments (pinnae) and the peculiar furrow of the stones, characteristic of dates.

Fig. 113.
Loroma amethystina Cook, Gillespie Estate, Montecito, California. Reproduced by permission from *Trees and Shrubs of California Gardens* by C. F. Saunders.

KEY TO THE SPECIES

Trunk 1½ to 5 feet in diameter; leaves 15 to 20 feet long.
  Leaves stiff, erect-ascending, curved, pale green or glaucous
                                       1. *P. dactylifera.*
  Leaves spreading, strongly arched and drooping, dark green
                                       2. *P. canariensis.*
Trunk 4 to 18 inches in diameter.
  Leaves 5 to 7 feet long, stiff and rigid............................3. *P. reclinata.*
  Leaves 1 to 3 feet long, rather flaccid.............................4. *P. roebelini.*

### 1. Phoenix dactylifera L. DATE PALM. Fig. 114.

A large feather palm, usually suckering at the base, 25 to 75 feet high, with an erect or often slightly bent trunk and a dense crown of leaves upward-arching from the upper part but more or less down-curving from the lower part. Leaf-blades 15 to 20 feet long, usually grayish glaucous. Fruit an ellipsoidal drupe, 1 to 3 inches long, edible.

Native probably to northern Africa or western Asia. Extensively cultivated for the commercial dates in Riverside and Imperial counties, California. Occasionally planted as an ornamental tree in the warmer parts of California.

2. **Phoenix canariénsis** Chaub. CANARY ISLAND DATE PALM. Fig. 115.

A large spreading feather palm, with an erect trunk 10 to 50 feet high and 2 to 5 feet in diameter, covered

Fig. 114. Phoenix dactylifera L., Santa Barbara, California.

Fig. 115. Phoenix canariensis Chaub., Oakland, California.

at least near the summit with the bases of the old petioles. Leaf-blades 15 to 20 feet long, spreading, strongly arched and drooping in age; pinnae 12 to 16 inches long, 1 to 2 inches wide, light green; petioles short, with long stiff spine-like reduced pinnae. Fruit an ovoid yellowish drupe, ½- to ¾-inch long, not edible.

Native to the Canary Islands. Extensively cultivated in central and southern California and less commonly in northern California. Not seen in Oregon and farther north. It is a very hardy and rapid-growing palm.

**3. Phoenix reclinàta** Jacq. CAPE PALM. SENEGAL DATE PALM. Fig. 116.

A slender feather palm, with a trunk 10 to 25 feet high and 8 to 18 inches in diameter. Leaf-blades 5 to 7 feet long, dark green, erect-spreading and recurved downward at the ends; pinnae rigid, with stiff points. Fruit ovoid or ellipsoidal, ½- to ¾-inch long, brownish, not edible.

Native to Africa. Occasionally planted as a specimen tree in California.

**4. Phoenix ròebelini** O'Brien. ROEBELIN PALM. Fig. 117.

A dwarf feather palm, with an erect trunk 3 to 6 feet high and 4 to 8 inches in diameter, often swollen at the base. Leaf-blades 1 to 3 feet long, shining green and somewhat glaucous, gracefully curving and drooping; pinnae 6 to 8 inches long, curved downward, long-pointed; petioles 1 to 2½ feet long, with slender weak spines. Fruit oblong-ovoid, about ½-inch long, borne in a many-branched cluster with a peduncle about 1 foot long.

Native to southeastern Asia. Grown in the warmer coastal parts of California as a lawn plant, also used as a tub plant.

Fig. 116. Phoenix reclinata Jacq., Santa Barbara, California.

**9. Roystònea** O. F. Cook. ROYAL PALMS

(Named in honor of General Roy Stone, an American engineer who carried on noted work in Puerto Rico.)

Three or 4 species of the American tropics.

**1. Roystonea oleràcea** O. F. Cook. CABBAGE PALM. Fig. 118.
*Oreodoxa oleracea* Mart.

A tall feather palm, with a cylindrical trunk from a swollen base, bearing erect and horizontally spreading leaves which form a crown more or less flat on the under side, and an erect spear-like unfolding bud at the top. Leaf-blades 4 to 8 feet long; pinnae 1½ to 2 feet long, ¾-inch to 1½ inches wide, borne in a single row or tier on either side of the unarmed rachis; petioles 2 to 4 feet long, unarmed. Flowers monoecious, borne in erect-spreading clusters at the bases of the sheathing petioles and thus below the crown. Fruit an ovoid drupe, ½- to ⅝-inch long.

Native in the West Indies. Occasionally cultivated as a specimen tree in southern California.

### 10. Trachycárpus Wendl.

(From the Greek *trachys*, harsh or rough, and *karpos*, fruit.)

About 5 species, native to eastern Asia and Japan These palms are often erroneously called *Chamaerops*.

### 1. Trachycarpus excélsa Wendl. WINDMILL PALM. FORTUNE PALM. Fig. 119.

*T. fortunei* Wendl. *Chamaerops excelsa* Thunb. *C. fortunei* Hook.

Fig. 117. Phoenix roebelini O'Brien, Whittier, California.

A slender fan palm, 10 to 30 feet high, with the trunk apparently enlarging from the base upward and densely covered with hair-like fibers and old leaf-bases. Leaf-blades nearly round, 2 to 4 feet across, divided to the middle or almost to the base into numerous stiff segments not drooping at the ends; petioles very slender, 2 to 3 feet long, ½-inch to 1 inch wide, with small irregular blunt points, but not spinose. Flowers monoecious, in clusters borne among the leaves. Fruit a small 3-lobed drupe, purplish, usually dry.

Native to central and eastern China, and probably to Japan. Windmill Palm is a very popular fan palm, extensively planted along streets, in parks, and in gardens throughout California. No specimens have been seen growing north of Crescent City and Redding, California. It is the hardiest of all palms planted on the Pacific Coast, but it does not stand prolonged freezing.

Some authors think the more robust plants with the leaves divided almost to the base worthy of varietal rank, but much variation occurs in these characters and therefore in this manual it seems best not to recognize the varieties.

Fig. 118. Roystonea oleracea O. F. Cook, Whittier, California.

## 11. **Washingtònia** Wendl.
### WASHINGTON PALM

(Named in honor of George Washington.)

Large fan palms, the trunks clothed with the deflexed old leaves and the remains of the sheaths and petioles, becoming smooth with age. Leaves large, erect-spreading, becoming deflexed; petioles long, stout, usually spiny along the margins. Flowers perfect, borne in long branching clusters among the leaves. Fruit a small drupe-like berry, black, ellipsoidal, with a thin dry sweet pulp.

Three species, native to Mexico, Arizona, and southern California. Much confusion exists in the classification of the forms of *Washingtonia*. The filaments along the leaf-margins, the size, length, and armed condition of the petioles, and the size of trunk vary to such a degree that they are of questionable value as species-criteria. Two types are usually seen in cultivation—one, typically with long stout well-armed petioles and robust trunks, the other with shorter more slender petioles and taller narrower trunks. Occasionally a specimen will be seen in cultivation in which the petioles are almost unarmed. This apparently is **Washingtonia filifera** Wendl. var. **microsperma** Beccari. *Washingtonia filifera* Wats. and *Washingtonia gracilis* Parish are extensively grown as avenue and specimen trees throughout central and southern California. *Washingtonia sonorae* Watson, native of western Sonora and southern Lower California, has not been seen in cultivation.

KEY TO THE SPECIES

Trunks stout, 2 to 3½ feet in diameter; leaf-blades with many thread-like filaments from the margins of the segments; petioles stout, 3 to 5 feet long..........................................................................1. *W. filifera.*

Trunks slender, 1 to 2 feet in diameter; leaf-blades almost devoid of thread-like filaments; petioles slender, 2 to 3 feet long
2. *W. gracilis.*

1. **Washingtonia filifera** Wats., not Wendl. CALIFORNIA WASHINGTON PALM. CALIFORNIA FAN PALM.
Fig. 120.

*W. filifera* var. *robusta* Parish. *W. robusta* Wendl. *W. filamentosa* O. Kuntze.

A tall robust fan palm, with a stout columnar trunk 20 to 70 feet high and 2 to 3 feet in diameter, enlarged at the base. Leaf-blades 3 to 6 feet broad, with 40 to 70 folds, deeply slashed ½ to ⅔ of the distance to the base, the margins of the divisions with many thread-like filaments; petioles stout, 3 to 5 feet long, 1 to 3 inches wide, armed along the margins with stout hooked spines. Flowering stems 8 to 10 feet long. Fruit ovoid, about ⅜-inch long, black.

Occurs in a few localities along the western and north-western borders of the Colorado Desert and the extreme

Fig. 119. Trachycarpus excelsa Wendl., Whittier, California.

southern Mohave Desert. One very fine grove occurs in Palm Canyon, Riverside County, California, and is considered one of the attractions of the State. Extensively cultivated along streets and in parks in California. No specimens have been seen growing farther north than Cloverdale on the Redwood Highway and Redding on the Pacific Highway.

2. **Washingtonia grácilis** Parish. MEXICAN WASHINGTON PALM. Fig. 121.

*W. robusta* Hort., not Wendl.

Fig. 120. Washingtonia filifera Wats., Oakland, California.

A tall fan palm, with a slender columnar trunk 40 to 100 feet high and 1 to 2 feet in diameter. Leaf-blades 2½ to 5 feet (usually about 3 feet) broad, with 70 to 80 folds, slashed about ⅓ to ½ of the distance to the base, the segments almost devoid of marginal thread-like filaments; petioles 2 to 3 feet long, armed along the margins with short hooked spines. Flowers and fruit similar to those of *W. filifera*.

Native probably to Lower California. Extensively cultivated with

Fig. 121. Washingtonia gracilis Parish, Anaheim, California.

the California Fan Palm, from which it can be distinguished by its more slender trunk, shorter petioles, and fewer thread-like filaments along the leaf-segments.

# III. BROAD-LEAVED TREES (Dicotyledons)

## Casuarinaceae. Casuarina Family

### 1. Casuarina L. BEEFWOOD. CASUARINA

(Probably Casuarius-like, referring to the resemblance of the long "weeping" branches to the drooping feathers of the Cassowary.)

Slender-branched leafless trees, with usually jointed pale green branchlets which resemble those of the Scouring-Rush (*Equisetum*). Leaves reduced to whorls of small brownish tooth-like scales at the joints of the branchlets. Flowers inconspicuous, monoecious or dioecious; the staminate in small slender cylindrical terminal spikes; the pistillate in small dense heads borne in the axils of the branches. Fruit a dry cone-like structure, consisting of hard persistent bracts subtending the simple pistils, each pistil becoming a thin-winged nutlet held between 2 bracts which open when mature as do the valves of a capsule.

About 25 species, native to Australia and the Pacific Islands. In general appearance, the casuarinas may be mistaken for pines, but their green "foliage" consists of branchlets instead of needle-like leaves. In Australia the casuarinas are called oaks. They thrive best in mild climates.

Fig. 122. Casuarina stricta Dry. Left, base of branchlet, × 1. Center, branchlet with staminate cone, × 1. Upper right, fruiting cone, × 1. Lower right, whorl of leaves at the node, × 5.

KEY TO THE SPECIES

Reduced tooth-like leaves of the branchlets 8 to 14.
    Cones ¾-inch to 1¼ inches long; internodes ½- (¼-) inch to 1 inch long....................1. *C. stricta*.
    Cones about ⅓-inch long; internodes about ¼-inch long; teeth usually 10.................2. *C. cunninghamiana*.
Reduced tooth-like leaves of the branchlets 6 to 8; cones ½- to ¾-inch long, the valves pubescent; internodes about ¼-inch long..3. *C. equisetifolia*.

### 1. Casuarina stricta Dry. SHE-OAK. Fig. 122.

A small tree, 10 to 30 feet high, with drooping branches. Branchlets with internodes ½- (¼-) inch to 1 inch long, and with about 10 teeth at the joints. Cones ovoid, ¾-inch to 1¼ inches long.

Native in Australia and Tasmania. Occasionally planted in California parks and gardens.

Fig. 123.
Casuarina
cunninghamiana
Miq.

Branchlet, × 2.
Whorl of leaves
at the node,
× 6.

**2. Casuarina cunninghamiàna** Miq. CUNNINGHAM BEEFWOOD. Fig. 123.

A tall tree, 30 to 70 feet high, with spreading or drooping branches and numerous slender branchlets. Branchlets with internodes ¼-inch or less long and with 8 to 10 teeth at the joints. Cones globular, ⅓-inch or less broad, the valves glabrous.

Native to Queensland and New South Wales. Commonly used in southern California in parks and gardens, and as a roadside tree.

**3. Casuarina equisetifòlia** L. HORSETAIL TREE. COMMON BEEFWOOD. Fig. 124.

Fig. 124.
Casuarina
equisetifolia L.

Branchlet, × 2.
Whorl of teeth
at the node,
× 6.

A tall narrow tree, 40 to 70 feet high, with spreading or drooping branches and scant gray-green terete branchlets. Branchlets with internodes about ¼-inch long and with 6 to 8 teeth at the joints. Cones subglobose, about ½-inch long, the valves pubescent.

Native to Australia. Cultivated in California gardens and parks. Sometimes used for reforestation and as an avenue tree.

## Salicaceae. Willow Family

### 1. Pópulus L. POPLAR. COTTONWOOD

(Ancient classical name of obscure derivation, probably from the root *pal*, to shake, in reference to the shaking or quivering of the leaves of many species.)

Deciduous trees with light colored rather soft wood and often resinous buds with several imbricated scales. Leaves simple, alternate, usually with long petioles. Flowers small, dioecious, in catkins, each flower subtended by a fringed bract and a cup-shaped disk, without a perianth. Fruit a 2- to 4-valved capsule with numerous hairy seeds.

About 30 species, which include the aspens, poplars, and cottonwoods, restricted to the northern hemisphere. Four of the species inhabit the Pacific Coast region. Five introduced species or hybrids are cultivated as ornamental trees and for windbreaks in our region. The Lombardy Poplar (*Populus nigra* var. *italica*), Carolina Poplar (*Populus canadensis*), and White Poplar (*Populus alba*) are the species most commonly seen in cultivation.

The identification of the numerous forms of poplars is difficult because of the variation which occurs in the size and shape of leaves among closely related species, among trees of the same species, and upon different parts of the same tree. Much of this variation has originated from hybridization among our native species and those of Europe and Asia. Many of these forms are reproduced by cuttings and are extensively cultivated throughout the United States. They frequently escape from cultivation and become naturalized. Owing to their habits of suckering, breaking in the wind, shedding cottony covered seeds, and to their short life-span, the poplars should not be planted where other more desirable trees are available.

### KEY TO THE SPECIES

In selecting leaves for identification, material should be taken from mature lateral branches and never from suckers or fast-growing new shoots.

Leaves white- or gray-tomentose beneath, palmately 3- or 5-lobed or toothed..............................................................................................1. *P. alba.*
Leaves glabrous on both surfaces or slightly pubescent on the veins beneath, pinnately veined, not lobed.
  Petioles round in cross section; leaves distinctly glaucescent, whitish, or rusty beneath.
    Leaves 4½ to 6 inches long.........................................2. *P. candicans.*
    Leaves 2½ to 4¾ inches long.
      Leaves broadly ovate; capsules pubescent..........3. *P. trichocarpa.*
      Leaves ovate to ovate-lanceolate; capsules glabrous
                                    4. *P. tacamahacca.*
  Petioles flattened (sometimes almost circular); leaves often paler beneath but not distinctly glaucous or whitish.
    Leaves 4 to 7 inches long; winter-buds very resinous
                                  5. *P. balsamifera.*
    Leaves 1 to 4 inches long.
      Trees very slender, with ascending or appressed branches forming a narrow columnar crown; leaves broadly cuneate at base
                        6. *P. nigra* var. *italica.*
      Trees of pyramidal or spreading habit.
        Basal pair of lateral veins prominent, arising from the summit of the petiole; leaves without narrow translucent margins.
          Leaves 1 to 2½ inches long, finely crenate-dentate to almost entire, not woolly beneath when young
                        7. *P. tremuloides.*
          Leaves 2 to 4 inches long, coarsely sinuate-dentate, densely woolly beneath when young..............8. *P. grandidentata.*

Basal pair of lateral veins not more prominent than others, usually arising above attachment of petiole to the blade; leaves with narrow translucent margins.

Leaves yellowish green, almost alike on both surfaces, coarsely crenate-dentate; native species..9. *P. fremonti.*

Leaves green above, paler beneath, crenate-serrate; introduced species..............................................10. *P. canadensis.*

1. **Populus álba** L. WHITE POPLAR. Fig. 125.

A medium-sized tree, 40 to 60 feet high. Leaf-blades ovate to roundish, 2 to 5 inches long, 1½ to 3½ inches broad, somewhat palmately 3- or 5-lobed or toothed, dark green and glabrous above, very white-tomentose beneath; petioles ¾-inch to 1½ inches long, tomentose.

Native in Europe and Asia. Occasionally cultivated as a park, garden, and street tree for the contrast in color of leaf-surfaces.

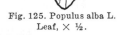

1*a*. Var. **pyramidàlis** Bunge.

*P. bolleana* Carr.

Fig. 125. Populus alba L. Leaf, × ½.

A form of columnar habit and with strongly lobed leaves.

2. **Populus cándicans** Ait. BALM OF GILEAD.

*P. balsamifera* var. *candicans* Gray.

A tree, 30 to 60 feet high, with stout spreading branches forming a broad crown. Leaf-blades broadly triangular-ovate, 4½ to 6 inches long, 3 to 4 inches wide, heart-shaped at the base, dark green and glabrous or slightly pubescent above, whitish and sparingly pubescent (at least on the veins) beneath, the margins crenate-serrate; petioles rounded, 1¼ to 2½ inches long, pubescent. Buds large, resinous, very aromatic.

Only the pistillate trees are known. Probably of hybrid origin and confused with *P. tacamahacca.* Occasionally planted as a street and park tree.

Fig. 126.
Populus trichocarpa T. & G.
Leaf, × ½.

3. **Populus trichocárpa** T. & G. WESTERN BALSAM POPLAR. BLACK COTTONWOOD. Fig. 126.

A tall tree, 40 to 100 feet high, with ascending or spreading branches forming a rather broad crown. Leaf-blades triangular-ovate, 2½ to 4¾ inches long (rarely longer), 2 to 2½ inches wide, truncate or rounded at the base, glossy and green above, whitish or rusty beneath, crenately serrate; petioles rounded, 1¼ to 2 inches long. Capsules pubescent.

Native to the Pacific Coast from southern Alaska to southern California, inhabiting living stream banks and valleys of the mountainous regions from near sea level to about 8000 feet elevation.

4. **Populus tacamahácca** Mill. BALSAM POPLAR. Fig. 127.

*P. balsamifera* Du Roi and authors, not L.

A tall tree, 40 to 90 feet high, with ascending branches. Leaf-blades ovate to ovate-lanceolate, 3 to 5 inches long, 1½ to 3 inches wide, rounded at the base, dark green and glabrous above, whitish or pale green beneath, finely crenate-serrate; petioles rounded, 1¼ to 2 inches long. Capsules glabrous.

Native from Labrador to Alaska, south to the northern United States. Occasionally cultivated. This species has been confused with *P. balsamifera* L., but differs from it in having rounded petioles, shorter leaves, and whitish lower leaf-surfaces.

Fig. 127.
Populus
tacamahacca Mill.
Leaf, × ½.

5. **Populus balsamífera** L. COTTONWOOD. Fig. 128.

*P. deltoidea* Marsh, in part. *P. monilifera* Ait. *P. angulata* Michx.

A large tree, 40 to 80 feet high, with massive spreading branches. Leaf-blades ovate, 4 to 7 inches long, 3 to 5 inches wide, heart-shaped or truncate at the wide base, glabrous, green and glossy above, paler beneath, finely crenate-serrate; petioles flattened, 2 to 4 inches long, with glands at the summit. Buds very resinous.

Native from southeastern Canada and the northeastern United States south and west to Texas. Occasionally planted as a park or street tree because of its rapid growth and abundant foliage.

6. **Populus nìgra** var. **itálica** Du Roi. LOMBARDY POPLAR. Fig. 129.

A slender columnar tree, 40 to 90 feet high,

Fig. 128. Populus
balsamifera L.
Leaf, × ¼.

with almost upright branches. Leaf-blades rhombic-ovate, 2 to 3½ inches long, 1¾ to 3 inches wide, broadly cuneate to truncate at the base, tapering or abruptly acuminate, glabrous, finely crenate-toothed; petioles flattened, 1¼ to 3 inches long.

Said to have originated as a staminate offshoot from *P. nigra* var. *typica* of the plains of Lombardy. A striking and picturesque tall slender tree, frequently planted along avenues, as a windbreak, and as a lawn tree. Suckers abundantly.

Fig. 129. Populus nigra
var. italica Du Roi.
Leaf, × ½.

Fig. 130. Populus
tremuloides Michx.
Leaf, × ½.

### 7. Populus tremuloìdes Michx. QUAKING ASPEN. Fig. 130.

A small slender tree, 10 to 60 feet high, with greenish white bark (dark on old trunks) and quaking foliage. Leaf-blades round-ovate, 1 to 2½ inches long and about as wide, truncate or rounded to rarely cuneate or heart-shaped at the base, abruptly tipped, glabrous and green above, paler beneath, finely crenate-serrate to almost entire; petioles flattened, 1 to 3 inches long.

Native in the colder mountainous regions of North America. On the Pacific Coast, the Quaking Aspen extends from Alaska southward to British Columbia, Washington, Oregon, California, and Lower California. In California it usually grows on the borders of mountain meadows, lakes, and moist benches. The fluttering leaves and smooth grayish green or almost white bark attract the attention of mountain travelers.

### 8. Populus grandidentàta Michx. LARGETOOTH ASPEN. Fig. 131.

A tall tree, 40 to 70 feet high, with smooth dull gray bark tinged with green becoming slightly fissured and brownish near the base on old trunks, and stout young branchlets coated with a dense heavy tomentum. Leaf-blades broadly ovate to suborbicular, 2 to 4 inches long and about as wide, rounded or truncate at the broad base, dark

green above, white-woolly beneath when young, becoming glabrous, coarsely sinuate-dentate, the midrib yellowish, the basal pair of lateral veins very prominent; petioles flattened, 1½ to 2½ inches long.

Native from Nova Scotia westward to Minnesota, southward to Iowa, Tennessee, and North Carolina. Occasionally cultivated as specimen trees. Closely related to the Quaking Aspen, but the leaves larger and more coarsely toothed.

9. **Populus frèmonti** Wats. FREMONT COTTONWOOD. Fig. 132.

Fig. 131. Populus grandidentata Michx. Leaf, × ½.

A large tree, 40 to 100 feet high, with grayish bark, spreading branches, and yellowish green foliage. Leaf-blades broadly triangular-ovate, 1½ to 3 inches long, 2 to 4 inches broad, heart-shaped or truncate to slightly wedge-shaped at the base, glabrous, yellowish green and glossy above, paler beneath, coarsely crenate; petioles flattened, 1½ to 3 inches long.

Native in California in the interior and North Coast Range valleys, foothills of the Sierra Nevada, the South Coast Ranges, and mountains of southern California, and eastward to Nevada, Arizona, and New Mexico. This tree is occasionally planted in parks and along highways. **P. arizonica** Sarg., **P. wislizeni** Sarg., and **P. macdougali** Rose, species closely related to one another, are sometimes found in the southern part of California.

10. **Populus canadénsis** Moench. CAROLINA POPLAR. Fig. 133.

*P. carolinensis* Hort. *P. caroliniana* Hort. *P. deltoides* var. *carolinensis* Bailey. *P. canadensis* var. *eugenei* Schelle.

A tall tree, 40 to 100 feet high, with many strongly ascending or spreading branches forming a columnar or pyramidal crown. Leaf-blades triangular-ovate, 2½ to 4 inches long, 2 to 3½ inches wide, truncate or slightly wedge-shaped at the base, glabrous, bright green above, paler beneath, crenately serrate with mostly incurved teeth and slightly ciliate; petioles flattened, 1 to 1¾ inches long. Buds slender, not viscid or seldom slightly so.

Supposedly a hybrid between *P. balsamifera* and *P. nigra*, occurring only as staminate trees. Frequently planted as shade trees, along streets, and in parks. Suckers badly, breaks easily, and roots penetrate sewers. It is therefore not recommended.

2. **Sàlix** L. WILLOW
(The classical name of the willow, from the Celtic

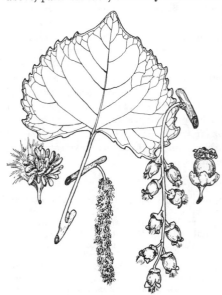

Fig. 132. Populus fremonti Wats.
Left to right: Staminate flower, × 4. Leaf, × ½. Staminate catkin, × ½. Fruiting catkin, × ½. Pistillate flower, × 4.

*sal*, near, and *lis*, water, in reference to its place of growth.)

Trees or shrubs with single bud-scales. Leaves simple, alternate, short-petioled, usually deciduous. Flowers small, dioecious, in catkins, each flower without a perianth and subtended by a scale-like bract; stamens 1 to 10 (commonly 2 or 3) on each bract; pistil 1. Fruit a 2- to 4-valved capsule with numerous hairy seeds.

Nearly 300 species of willows are now recognized, of which about 50 are native on the Pacific Coast. They inhabit stream banks and moist ground from sea level to alpine slopes of the highest mountains. Seven of these species have trunks

Fig. 133.
Populus canadensis Moench.
Leaf, × ½.

of a size, and grow to a height, that for practical purposes require them to be classed as trees. Nine other species are occasionally tree-like and are listed after the descriptions of the species which follow. Many willows produce several trunks from a common root-crown and when young are shrub-like.

Willows, as a group, are easily distinguished from other trees, but the various species are often difficult to identify because of the variations in foliage characters. These variations have resulted from adaptations to habitat conditions and from the unusual number of natural hybrids. The Weeping Willow (*Salix babylonica*), cultivated for its habit of growth, and the Goat Willow (*Salix caprea*), grown for its large showy catkins which appear in early spring, are the two species usually planted as ornamentals. Several of the native species are used for holding stream banks and highway embankments, and to prevent the denuding of soil areas.

### KEY TO THE SPECIES

Leaves whitish, felty-tomentose, or short-silky to rarely glabrate beneath.
  Leaves entire or obscurely serrulate, oblanceolate to obovate.
    Leaves 1¼ to 2 (or 4) inches long, ½-inch to 1½ inches wide, thin, glabrate to short-silky beneath; stamens 2
           1. *S. scouleriana*.
    Leaves 2 to 5 inches long, 1 to 3 inches wide, thick, densely white-tomentose beneath; stamen 1...............................2. *S. sitchensis*.
  Leaves irregularly serrate, ovate to oblong-orbicular, 2 to 5 inches long.................................................................................3. *S. caprea*.
Leaves glabrous (rarely pubescent), green or glaucous beneath.
  Trees of "weeping" habit.
    Leaves not curled.................................................................4. *S. babylonica*.
    Leaves curled as in a spiral.............4a. *S. babylonica* var. *annularis*.
  Trees without "weeping" habit.
    Petioles with warty glands near base of the blades
           5. *S. lasiandra*.
    Petioles without warty glands.
      Leaves linear-lanceolate, ½-inch or less wide, gray-green on both surfaces.................................................6. *S. nigra* var. *vallicola*.
      Leaves lanceolate to ovate-lanceolate, or often oblanceolate to obovate, ⅓-inch to 1¼ inches wide, glabrous and shining above, paler and glaucous beneath.
        Leaves usually broader above the middle, the margins nearly entire and somewhat revolute; catkins sessile
           7. *S. lasiolepis*.
        Leaves broader below the middle (or broader above when developing), the margins finely serrulate; catkins peduncled.
          Petioles puberulent, less than ½-inch long....8. *S. laevigata*.
          Petioles glabrous, ½- to ¾-inch long......9. *S. amygdaloides*.

1. **Salix scouleriàna** Barr. NUTTALL WILLOW. SCOULER WILLOW. Fig. 134.

*S. nuttalli* Sarg. *S. flavescens* Nutt.

A large shrub or small tree, 4 to 30 feet high, with dull gray to very dark bark. Leaf-blades mostly oblanceolate to obovate, 1¼ to 2½ (or 4) inches long, ½-inch to 1½ inches wide, thin, yellow-green

and glabrous above, short-silky to almost glabrous beneath, entire to finely and obscurely serrulate; petioles ¼- to ½-inch long. Catkins appearing before the leaves, 1 to 2 inches long, very showy.

Nuttall Willow is one of the commonest species on the Pacific Coast. It ranges from southern Alaska to southern California, usually occurring in the Transition and Canadian life zones. In California, it occurs from Modoc and Siskiyou counties southward to Tulare County at 4000 to 10,000 feet elevation, in the Coast Ranges near the coast from Del Norte County southward to Alameda and Monterey counties, and in the San Antonio, San Jacinto, and San Bernardino mountains of southern California.

Fig. 134.
Salix scouleriana Barr.
Leaves, × ½.

2. **Salix sitchénsis** Sanson. SATIN WILLOW. SILKY WILLOW.

A shrub or small tree, 6 to 25 feet high. Leaf-blades cuneate-obovate or oblanceolate to oblong-obovate, 2 to 5 inches long, 1 to 3 inches wide, thickish, becoming thinner in age, dark green and glabrous above, densely white-tomentose beneath, and entire or obscurely denticulate; petioles stout, rarely to ½-inch long. Catkins appearing with the leaves, 1½ to 3 inches long.

Native to stream banks from southern Alaska to southwestern Oregon. The variety **còulteri** Jepson, VELVET WILLOW, which is scarcely separable from the species, extends from Santa Barbara northward along the coast to southern Oregon where it passes into the species. The leaf-blades are typically more rounded than cuneate at the base.

Fig. 135.
Salix caprea L.
Leaf, × ½.

3. **Salix cáprea** L. GOAT WILLOW. Fig. 135

A small tree, 15 to 30 feet high, with erect or spreading branches and brown or purplish branchlets tomentose when young. Leaf-blades broadly ovate to oblong-orbicular, 2 to 5 inches long, 1 to 1¾ inches wide, thick, deep green and glabrous above, felty-tomentose beneath, irregularly serrate to almost entire; petioles ⅜- to ⅝-inch long.

Catkins appearing before the leaves, 1½ to 2¼ inches long, the staminate very showy.

Native in Europe and Asia. Frequently cultivated for its large staminate catkins flowering in early spring.

**4. Salix babylónica** L. WEEPING WILLOW. Fig. 136.

A tall tree, 30 to 70 feet high, with rough bark and long slender drooping branchlets. Leaf-blades oblong-lanceolate, 3 to 6 inches long, ½- to ⅞-inch wide, light green, glossy, and glabrous above, glaucous beneath, finely serrate; petioles ⅛- to ⅜-inch long. Catkins appearing with the leaves, slender, about 1 inch long.

Native to China. Linnaeus thought it native to the Babylonian region, hence its species-name. Much cultivated for its "weeping" habit of growth.

*4a.* Var. **annulàris** Forbes. RINGLEAF WEEPING WILLOW. Fig. 137.

A large graceful tree, 30 to 50 feet high, with drooping branches. Leaves curled into a spiral, narrower than in the species. Not so commonly cultivated, but as desirable as the species.

**5. Salix lasiándra** Benth. YELLOW WILLOW. BLACK WILLOW. Fig 138.

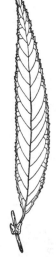

Fig. 136.
Salix baby-
lonica L.
Leaf,
× ½.

A medium-sized tree, 15 to 45 feet high, with rough brown bark, ascending or spreading branches, and yellow one-year-old branchlets. Leaf-blades oblong-lanceolate, 3 to 6 inches long, ¾-inch to 1¼ inches wide, glabrous and dark green above, glaucous beneath, finely glandular-serrulate; petioles ¼- to ¾-inch long, distinctly warty-glandular near the base of the blades. Catkins appearing with and after the leaves, 1¼ to 3 inches long.

Native along stream banks from British Columbia southward to southern California and New Mexico. In California it usually occurs along permanent streams

Fig. 137.
Salix babylonica
var. annularis
Forbes.
Leaves. × ½

in the Great Valley, the Coast Ranges, and the Sierra Nevada up to 4000 feet elevation, or to higher elevations in the mountains of southern California.

6. **Salix nìgra** var. **vallícola** Dudley. BLACK WILLOW. Fig. 139.
*S. gooddingi* Ball.

A medium-sized tree, 20 to 40 feet high, with dark rough bark, spreading or ascending branches, and yellowish branchlets. Leaf-blades linear-lanceolate, 2¼ to 4 inches long, ½-inch or less wide, grayish green and glabrous on both surfaces, the young leaves usually pubescent, finely glandular-denticulate; petioles ⅛- to ⅜-inch long. Catkins appearing with the leaves, 1½ to 2½ inches long.

Native along stream banks in the Great Valley, the Sierra Nevada foothills up to 1500 feet elevation, southern California, and Lower California, eastward to Nevada, Arizona, New Mexico, and Texas.

7. **Salix lasiólepis** Benth. ARROYO WILLOW. Fig. 140.

A large shrub or small tree, 10 to 30 feet high, usually with clustered stems and smooth dark gray bark. Leaf-blades oblanceolate, 2¼ to 5 inches long, ⅜-inch to 1¼ inches wide, dark green and glabrous above, glabrate or glaucous to pubescent beneath, entire or seldom obscurely serrulate and somewhat revolute; petioles ⅛- to ⅝-inch long. Catkins appearing before the leaves, 1¼ to 2½ inches long, sessile.

Native along dry stream beds or permanent water courses from eastern Washington to California, Arizona, western Nevada, western New Mexico, and Lower California. In California, Arroyo Willow occurs throughout the Coast Ranges, the Great Valley, the Sierra Nevada foothills up to 2500 feet elevation or to nearly 4000 feet in the southern part, and in southern California.

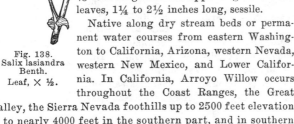

Fig. 138.
Salix lasiandra
Benth.
Leaf, × ½.

Fig. 139.
Salix
nigra var.
**vallicola**
Dudley.
Leaf,
× ½.

8. **Salix laevigàta** Bebb. RED WILLOW. SMOOTH WILLOW. Fig. 141.

A medium-sized tree, 20 to 40 feet high, with dark rough trunk-bark and reddish brown bark on the young branchlets. Leaf-blades broadly lanceolate, 2½ to 6 inches long, ¾-inch to 1⅛ inches wide, rounded or subcordate at the base, glabrous, shining, and green above, glaucous beneath, finely serrulate; petioles ¼- to ⅜-inch long, puberulent. Catkins appearing with the leaves, 1¼ to 4 inches long, peduncled, the staminate showy and bent or flexuous.

Native along stream banks from extreme southern Oregon southward to southern California, east to Arizona, Utah, and Nevada. In California, Red Willow occurs in the foothills of the Sierra Nevada, in the Great Valley, the Coast Ranges, and southern California, from sea level to 4000 feet elevation. Red Willow and Yellow Willow often are found growing together.

Fig. 140. Salix lasiolepis Benth.
Left to right: Pistillate catkin, × ½.
Pistillate flower, × 2. Leaves, × ½.
Staminate flower, × 2. Staminate catkin, × ½.

9. **Salix amygdaloìdes** Anderss. PEACHLEAF WILLOW. Fig. 142.

A tree, 10 to 40 feet high, with somewhat drooping slender yellow young branches and yellowish green foliage. Leaf-blades lanceolate to ovate-lanceolate, 2 to 4½ inches long, ¾-inch to 1¼ inches wide, light green and glabrous above, pale and glaucous beneath, finely serrulate; petioles slender, ½- to ¾-inch long, glabrous. Catkins appearing with the leaves, 2 to 3 inches long, peduncled.

Native along stream banks from southeastern British Columbia southward to Washington and Oregon east of the Cascade Mountains. It extends eastward to the Mississippi Valley and through the Great Lakes basin to New York.

OTHER TREE-LIKE SPECIES NATIVE ON THE PACIFIC COAST

10. **Salix alaxénsis** Cov. FELTLEAF WILLOW. Coast of Alaska.

11. **Salix amplifòlia** Cov. BROADLEAF WILLOW. Alaska.

12. **Salix bebbiàna** Sarg. BEBB WILLOW. Alaska and Coast Ranges of British Columbia.

Fig. 141.
Salix laevigata
Bebb.
Leaf, × ½.
Leaf-margin, × 5.

13. **Salix hookeriàna** Barr. COAST WILLOW. Vancouver Island and along the coast of Washington and Oregon.

14. **Salix mackenziàna** Barr. *S. cordata* var. *mackenziana* Hook. MAC-KENZIE WILLOW. British Columbia, Washington, Oregon, and northern California.

Fig. 142.
Salix amygdaloides Anderss.
Leaf, × ½.

15. **Salix melanópsis** Nutt. DUSKY WILLOW. British Columbia south to southern California.

16. **Salix pìperi** Bebb. DUNE WILLOW. PIPER WILLOW. Along the coast of Washington, Oregon, and Del Norte County, California.

17. **Salix sessilifòlia** Nutt. SOFTLEAF WILLOW. British Columbia, the Puget Sound region, south to the Willamette and Umpqua rivers in Oregon.

18. **Salix hindsiàna** Benth. *S. sessilifolia* var. *hindsiana* Anderss. VALLEY WILLOW. Fig. 143.

Fig. 143.
Salix hindsiana Benth.
Leaf,
× ½.

Very common along river banks and flood beds in Lower California, California, and southwestern Oregon.

## Myricaceae. Sweetgale Family

### 1. Myrìca L. WAX-MYRTLE

(From the Greek *myrike,* probably first applied to the tamarisk.) About 30 to 40 species, occurring in the temperate and subtropical parts of both hemispheres.

1. **Myrica califórnica** Cham. CALIFORNIA WAX-MYRTLE. Fig. 144.

A large evergreen shrub or small tree, 10 to 35 feet high, with slender ascending branches and smooth gray or light brown bark. The leaves simple, alternate; blades oblong to oblanceolate, 2 to 4½ inches long, ½- to ¾-inch wide, glabrous, dark green and glossy above, slightly paler beneath, remotely serrate or almost entire, narrowed at the base to a petiole ⅛- to ⅜-inch long. Flowers inconspicuous, monoecious, in short catkins. Fruit a small globose dark purplish berry-like nut, about ¼-inch in diameter, coated with a granulated waxy bloom, borne in short clusters.

Native in canyons and on moist hill slopes of the coastal region from

Washington south to the Santa Monica Mountains of southern California. The evergreen glossy foliage and bushy habit make the California Wax-Myrtle a fine plant for cultivation in moist habitats.

## Juglandaceae. Walnut Family

### 1. Hicòria Raf. HICKORY

*Carya* Nutt.

(From the aboriginal name *Hicori*.)

Deciduous trees with smooth or shaggy bark. Leaves once-pinnately compound, alternate. Flowers small, monoecious; the staminate in slender drooping catkins; the pistillate in 2- to 6-flowered clusters,

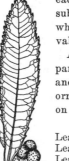

each with a perianth-like involucre. Fruit an ovoid, subglobose, or ellipsoidal nut, enclosed in a husk which becomes hard and woody and splits into 4 valves at maturity.

About 15 species, chiefly native in the eastern part of the United States, one in southern China, and one in Mexico. Grown for the edible nuts and as ornamentals. Occasionally planted as specimen trees on the Pacific Coast.

KEY TO THE SPECIES

Leaflets 5 (rarely 7)..................................1. *H. ovata.*
Leaflets 7 or 9..........................................2. *H. alba.*
Leaflets 11 to 15......................................3. *H. pecan.*

### 1. Hicoria ovàta Britt. SHAGBARK HICKORY.

*Carya ovata* K. Koch.

Fig. 144.
Myrica californica
Cham.

Leaf, fruit-cluster,
× ½.

A large tree, 60 to 100 feet high, with shaggy bark. Leaves 8 to 14 inches long; leaflets 5 (rarely 7), oblong-lanceolate (or the upper obovate), 4 to 6 inches long, 2 to 3 inches wide, dark green and glabrous above, paler underneath, serrate except near the base, sessile, the larger terminal leaflet decurrent on a slender petiolule. Bud-scales imbricated. Fruit 1¼ to 2½ inches long, the husk thick and splitting to the base into 4 valves, the nut ovoid, usually 4-ridged, with a thin nearly white shell. Seeds sweet.

Native in the central and eastern United States. Occasionally cultivated as a specimen tree.

### 2. Hicoria álba Britt. MOCKERNUT. Fig. 145.

*Carya alba* K. Koch.

A large tree, 60 to 80 (or rarely 100) feet high, with rough deeply furrowed but not shaggy bark. Leaves 8 to 12 inches long; leaflets

7 to 9, oblong-lanceolate or obovate to lance-obovate, the upper ones 5 to 8 inches long, 2½ to 4 inches wide, the lower ones much smaller, dark yellow-green and glossy above, paler beneath, finely serrate, sessile, very fragrant when crushed. Bud-scales imbricated. Fruit broadly ellipsoidal or obovoid, 1½ to 2½ inches long, the husk thick and splitting nearly to the base, the nut variable in shape, angled, with a thick reddish brown shell. Seeds sweet.

Native to the eastern United States. Occasionally cultivated as a specimen tree.

Fig. 145. Hicoria alba Britt. Leaf, fruit, × ⅓.

### 3. Hicoria pècan Britt. PECAN.
*Carya pecan* Engelm. & Graebn.

A large tree, 70 to 150 (or 180) feet high, with light brown bark deeply and irregularly fissured. Leaves 12 to 18 inches long; leaflets 11 to 15, oblong-lanceolate or lanceolate and more or less falcate, 4 to 7 inches long, 1 to 3 inches wide, dark yellow-green and in age usually glabrous above, paler and glabrous or pubescent beneath, coarsely serrate, sessile or very short-stalked. Bud-scales valvate. Fruit oblong-ovoid, 1½ to 2½ inches long, the husk thin, the nut ovoid to narrow-ellipsoidal, smooth, reddish brown, thin-shelled. Seeds very sweet.

Native in the Mississippi Valley from Indiana south to Texas. Cultivated in the southern part of its range for the edible nuts. Occasionally cultivated in southern California as a specimen tree.

### 2. Júglans L. Walnut

(From the Latin *Jovis*, Jupiter, and *glans*, nut; the classical name of the walnut tree.)

Deciduous trees or large shrubs with resinous aromatic bark and foliage. Leaves once-pinnately compound, alternate. Flowers small, monoecious, appearing after the leaves; the staminate flowers in long

Fig. 146. Juglans regia L. Leaf, × ⅛. Nut, × ⅔.

drooping catkins, each flower composed of 4 to 40 stamens and a 3- to 6-lobed calyx in the axil of and adnate to a bractlet; the pistillate few, on short terminal erect spikes, each flower composed of a 1- to 3-celled inferior ovary and a 4-lobed calyx. Fruit a drupaceous nut with a husk either indehiscent or dehiscent and breaking into 2 parts, exposing the thick- or thin-shelled nut. Seed 1, deeply 2-lobed and again irregularly lobed to fit into the interior sculpturings of the nut.

About 11 species, native in the temperate regions of the northern hemisphere and extending into the mountains of tropical America and the Andes. Two species are native on the Pacific Coast.

1. **Juglans règia** L. ENGLISH WALNUT. PERSIAN WALNUT. Fig. 146.

A large tree, 40 to 70 feet high, with smooth gray bark and spreading branches forming a broad round crown. Leaves 8 to 16 inches long; leaflets usually 7 or 9, oblong or oblong-ovate, 2 to 5 inches long, 1 to 2½ inches wide, nearly glabrous and entire, the terminal leaflet larger than the lateral ones; petiolules about ⅛-inch long. Fruit globular, about 1½ inches in diameter, green, glabrous, the nut rather thin-shelled. Seeds sweet.

Native in southeastern Europe, the Himalayas, and China. Many varieties are in cultivation for their edible nuts. Also planted for street trees.

2. **Juglans hìndsi** Rehd. CALIFORNIA BLACK WALNUT. Fig. 147.

*J. californica* var. *hindsi* Jepson.

A large tree, 30 to 70 feet high, with a trunk unbranched for 10 to 20 feet, gray-brown bark furrowed into longitudinal fissures, and a broad round-topped crown. Leaves 9 to 14 inches long; leaflets 15 to 23, ovate-lanceolate to lanceolate, 2½ to 5 inches long; ¾-inch to 1½ inches wide, bright green and glabrous above, paler and villose-pubescent on the main veins beneath, serrate, almost sessile. Fruit 1¼ to 2 inches in diameter, the nuts almost smooth, somewhat flattened at the ends. Seeds small, sweet.

Fig. 147.
Juglans hindsi
Rehd.
Nut, husk
removed, × ½.

Native in the warmer valleys about the old Indian habitations in only a few specific locations of central California. Often cultivated as a shade tree and commonly used as a stock upon which the English Walnut is grafted because its roots are better adapted to the local soil conditions.

3. **Juglans califórnica** Wats. SOUTHERN CALIFORNIA-BLACK WALNUT.

A round-headed tree, 15 to 30 (or 50) feet high, usually branching from the ground and appearing shrub-like. Leaves 6 to 10 inches long, with glandular-pubescent petioles; leaflets 11 to 15 (rarely to 19), oblong-lanceolate, 1 to 2½ inches long, ⅓- to ¾-inch wide, glabrous, finely serrate, sessile. Fruit ⅓- to ¾-inch in diameter, the nut grooved with longitudinal furrows. Seeds small, sweet.

Native to coastal southern California from Santa Barbara and the Ojai Valley to the Santa Ana Mountains, eastward to the foothills of the San Bernardino Mountains. Rarely cultivated.

Fig. 148. Juglans nigra L.
Leaf, × ⅙. Nut, × ½.

4. **Juglans cinèrea** L. BUT-TERNUT.

A large tree, 50 to 80 feet high, with rough ridged gray bark on old trunks which usually divide 20 to 30 feet above the ground, and glandular villous young branchlets. Leaves 12 to 24 inches long, with stout pubescent petioles; leaflets 11 to 17, oblong-lanceolate, 2 to 3 inches long, 1½ to 2 inches wide, pubescent on both surfaces (sticky-pubescent when young), finely serrate except near the base, sessile. Fruit oblong-ovoid, 1½ to 2½ inches long, covered with a sticky rusty pubescence, the nut ovoid and deeply sculptured into irregular plates between narrow ribs. Seeds sweet, very oily, becoming rancid.

Native in the north central and eastern United States. Occasionally cultivated.

5. **Juglans nìgra** L. BLACK WALNUT. Fig. 148.

A large tree, 60 to 100 feet high, with dark brown rough bark on old trunks, often unbranched for 40 to 60 feet, and stout pale or rusty-hairy young branchlets becoming glabrous or nearly so. Leaves 12 to 24 inches long, with pubescent petioles; leaflets 15 to 23, ovate-lanceolate, 3 to 3½ inches long, 1 to 1½ inches wide, bright yellow-green and glabrous above, soft-pubescent beneath, serrate except near the base, sessile.

Fruit globose, 1½ to 2 inches in diameter, light yellow-green, seemingly glabrous but roughened by clusters of short papillae, the nut very hard and deeply fissured into irregular ridges and plates, dark brown to black. Seeds sweet, soon becoming rancid.

Native in the eastern and central United States. Occasionally cultivated.

### 3. **Pterocàrya** Kunth. WINGNUT

(From the Greek *pteron*, wing, and *karyą* nut, in reference to the winged nuts.)

Eight species, native in China, Japan, and western Asia.

### 1. **Pterocarya stenóptera** DC. CHINESE WINGNUT. Fig. 149.

A deciduous tree, 30 to 60 feet high. Leaves pinnately compound, alternate, 6 to 12 inches long, with

Fig. 149. Pterocarya stenoptera DC. Fruit cluster, × ⅛. Single fruit, × ⅔. Leaf, × ⅓.

a winged and pubescent rachis; leaflets 11 to 23, oblong, 2 to 4 inches long, ¾-inch to 1¾ inches wide, glabrous above, pubescent beneath (at least on the midrib), serrate, sessile. Flowers small, monoecious, in drooping catkins appearing with the leaves. Fruit a small 1-seeded nut, with an oblong wing, borne in pendulous racemes often 1 foot long.

Native to China. Occasionally cultivated in southern California.

## Betulaceae. Birch Family

### 1. **Álnus** Hill. ALDER

(The classical name of the alder.)

Trees or shrubs. Leaves simple, alternate, deciduous. Flowers small, monoecious, in catkins appearing before the leaves; the staminate catkins pendulous and clustered at the ends of the branchlets; the pistillate erect, usually clustered near the base of the staminate catkins, becoming cone-like when mature. Fruit a flat 1-seeded nutlet, borne between the bractlets and persistent scales of the small woody cone.

About 18 species, native in the northern hemisphere and extending southward to the Andes in South America. Four species are native on the Pacific Coast. The alders inhabit stream banks, swampy regions, and high mountain slopes. They are rarely cultivated.

## KEY TO THE SPECIES

Fruiting peduncles slender, as long as (most of them longer than) the cones, bearing at least one leaf toward the base...........1. *A. sinuata.*
Fruiting peduncles stout, shorter than the cones, leafless.
  Leaves coarsely toothed, the teeth finely serrate.
    Leaf-margins with a narrow underturned edge; cones ¾-inch to 1⅛ inches long; coastal species...............................2. *A. rubra.*
    Leaf-margins without a narrow underturned edge; cones ⅜- to ⅝-inch long; high montane species.........................3. *A. tenuifolia.*
  Leaves finely serrate, with small glandular teeth (seldom coarsely double-serrate); cones ½- to ¾-inch long; species of low altitudes.....................................................................................4. *A. rhombifolia.*

Fig. 150. Alnus sinuata Rydb. Fruit-cluster, leaf, × ½.

1. **Alnus sinuàta** Rydb. THINLEAF ALDER. SITKA ALDER. Fig. 150.

*A. viridis* var. *sinuata* Regel. *A. sitchensis* Sarg.

A slender shrub or small tree, 5 to 40 feet high, with short nearly horizontal branches forming a narrow crown. Leaf-blades thin, ovate, 2 to 6 inches long, 1½ to 3½ inches wide, yellowish green above, pale and glossy beneath, glabrous, or hairy along the midrib and in the axils of the main veins, doubly glandular-serrate; petioles ½- to ¾-inch long. Cones about ½-inch long, on slender peduncles usually leafy at the base and longer than the cones. Nutlets oval, about as wide as the wings.

Native from the coast of Alaska south through British Columbia, Washington, and Oregon to Del Norte County (1 mile southeast of Crescent City) and the high mountains of northern California, east to Alberta and the Rocky Mountains of western Montana and Wyoming.

2. **Alnus rùbra** Bong. RED ALDER. Fig. 151.

*A. oregona* Nutt.

A medium-sized or large tree, 40 to 80 feet high, with slender spreading or somewhat pendulous branches, gray or whitish outer bark, and red-brown inner bark. Leaf-blades ovate to elliptic-ovate, 2½ to 6 inches long, 1½ to 3 inches wide, dark green and glabrous or slightly pubescent above, commonly rusty-pubescent beneath, doubly and coarsely glandular-serrate, with a narrow underturned edge; petioles ½-inch to 1¼ inches long. Stamens usually 4. Cones ¾-inch to 1⅛ inches long, on stout peduncles shorter than the cones. Nutlets ovate to nearly circular, much wider than the membranous wings.

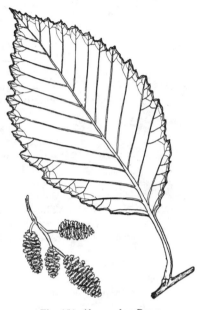

Fig. 151. Alnus rubra Bong.
Fruit-cluster, leaf, × ½.

Native on coastal stream banks and shore flats from Alaska southward to the mountains of San Luis Obispo County, California, often extending inland along the streams for 10 to 20 miles.

3. **Alnus tenuifòlia** Nutt. MOUNTAIN ALDER. Fig. 152.

A shrub or small tree, 6 to 25 feet high, with a narrow trunk and smooth gray or reddish brown bark. Leaf-blades ovate-oblong to roundish, 2 to 4 inches long, 1 to 2½ inches wide, dark green and glabrous above (pubescent when young), pale yellowish green and glabrous or finely pubescent beneath, coarsely toothed and again finely serrate; petioles ¼-inch to 1 inch long. Cones ⁵⁄₁₆- to ½-inch long, on stout leafless peduncles shorter than the cones. Nutlets nearly circular, surrounded by a very narrow thin membranous border.

Native from the Yukon Territory and British Columbia south in the Rocky Mountains, the Cascades, and the Sierra Nevada to New Mexico and Lower California. This is the common alder of stream banks and high mountain meadows in the Sierra Nevada and Cascade Mountains.

**4. Alnus rhombifòlia** Nutt. WHITE ALDER. Fig. 153.

A tree, 40 to 100 feet high, with a tall straight trunk, ascending or spreading branches pendulous at the ends, and whitish or gray-brown bark irregularly broken into plates on old trunks. Leaf-blades ovate, 2

to 3½ inches long, 1½ to 2 inches wide, dark green and glossy above, paler and pubescent beneath, finely and irregularly glandular-serrate (occasionally double-serrate), not turned under along the margins; petioles ½- to ¾-inch long. Stamens usually 2 or 3. Fruiting cones ⅜- to ⅝-inch long, on leafless peduncles shorter than the cones. Nutlets broadly ovate, with a thin narrow margin.

Native along stream banks from British Columbia southward to Idaho, Washington, Oregon, and California. In California, White Alder is the common alder of the Coast Range valleys (except close to the coast from Marin County to Humboldt County), the lower western slope of the

Fig. 152. Alnus tenuifolia Nutt.
Leaf, staminate catkins,
fruiting cone, × ½.

Sierra Nevada, the Great Valley, and the lower mountains of southern California. It extends from near sea level to about 7000 feet elevation.

### 2. Bétula L. BIRCH

(The classical name of the birch tree.)

Mostly short-lived trees or shrubs, with smooth resinous light gray or whitish bark having large transverse lenticels and commonly separating into thin papery plates or strips. Leaves simple, alternate, deciduous, toothed. Flowers small, monoecious, in catkins appearing before or with the leaves; the staminate catkins appearing in late summer and early fall and remaining more or less erect during the winter, becoming long and pendulous in the spring; the pistillate catkins erect or spreading on the ends of short lateral spur-like branchlets, becoming cone-like

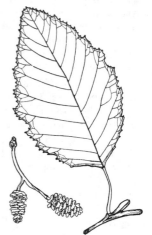

Fig. 153.
Alnus rhombifolia Nutt.
Fruiting cones, leaf, × ½.

in fruit, their scales 3-lobed, closely imbricated, falling with the winged nutlets when ripening in the autumn.

About 30 species, confined to the north temperate and arctic zones. Two tree-species and several varieties are native to the Pacific Coast. The birches are conspicuous and generally well-known trees because of their thin whitish or gray bark, often drooping branches, and numerous catkins which usually appear before the leaves. They are widely planted for ornamental purposes and are very hardy in colder climates. The European White Birch, with its several forms, is the most extensively cultivated species.

KEY TO THE SPECIES

Petioles ¼- to ½-inch long; bark brown, not exfoliating in thin layers; the lateral lobes of the fruiting bracts ascending, shorter than the middle lobe..........................................................................1. *B. fontanalis.*
Petioles ½-inch to 2 inches long; bark white, exfoliating in thin layers.
    Branchlets not resinous-glandular (or only slightly so when young), densely pubescent at least when young; lateral lobes of the fruiting bracts ascending or rarely widely spreading, shorter than the middle lobe.
        Leaves rounded, truncate, or subcordate at base, 1½ to 4 inches long..........................................................................2. *B. papyrifera.*
        Leaves cuneate or truncate at base, 1 to 2 inches long
                3. *B. alba,* form *pubescens.*
    Branchlets resinous-glandular, glabrous; lateral lobes of the fruiting bracts widely spreading and recurved, usually larger than the middle lobe.
        Leaves broadly rhombic-ovate, short-acuminate to acute..3. *B. alba.*
        Leaves triangular-ovate, long-acuminate.................4. *B. populifolia.*

1. **Betula fontanàlis** Sarg. SPRING BIRCH. BLACK BIRCH. MOUNTAIN BIRCH. Fig. 154.

A slender tree, 10 to 25 feet high, or sometimes shrub-like, with shining red-brown bark not separable into layers. Leaf-blades mostly round-ovate, 1 to 2 inches long, ⅜-inch to 1 inch wide, acute at the apex, cuneate or rounded at the base, glabrous or nearly so, glandular-dotted beneath, sharply, coarsely, and often doubly serrate; petioles ¼- to ½-inch long. Fruiting catkins 1 to 1¼ inches long, on slender glandular peduncles; bracts glabrous or puberulent, ciliate on the margins, the lateral lobes erect and shorter than the middle lobe. Nutlets narrower than (rarely as wide as) the wings.

Fig. 154.
Betula fontanalis Sarg.
Bract from fruiting catkin, × 2½. Branchlet with leaves and pistillate catkin, × ½. Single fruit, × 2½.

Native from British Columbia southward in the Rocky Mountains to eastern Utah, Arizona, and northern New Mexico, eastward to the

Black Hills of South Dakota and the eastern foothills of the Rocky Mountains of Colorado, westward to eastern Washington, Oregon, and California in a few scattered localities in the Sierra Nevada from Tulare and Inyo counties, chiefly on the east slope, northward to Mount Shasta, thence westward through Siskiyou County to northern Humboldt County.

1*a*. Var. **pìperi** Sarg.

*B. piperi* Britt.

A tree, 30 to 60 feet high, with short spreading branches and with catkins usually longer and narrower than in the species.

Native in eastern Washington.

2. **Betula papyrífera** Marsh. CANOE BIRCH. PAPER BIRCH. Fig. 155.

A large tree, 40 to 70 feet high, with exfoliating white bark, spreading branches, and pubescent slightly glandular young branchlets. Leaf-blades ovate, 1½ to 4 inches long, 1½ to 2 inches wide, long-

acuminate, rounded, truncate, or subcordate at the base, dark green and glabrous above, lighter and usually glandular and pubescent at least on the veins beneath, with axillary tufts of hairs, coarsely and irregularly serrate; petioles ½-inch to 1 inch long, pubescent. Fruiting catkins 1¼ to 2 inches long; bracts ciliate on the margins, the lateral lobes ascending or widely spreading and shorter than the middle lobe. Nutlets much narrower than the wings.

Fig. 155. Betula papyrifera Marsh. Leaf, × ½.

Native to Canada and the northern United States. Several natural and horticultural varieties are occasionally cultivated.

2*a*. Var. **occidentàlis** Sarg. WESTERN PAPER BIRCH.

*B. occidentalis* Hook.

A large tree, 80 to 120 feet high, with a trunk 3 to 4 feet in diameter. Bark orange-brown or white. Lateral lobes of the fruiting bracts widely spreading. Nutlets nearly as wide as the wings.

Native from southwestern British Columbia and northwestern Washington, eastward to eastern Washington, northern Idaho, and northern Montana west of the Continental Divide.

2*b*. Var. **subcordàta** Sarg.

*B. subcordata* Rydb.

A small tree, 20 to 40 feet high. Leaves rounded to cordate at the base. Fruiting bracts puberulent, the middle lobe acute and slightly longer than the ascending or broadly spreading lateral lobes.

Native from British Columbia south to western Washington and northeastern Oregon (Minum River Valley), eastward through Idaho to northern Montana and Alberta, Canada.

2c. Var. **kenaìca** Henry. RED BIRCH.

Fig. 156.
Betula pen-
dula Roth.
Leaf, fruiting
catkin, × ½.

Leaves usually wedge-shaped at the base. Fruiting catkins about 1 inch long. Middle lobe of the fruiting bracts not much longer than the broad lateral lobes.

Native along the western coast of Alaska.

3. **Betula álba** L. EUROPEAN WHITE BIRCH.

A small tree, 20 to 60 feet high, with white bark, ascending, spreading, or drooping branches, and resinous-glandular branchlets. Leaf-blades ovate or rhombic-ovate, 1 to 2½ inches long, ¾-inch to 1½ inches wide, short-acuminate to acute at the apex, truncate or wedge-shaped to rounded at the base, glabrous or pubescent, the margins unequally serrate to deeply cut; petioles about 1 inch long. Fruiting catkins about 1 inch long; bracts glabrous or puberulent, the lateral lobes widely spreading and recurved, usually larger than the middle lobe. Nutlets narrower than the wings.

A native of Europe. European White Birch is extensively culti-

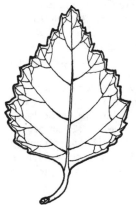

Fig. 157.
Betula pubescens Ehrh.
Leaf, × 1.

Fig. 158.
Betula pendula var. gracilis Rehd.
Leaf, × ¾.

vated in gardens and parks under many names which indicate the habit of growth and characteristics of foliage. The following forms are commonly seen in cultivation: (1) **B. péndula** Roth. (*B. alba* var.

*pendula* Hort.). Fig. 156. Branchlets pendulous. (2) **B. pubéscens** Ehrh. (*B. alba* L., in part). Fig. 157. Young branchlets densely pubescent. Leaves pubescent beneath. (3) **B. álba** var. **fastigiàta** Carr. (*B. pendula* var. *fastigiata* Koch). Branchlets straight, ascending, forming a columnar crown. Leaves very dark green. (4) **B. álba** var. **atropurpùrea** Lauche. (*B. pendula* var. *purpurea* Schneid.). Leaves purple, turning greenish. (5) **B. péndula** var. **grácilis** Rehd. (*B. alba laciniata gracilis pendula* Hort.). Fig. 158. Leaves deeply and irregularly cut or dissected.

4. **Betula populifòlia** Marsh. AMERICAN WHITE BIRCH. GRAY BIRCH. Fig. 159.

*B. alba* var. *populifolia* Spach.

A small tree, 20 to 30 feet high, with chalky white bark tardily separating into thin layers, and short resinous-glandular and more or less pendulous branchlets. Leaf-blades triangular-ovate, 2½ to 3¼ inches

long, 1½ to 2¼ inches wide, long-acuminate, truncate or slightly cordate or cuneate at the base, dark green and glossy above, light green beneath, coarsely and somewhat doubly serrate with spreading glandular teeth; petioles ¾-inch to 1¼ inches long, slender, covered with minute black glands. Fruiting catkins ¾-inch to 1¼ inches long; bracts puberulent, the lateral lobes widely spreading and recurved, usually larger than the middle lobe. Nutlets a little narrower than the wings.

Native from Nova Scotia south to Pennsylvania and Delaware. Cultivated in parks and gardens. The following horticultural forms are frequently seen: (1) **B. populifòlia** var. **laciniàta** Loud. Leaves pinnately lobed. (2) **B. populifòlia** var. **péndula** Loud. Branchlets pendulous. (3) **B. populifòlia** var. **purpùrea** Ellw. & Barry. Leaves purple when young, turning green.

Fig. 159. Betula populifolia Marsh. Bract of fruiting catkin, × 2½. Leaf, × ½. Single fruit, × 2½.

### 3. Carpìnus L. HORNBEAM

(The ancient Latin name.)

About 20 species, native to central and eastern Asia, Europe, and North America.

1. **Carpinus bétulus** L. EUROPEAN HORNBEAM. Fig. 160.

A small deciduous tree, 30 to 60 feet high, with smooth gray bark.

Leaves simple, alternate; the blades ovate or oblong-ovate, 2 to 4 inches long, 1 to 1¾ inches wide, glabrous except in the axils of the veins beneath, sharply and doubly glandular-serrate; petioles ¼- to ½-inch long. Flowers monoecious, in catkins; the staminate catkins pendulous, axillary; the pistillate slender, terminal on the new leafy branchlets. Fruit a small nutlet, subtended by a 3-lobed leafy bract, borne in pendulous light green fruiting catkins 3 to 5½ inches long. Bracts 1½ to 2 inches long, the middle lobe much longer than the 2 equal lateral ones.

Native to Europe, extending to Persia. Several garden forms varying in habit of growth and leaf characters are occasionally cultivated.

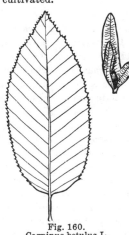

Fig. 160.
Carpinus betulus L.
Leaf, single fruit, × ½.

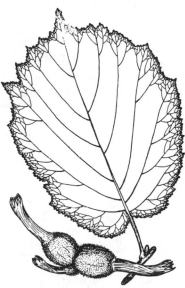

Fig. 161. Corylus maxima Mill.
Leaf, fruit, × ½.

#### 4. Córylus L. HAZEL

(From the Greek *korys*, a hood or helmet, in reference to the involucre covering the nut.)

About 15 species, native to North America, Europe, and Asia.

#### 1. Corylus máxima Mill. GIANT FILBERT. Fig. 161.

A large deciduous shrub or small tree, 15 to 30 feet high, with glandular-hairy young branchlets and petioles. Leaves simple, alternate; the blades round ovate to broadly obovate, 3 to 4½ inches long, 2 to 3½ inches wide, cordate at the base, more or less pubescent on both surfaces, slightly lobed, finely and doubly serrate; petioles ½- to ¾-inch long.

Flowers small, monoecious, appearing before the leaves; the staminate in long pendulous catkins; the pistillate few to several in scaly buds, with only the red styles protruding. Fruit a subglobose nut, enclosed by a leafy involucre extending as a tube beyond the nut and lobed and fringed at the apex.

Native to southeastern Europe. Occasionally grown in gardens. Hybrids between this species and **C. avellana** L. are cultivated in Oregon (vicinity of Portland and in the Willamette Valley) and in northern California for the edible nuts. The variety **purpùrea** Rehd. has dark purple leaves.

## Fagaceae. Beech Family

### 1. Castànea Mill. CHESTNUT

(From the Greek name of a city in Thessaly.)

Trees or shrubs. Leaves simple, alternate, deciduous, straight-veined, toothed. Flowers inconspicuous, monoecious; the staminate in erect or

drooping cylindrical catkins; the pistillate usually 3 together in a prickly involucre, the involucres few on the lower part of the staminate catkins or rarely in separate catkins. Fruit a large polished brown nut, 1 to 3 (or 5) in a prickly involucre splitting at maturity into 2 to 4 valves, maturing the first year.

About 10 species, restricted to the temperate regions of the northern hemisphere, but none native on the Pacific Coast. Chestnut trees in the eastern part of the United States have been almost exterminated by the attacks of a fungus, *Endothia parasitica* Anders.

KEY TO THE SPECIES

Leaves glabrous, wedge-shaped or some
　　subcordate at base; nut 1 inch or
　　less broad......................1. *C. dentata.*
Leaves tomentose beneath, at least when
　　young, often rounded at base; nut
　　1 inch or more broad......2. *C. sativa.*

Fig. 162.
Castanea dentata Borkh.
Leaf, fruit, × ⅛.

**1. Castanea dentàta** Borkh. AMERICAN CHESTNUT. Fig. 162.

　　*C. americana* Raf.

A large tree, 50 to 90 feet high. Leaf-blades oblong-lanceolate, 5 to 9 inches long, 1½ to 2½ inches wide, thin, glabrous, wedge-shaped or truncate to cordate at the base, coarsely serrate, the teeth usually tipped with ascending bristle-like appendages; petioles about ½-inch long, puberulent. Involucral burs about 2 inches across, enclosing 2 or 3 flattened nuts 1 inch or less broad.

Native to the northeastern United States. Occasionally cultivated in parks and gardens.

2. **Castanea sativa** Mill. SPANISH CHESTNUT.

*C. vesca* Gaertn.

A large tree, 30 to 70 feet high. Leaf-blades 6 to 9 inches long, about 2 inches wide, usually rounded at the base, glabrous above, tomentose beneath at least when young, coarsely serrate; petioles about ½-inch long. Involucral burs very prickly, about 2 inches across, enclosing 1 to 3 nuts 1 inch or more broad.

Native to south Europe, north Africa, and south Asia. Occasionally cultivated.

2. **Castanópsis** Spach. CHINQUAPIN

(From the Greek *kastanea*, chestnut, and *opsis*, resemblance, in reference to the resemblance to the chestnut tree.)

About 30 species, native to Asia, one on the Pacific Coast.

1. **Castanopsis chrysophýlla** DC. GIANT CHINQUAPIN. Fig. 163.

A forest tree, 50 to 125 feet high, with thick bark broken into longitudinal furrows and ridges, and stout spreading branches forming a broad round-topped or conical crown. Leaves simple, alternate, evergreen; the blades thick and leathery, oblong or oblong-lanceolate, 2 to 5½ inches long, ½-inch to 1½ inches wide, tapering to both ends, dark green and glabrous above, golden-tomentose beneath, becoming pale yellow, entire; petioles ¼- to

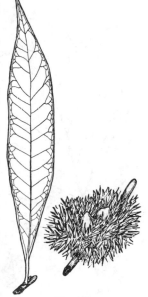

Fig. 163.
*Castanopsis chrysophylla* DC.
Leaf, fruit, × ½.

½-inch long. Flowers inconspicuous, monoecious; the staminate numerous, in erect elongated sometimes branching catkins; the pistillate 1 to 3 in an involucre, the involucres at the base of the staminate catkins or in shorter separate catkins. Fruit a globose or ovoid nut, 1 to 3 in a spiny involucre, maturing the second year.

Native on the mountain slopes of the outer Coast Ranges of Mendocino County, California, north to northwestern Siskiyou County and along the western slope of the Cascade Mountains to Skamania County, Washington.

1*a*. Var. **mìnor** Benth. GOLDEN CHINQUAPIN.

A small tree or large shrub with ascending or spreading branches forming a compact crown. Leaf-blades 2 to 3 inches long, usually folded upward along the midrib and becoming trough-like, very dark green above, golden brown beneath.

This variety occurs in small groups on the rocky ridges in the mountains from Monterey County northward to Mendocino County, California.

### 3. **Fàgus** L. BEECH

(The ancient Latin name of the beech.)

Eight species restricted to the northern hemisphere, one in the central and eastern United States and 7 in Europe. Beeches are valuable ornamental and timber trees with edible nuts.

1. **Fagus sylvática** var. **purpùrea** Ait. PURPLE BEECH. COPPER BEECH. Fig. 164.

*F. sylvatica* var. *atropunicea* West. *F. sylvatica* var. *riversi* Rehd.

A small tree, 15 to 30 feet high, with ascending or spreading branches forming a round-topped crown. Leaves simple, alternate and clustered at the ends of short lateral branchlets, deciduous, purple, or copper-colored; the blades ovate to elliptic, 2 to 4 inches long, 1 to 2 inches wide, distinctly straight-veined, glabrous when

Fig. 164. Fagus sylvatica var. purpurea Ait.
Leaf, fruit, × 1.

mature, silky hairy on the margins and on the veins beneath when young, finely serrate; petioles ¼- to ⅜-inch long. Flowers small, monoecious, appearing with the leaves; the staminate in slender-peduncled heads; the pistillate usually 2, surrounded by numerous bracts. Fruit a smooth triangular brown nut, 1 or 2 enclosed in a prickly bur about ¾-inch long, separating at maturity into 4 valves.

Cultivated as a street or park tree. The species **F. sylvática** L., EUROPEAN BEECH, and **F. americàna** Sweet, AMERICAN BEECH, are occasionally cultivated as specimen trees. They are large spreading trees with green leaves similar to the variety above.

### 4. Lithocárpus Bl.

(From the Greek *lithos*, stone, and *karpos*, fruit, in reference to the hard shell of the nut.)

A genus intermediate between *Quercus* and *Castanea*. About 100 species, native to southern Asia, one on the Pacific Coast.

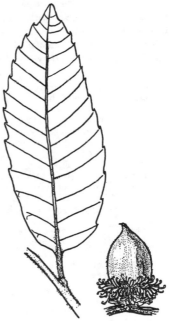

### 1. Lithocarpus densiflòra Rehd.
TANBARK-OAK. TAN-OAK. Fig. 165.

*Quercus densiflora* H. & A. *Pasania densiflora* Örst.

An evergreen forest tree, 50 to 75 (or 150) feet high, with thick furrowed bark, ascending or spreading branches, and tomentose young branchlets forming a narrow columnar or round-topped crown. Leaves simple, alternate; the blades thick and leathery, oblong or oblong-ovate, 2 to 5 inches long, ¾-inch to 2½ inches wide, strongly straight-veined, glabrous or with scattered pubescence above, white- or rusty-tomentose beneath, becoming glabrous, dentate with thickened teeth; petioles ½- to ¾-inch long, tomentose. Flowers inconspicuous, monoecious; the staminate numerous, in erect ill-smelling catkins 2½ to 4 inches long; the pistillate solitary in the axils of involucral bracts; the involucres few at the base of some

Fig. 165. Lithocarpus densiflora Rehd. Leaf, fruit, × ¾.

staminate catkins. Fruit an ovoid nut, ¾-inch to 1 inch long, surrounded at the base by an involucral cup beset with numerous slender scales and fascicled hairs, maturing the second year.

Native in the Coast Ranges from the Santa Inez Mountains of southern California north to the Umpqua River, Oregon, and on the western slope of the Sierra Nevada up to 4000 feet elevation, from Mariposa County north and westward to Siskiyou County, California. Most abundant in the redwood belt. Occasionally cultivated in parks and gardens. The bark is used for tanning leather.

### 5. Quércus L. OAK

(From the Celtic *quer*, fine, and *cuez*, a tree, signifying a beautiful tree.)

Deciduous or evergreen trees or shrubs. Leaves simple, alternate. Flowers inconspicuous, monoecious; the staminate in drooping catkins; the pistillate solitary, in many-bracted involucres. Fruit an acorn.

A large genus of more than 200 species, native in the colder and temperate regions of the northern hemisphere and the mountains of the tropics. Sixteen species with several named varieties are native to the Pacific Coast, 11 of which are trees and 5 are shrubs. Many species are cultivated for their ornamental foliage, which in the deciduous species often becomes brilliantly colored in the autumn.

#### KEY TO THE SPECIES

I. Leaves not lobed, evergreen except in *Q. douglasi.*

Leaves green and glabrous on both surfaces except for a few axillary tufts of hairs beneath in *Q. agrifolia.*

Leaves often convex above, broadly elliptical or roundish; acorns maturing the first year..............................................1. *Q. agrifolia.*

Leaves plane, oblong or ovate to ovate-lanceolate; acorns maturing the second year.................................................2. *Q. wislizeni.*

Leaves dull, whitish, or grayish tomentose beneath.

Leaves bluish green above; acorns maturing the first year.

Leaves deciduous; acorn-cup very shallow, ¼ to ⅓ as high as the nut.....................................................................3. *Q. douglasi.*

Leaves evergreen; acorn-cup almost ½ as high as the nut

4. *Q. engelmanni.*

Leaves not bluish green above.

Bark thick and corky, deeply furrowed; acorns maturing the first year.......................................................................5. *Q. suber.*

Bark not thick and corky; smooth or scaly.

Acorns maturing the first year; introduced species....6. *Q. ilex.*

Acorns maturing the second year; native species.

Leaves ¾-inch to 2¼ (or 4) inches long, the lateral veins not prominent or parallel.........................7. *Q. chrysolepis.*

Leaves 2 to 4 inches long, the lateral veins prominent and nearly parallel.............................................8. *Q. tomentella.*

II. Leaves distinctly lobed, deciduous except in *Q. morehus.*

Lobes with spinose- or bristle-tipped teeth.

Leaves shallowly lobed, the lobes spinose-tipped, almost evergreen

9. *Q. morehus.*

Leaves deeply lobed, the lobes bristle-tipped, deciduous.

Leaves dull green above, whitish or gray-tomentulose or nearly
glabrous beneath.
Leaves 3- to 7-lobed, the lobes often curved.............10. *Q. rubra.*
Leaves 7- to 11-lobed..................................................11. *Q. borealis.*
Leaves bright green and glossy above.
Leaves glabrous beneath, with small or conspicuous axillary
tufts of hairs.
Acorn-cup saucer-shaped, about ⅓ as high as the nut; the
nut about ½-inch long; leaves with conspicuous axil-
lary tufts of hairs.......................................12. *Q. palustris.*
Acorn-cup top- or cup-shaped, ⅓ to ½ as high as the nut;
the nut ovoid, ½-inch to 1 inch long; leaves with small
axillary tufts of hairs..................................13. *Q. coccinea.*
Leaves minutely pubescent beneath, at least when young, with-
out axillary tufts of hairs; nut oblong-ovoid or ellipsoidal,
1 to 1⅛ inches long........................................14. *Q. kelloggi.*
Lobes not spinose- or bristle-tipped (or sometimes bristle-tipped in
*Q. macdonaldi*).
Leaves 1 to 3 inches long.
Leaves bluish green and puberulent above, shallowly lobed or
coarsely and unequally few-toothed to entire....3. *Q. douglasi.*
Leaves green and glabrous above; species of the coastal islands
15. *Q. macdonaldi.*
Leaves 2½ to 10 inches long.
Leaves 4 to 10 inches long, grayish tomentulose beneath; upper
scales of acorn-cup with awn-shaped tips forming a fringe
around the nut..............................................16. *Q. macrocarpa.*
Leaves 2½ to 4 (or 6) inches long, pale or pubescent beneath;
upper scales of acorn-cup not awn-tipped.
Leaves cuneate or rounded at base; native species.
Leaves never more than 3 inches broad; nut 1¼ to 2 inches
long, tapering to apex; acorn-cup ½- to ¾-inch deep
17. *Q. lobata.*
Leaves often 3 to 5 inches broad; nut ¾-inch to 1 inch long,
rounded at apex; acorn-cup ½-inch or less deep
18. *Q. garryana.*
Leaves auriculate or truncate at base; introduced species
19. *Q. robur.*
1. **Quercus agrifòlia** Née. CALIFORNIA LIVE OAK. COAST LIVE OAK.
Fig. 166.
An evergreen tree, 30 to 75 feet high, with spreading branches from
a trunk dividing a few feet above the ground forming a broad round
crown. Leaf-blades stiff, leathery, oval or broadly elliptical to almost
round, 1 to 3 inches long, ¾-inch to 2 inches wide, dark green and
glabrous above, glossy, often convex, paler beneath and usually with
axillary tufts of hairs, the margins finely to coarsely spine-tipped or
rarely entire; petioles ¼- to ½-inch long. Acorns maturing the first
year; the cup broadly top-shaped, enclosing only the base or about ⅓
of the nut, with thin closely imbricated scales; nut slender, pointed,
1 to 1½ inches long.

Native on the lower mountain slopes, rocky hills, and on valley flats of the Coast Ranges from Sonoma and Napa counties southward to the mountains of southern California, the coastal islands, and in Lower California to the San Pedro Mártir Mountains. Frequently cultivated in parks and large gardens for the handsome dark green foliage and broad symmetrical crown. Often known as Encina or Encina Oak. Dr. Albert Kellogg (Appendix, *Mineralogist's Report of California*, p. 67, 1882)

Fig. 166. Quercus agrifolia Née.
Leaves, fruit, × ¾.

thought the spelling of this species should have been *aquifolia*. He states, "This *agrifolian* oak—probably a printer's immortalized mistake for *aquifolia*, or Hollyleaf Oak (?)."

**2. Quercus wislizèni** A. DC. INTERIOR LIVE OAK. Fig. 167.

An evergreen tree, 25 to 75 feet high, with stout spreading branches forming a round-topped crown. Leaf-blades stiff, leathery, oblong, ovate to ovate-lanceolate, or elliptical, 1 to 3 (or 4) inches long, ¾-inch to 1¾ inches wide, glabrous, dark green and glossy above, pale yellowish green beneath, entire or spiny-toothed, plane; petioles ¼- to ¾-inch long. Acorns maturing the second year; the cup top-shaped or hemispheric, enclosing ¼ to ½ of the nut, with thin lanceolate closely imbricated scales; nut slender, oblong, pointed, 1 to 1½ inches long.

Native in California, on the lower mountain slopes and foothills of the Sierra Nevada from Kern County northward to Mount Shasta, south-

Fig. 167. Quercus wislizeni A. DC.
Leaves, × ½. Fruit, × 1.

ward in the Great Valley and inner Coast Range to Napa County, south of which it occurs as a shrub in the chaparral. Occasionally cultivated.

3. **Quercus doúglasi** H. & A. BLUE OAK. Fig. 168.

A small to medium-sized deciduous tree, 20 to 60 feet high, with light gray bark checked into thin scales, and short stout spreading branches forming a round-topped crown. Leaf-blades rather firm and rigid, oblong to obovate in outline, 1½ to 4 inches long, ¾-inch to 2 inches wide, shallowly and irregularly lobed or few-toothed to entire, minutely pubescent, bluish green above, pale beneath; petioles ¼- to ½-inch long. Acorns maturing the first year; the cup shallow, ¼ to ⅓ as high as the nut, with small thickened usually pubescent scales; nut quite variable in shape, commonly ovoid, ¾-inch to 1¼ inches long, often narrowed at the base.

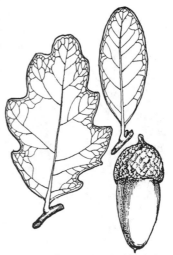

Fig. 168. Quercus douglasi H. & A.
Leaves, fruit, × ¾.

Native in California, forming scattered groves along the lower dry mountain slopes and rocky foothills of the middle and inner Coast

Ranges, the Sierra Nevada, and the Sierra Liebre of Los Angeles County. This tree is rarely cultivated.

4. **Quercus éngelmanni** Greene. MESA OAK. ENGELMANN OAK. Fig. 169.

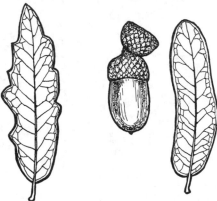

Fig. 169. Quercus engelmanni Greene.
Leaves, fruit, × ¾.

An evergreen tree, 20 to 50 feet high, with spreading branches forming a round-topped crown. Leaf-blades stiff, leathery, oblong to

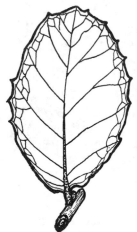

Fig. 170. Quercus suber L.
Leaf, × 1.

obovate, 1 to 3 inches long, ½-inch to 1 inch wide, bluish green and glabrous or slightly pubescent above, pale and glabrous or pubescent beneath, entire or irregularly toothed; petioles ¼- to ½-inch long. Acorns maturing the first year; the cup top- or cup-shaped, enclosing about ¼ to ⅓ of the nut, with tomentose scales which become tuberculate toward the base; nut subcylindric or ovoid, ¾-inch to 1 inch long.

Native to California, in the foothills between the coast and the mountains of San Diego County, northward to the slopes below the San Gabriel Mountains near Pasadena, and in Lower California.

5. **Quercus sùber** L. CORK OAK. Fig. 170. An evergreen tree, 20 to 50 feet high, with a short trunk dividing into several stout stems, thick corky and deeply furrowed bark, and spreading branches

often touching the ground and forming a compact round-topped crown. Leaf-blades leathery, broadly elliptical or oval to oblong-ovate, 1 to 2½ inches long, ¾-inch to 1¼ inches wide, dark green and glabrous above, prominently veined and grayish tomentose beneath, the margins with 4 or 5 pairs of short pickle-like teeth or rarely almost entire; petioles ¼- to ½-inch long. Acorns maturing the first year; the cup enclosing ⅓ to ½ the nut, with thick well-imbricated scales, usually with short and recurved tips; nut ovoid to oblong-ovoid, ½-inch to 1 inch long.

Native to southern Europe and northern Africa. Frequently cultivated in parks. The thick corky bark supplies the commercial cork.

**6. Quercus ilex** L. HOLLY OAK. HOLM OAK. Fig. 171.

An evergreen tree, 20 to 60 feet high, with gray nearly smooth bark and spreading or descending branches forming a large round-topped crown. Leaf-blades stiff, variable, usually ovate to lanceolate, 1¼ to 2½ inches long, ½-inch to 1 inch wide, acute or acuminate, rounded and somewhat uneven at the base, dark green and glabrous above, gray-tomentose beneath, irregularly and finely toothed or entire; petioles ¾- to ⅝-inch long, pubescent. Acorns maturing the first year; the cup top-shaped, enclosing about ½ of the nut, with thin closely appressed scales; nut ovoid, ¾-inch to 1¼ inches long.

Native to southern Europe. Occasionally planted in parks. Sometimes

Fig. 171. Quercus ilex L. Leaf, fruit, × 1.

confused with *Q. suber* L., Cork Oak, which has a thick corky bark.

**7. Quercus chrysólepis** Liebm. MAUL OAK. CANYON OAK. Fig. 172.

An evergreen tree, 25 to 50 feet high, or often shrub-like, with grayish bark and spreading branches forming a round-topped crown sometimes 100 feet across. Leaf-blades thick, leathery, elliptic-oblong or ovate, ¾-inch to 3 (or 4) inches long, ½-inch to 1¾ inches wide, green and glabrous above, yellowish tomentose beneath, becoming grayish tomentose or glaucous, entire or irregularly spiny-toothed, especially on young growth; petioles ¼- to ½-inch long. Acorns maturing the second year; the cup shallow, surrounding only the base of the nut, very thick-walled and broad-rimmed, with small scales densely covered by a fine felt-like often golden-yellow tomentum; nut ovoid or oblong-ellipsoidal, 1 to 1¼ inches long.

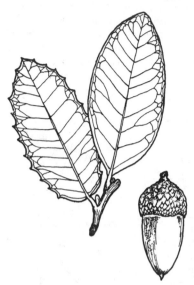

Fig. 172. Quercus chrysolepis Liebm.
Leaves, fruit, × ¾.

Native in mountain canyons and on moist ridges and flats of the Coast Ranges of southern Oregon and California, the lower and middle slopes of the western side of the Sierra Nevada, and the mountains of southern California, Lower California, Arizona, and New Mexico. Occasionally cultivated.

8. **Quercus tomentélla** Engelm. ISLAND OAK. Fig. 173.

A small round-headed evergreen tree, 20 to 40 feet high, with heavy tomentose young branchlets. Leaf-blades thick and leathery, elliptic to oblong-lanceolate, 2 to 4 inches long, 1 to 2 inches wide, dark green and glabrous above, or tomentose at first, pale and densely pubescent beneath, with prominent and nearly parallel lateral veins, the margins entire or crenately dentate with swollen teeth, often revolute; petioles about ¼- to ½-inch long, pubescent. Acorns maturing the second year; the cup hemispheric, enclosing ⅓ to ½ of the nut, with acute scales partly covered by a hoary tomentum; nut ovoid, rounded at the summit, ¾-inch to 1¼ inches long.

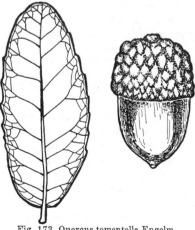

Fig. 173. Quercus tomentella Engelm.
Leaf, fruit, × ¾.

Native on Santa Catalina, San Clemente, Santa Cruz, Santa Rosa, and Guadalupe islands. Not occurring on the mainland.

9. **Quercus mórehus** Kell. ORACLE OAK. Fig. 174.

A small almost evergreen tree, 20 to 40 feet high, or often shrub-like. Leaf-blades oblong to elliptic, 2 to 4 inches long, 1 to 2 inches wide, with shallow spinose-tipped lobes, glabrous, dark green and glossy above,

Fig. 174. Quercus morehus Kell.
Leaves, fruit, × ¾.

paler beneath; petioles ¼- to ½-inch long. Acorns maturing the second year; the cup hemispherical, enclosing ⅓ to ½ the nut, with thin closely appressed scales; nut slender-ellipsoidal, ¾-inch to 1¼ inches long.

Native on the lower slopes and foothills of the Sierra Nevada, the Coast Ranges, and the San Bernardino Mountains of southern California, occurring in scattered locations and nowhere abundant.

10. **Quercus rùbra** L. SPANISH OAK. RED OAK. Fig. 175.

A deciduous tree, 50 to 70 feet high, with large spreading branches forming a round-topped crown. Leaf-blades elliptic-oblong to obovate, 4 to 10 inches long, 3 to 5 inches wide, deeply 3- to 7-lobed, the lobes with 1 to 3 bristle-like teeth, dark green and glabrous above, pale and grayish

or rusty-pubescent beneath; petioles 1 to 2 inches long. Acorns maturing
the second year; the cup saucer-shaped, enclosing ¼ to ⅓ of the nut,
with thin reddish brown scales; nut subglobose, ½-inch to 1 inch long.

Native to the eastern and southern United States. Occasionally
cultivated in parks.

### 11. Quercus boreàlis Michx. RED OAK.

A deciduous tree, 40 to 60 feet high. Leaf-blades oblong, 5 to 9
inches long, 3½ to 5 inches wide, 7- to 11-lobed to about halfway to

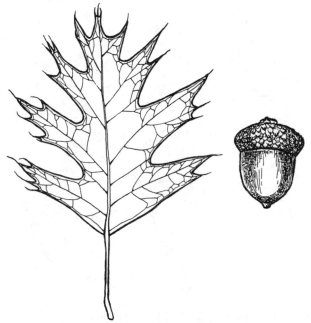

Fig. 175. Quercus rubra L.
Leaf, × ⅝. Fruit, × ¾.

the midrib, the sinuses rounded, the lobes with usually 3 or 4 bristle-
pointed teeth, dull green above, grayish or whitish or sometimes yellow-green and nearly glabrous beneath, but with axillary tufts of
brown hairs; petioles 1 to 2 inches long. Acorns maturing the second
year; the cup hemispheric, enclosing about ⅓ of the nut, with closely
appressed pubescent scales; nut ovoid, ¾-inch to 1 inch long.

Native to the eastern United States, north to Nova Scotia. Occasionally cultivated in parks.

12. **Quercus palùstris** L. PIN OAK. Fig. 176.

A large deciduous tree, 50 to 80 feet high. Leaf-blades oblong-elliptical to obovate, 4 to 6 inches long, 2 to 4 inches wide, deeply

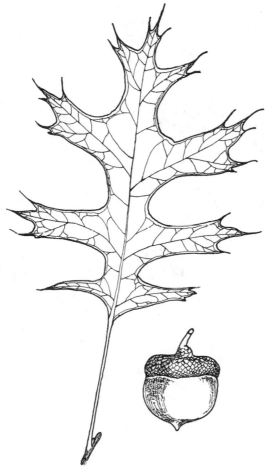

Fig. 176. Quercus palustris L.
Leaf, × ¾. Fruit, × 1½.

5- to 7-lobed, the lobes bristle-tipped, dark green, glossy, and glabrous above, paler and glabrous beneath, except for axillary tufts of hairs; petioles ½-inch to 2 inches long. Acorns maturing the second year;

the cup saucer-shaped, enclosing the base or ⅓ of the nut, with thin puberulent scales; nut hemispheric, about ½-inch long.

Native to the north central and eastern United States. Occasionally cultivated in parks.

13. **Quercus coccínea** Muench. SCARLET OAK. Fig. 177.

A large deciduous tree, 60 to 80 feet high, with bright green foliage turning a brilliant scarlet in autumn. Leaf-blades elliptic or oblong, 3

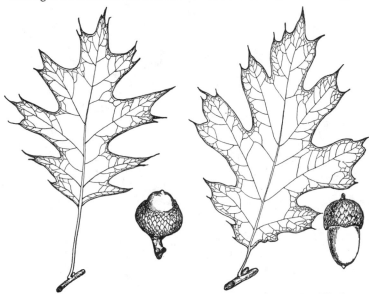

Fig. 177. Quercus coccinea Muench.
Leaf, fruit, × ½.

Fig. 178. Quercus kelloggi Newb.
Leaf, fruit, × ½.

to 6 inches long, 2 to 4 inches wide, deeply 7- or rarely 9-lobed, the lobes bristle-tipped, bright green and glabrous above, paler and glabrous beneath, or occasionally with axillary tufts of hairs; the petioles 1½ to 2½ inches long. Acorns maturing the second year; the cup top- or cup-shaped, enclosing ⅓ to ½ of the nut, with thin brownish puberulent scales; nut ovoid to hemispheric, ½-inch to 1 inch long.

Native to the eastern United States from Maine to Florida, west to Wisconsin, and south to Arkansas and Oklahoma. Occasionally cultivated in parks.

14. **Quercus kélloggi** Newb. CALIFORNIA BLACK OAK. Fig. 178.

A large deciduous tree, 30 to 80 feet high, with stout spreading branches forming a broad round-topped crown. Leaf-blades broadly elliptical to obovate, 4 to 8 inches long, 2½ to 4 inches wide, deeply 7- (or 5-) lobed, the lobes with 1 to 4 bristle-tipped teeth, bright green, glossy, and glabrous above, paler beneath, gray-tomentose on both surfaces when young; petioles 2 to 4 inches long. Acorns maturing the first year; the cup hemispheric, enclosing ⅓ to ½ of the nut, with thin closely imbricated puberulent scales; nut cylindric, rounded at the apex, 1 to 1⅛ inches long.

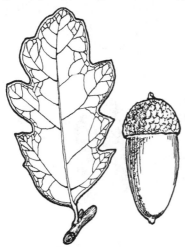

Native on lower and middle mountain slopes and the foothills from southwestern Oregon south in the Coast Ranges and

Fig. 179. Quercus macdonaldi Greene.
Leaf, fruit, × 1.

the Sierra Nevada to the mountains of southern California.

15. **Quercus macdónaldi** Greene. MAC-DONALD OAK. Fig. 179.

*Q. dumosa* var. *macdonaldi* Jepson.

A small deciduous tree, 15 to 40 feet high, with a dense rounded crown. Leaf-blades oblong to obovate, 1½ to 2¾ inches long, 1 to 1½ inches wide, with 5 to 9 obtuse or sharp-pointed lobes, rarely bristle-tipped, green and glabrous above, paler and pubescent beneath; petioles ¼- to ½-inch long. Acorns maturing the first year; the cup deeply cup-shaped to hemispheric, enclosing about ⅓ of the nut, with blunt tuberculate scales; nut cylindric, tapering at the apex, ¾-inch to 1¼ inches long.

Fig. 180.
Quercus macrocarpa Michx.
Leaf, fruit, × ⅓.

Native on Santa Catalina and Santa Cruz islands off the coast of southern California.

16. **Quercus macrocárpa** Michx. BUR OAK. MOSSYCUP OAK. Fig. 180.

A large deciduous tree, 70 to 150 feet high, with the main branches often originating 50 feet or more above the ground. Leaf-blades obovate to oblong, 4 to 10 inches long, 2½ to 5 inches wide, either shallowly or deeply 5- or 7-lobed, dark green and glabrous or sometimes pubescent above, grayish tomentulose beneath, the terminal lobe large, obovate, and with rounded secondary lobes; petioles ½-inch to 1 inch long. Acorns maturing the first season, variable in

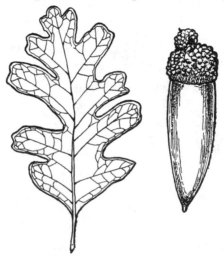

Fig. 181. Quercus lobata Née.
Leaf, fruit, × ¾.

form and size; the cup with hoary-tomentose well-imbricated scales, bearing near the rim awn-shaped tips forming a fringe around the nut; nut broadly ovoid or ellipsoidal, ¾-inch to 1¾ inches long.

Native in southeastern Canada and the north central and northeastern parts of the United States. Occasionally cultivated.

17. **Quercus lobàta** Née. VALLEY OAK. ROBLE. CALIFORNIA WHITE OAK. Fig. 181.

A large, graceful deciduous tree, 40 to 125 feet high, with thick checkered bark covered by light gray scales, and ascending or spreading and ultimately drooping branches. Leaf-blades oblong to obovate, 2½ to 4 (or 6) inches long, 1½ to 3 inches wide, deeply 7- to 11-lobed, the lobes often broader toward the apex, green and pubescent or glabrate above, paler and pubescent and with yellow

veins beneath; petioles ¼- to ½-inch long, pubescent. Acorns maturing the first year; the cup hemispherical, enclosing about ⅓ of the
nut, with scales especially warty near the base; nut long-conical, 1¼
to 2 inches long.

Native in the fertile parts of the Sacramento, San Joaquin, and
adjacent valleys, the foothills of the Sierra Nevada, the inner and
middle Coast Ranges, and extending south to the San Fernando Valley. This oak is the most characteristic native tree of the Great Valley

Fig. 182. Quercus garryana Dougl.
Leaf, fruit, × ¾.

of California. One very large specimen at Chico has been named the
Hooker Oak in honor of Sir Joseph Hooker, an eminent English botanist. Frequently cultivated in the warmer parts of California.

18. **Quercus garryàna** Dougl. OREGON OAK. Fig. 182.

A round-headed deciduous tree, 35 to 60 feet high. Leaf-blades leathery, obovate to oblong, 3 to 4 (or 6) inches long, 2 to 3 (or 5) inches
wide, deeply 5- to 9-lobed, the lobes rounded or unequally toothed, dark
green and glossy above, paler and finely pubescent beneath; petioles
½-inch to 1 inch long. Acorns maturing the first year; the cup shallow,
enclosing only the base or ⅓ of the nut, with pubescent thickened scales;
nut ovoid or subcylindric, ¾-inch to 1 inch long, rounded at the apex.

Native from Vancouver Island south to western Washington, Oregon, and the Coast Ranges of California to Marin County and in the Sierra Nevada to eastern Shasta County.

**19. Quercus ròbur** L. ENGLISH OAK. Fig. 183.

A large deciduous tree, 60 to 100 feet high, with spreading branches forming a broad round-topped crown. Leaf-blades obovate-oblong or obovate, 2½ to 5 inches long, 1½ to 3 inches wide, auriculate or truncate at the base, with 3 to 7 shallow rounded lobes on each side, glabrous and dark green above, pale bluish green beneath; petioles ⅛- to ¼-inch long. Acorns maturing the first year; the cup hemispherical, enclosing about ⅓ of the nut, with closely imbricated scales; nut ovoid to oblong-ovoid, about 1 inch long.

Native to Europe, northern Africa, and western Asia. A variable species with numerous forms in cultivation.

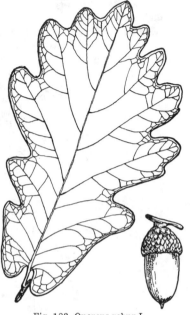

Fig. 183. Quercus robur L.
Leaf, fruit, × ¾.

## Ulmaceae. Elm Family
### 1. Céltis L. HACKBERRY

(The classical name for a species of lotus.)

Usually trees with smooth thin often warty bark. Leaves simple, alternate, deciduous (evergreen in some tropical species). Flowers inconspicuous, polygamo-monoecious, appearing with the leaves; the staminate fascicled; the pistillate and perfect flowers solitary or in small clusters in the axils of the upper leaves. Fruit a subglobose drupe with a thin outer coat and hard stone.

About 70 species, native in temperate and tropical regions.

KEY TO THE SPECIES

Leaves entire to irregularly serrate, ovate, conspicuously reticulate and usually glabrous beneath..........................................1. *C. douglasi.*
Leaves sharply serrate, elliptic-ovate to oblong-lanceolate, soft-pubescent and grayish green beneath..........................................2. *C. australis.*

**1. Celtis doúglasi** Planch. WESTERN HACKBERRY. PALOBLANCO. Fig. 184.

*C. mississippiensis* var. *reticulata* Sarg.

A small tree, 10 to 30 feet high, or often shrub-like, with roughish gray-brown bark. Leaf-blades very rough to the touch, ovate or ovate-lanceolate, 1 to 3½ inches long, ¾-inch to 1½ inches wide,

unequal and slightly heart-shaped at the base, conspicuously netted-veined beneath, often with 3 prominent veins, deep green above, lighter and pubescent on the veins beneath, entire or serrulate to coarsely serrate; petioles ¼-inch or less long, slightly pubescent. Drupes about ¼-inch in diameter, dark brown or orange when mature, on pubescent peduncles ¼- to ⅜-inch long.

Sparingly native in gravelly washes in the mountains bordering the deserts of southern California, more common eastward to Texas, south to Mexico, and north to eastern Washington and Oregon. Rarely cultivated.

Fig. 184.
Celtis douglasi Planch.
Branchlet with leaves
and fruit, × ½.

**2. Celtis austràlis** L. EUROPEAN HACKBERRY. Fig. 185.

A tree, 40 to 80 feet high, with spreading branches forming a round-topped crown. Leaf-blades elliptic-ovate to oblong-lanceolate, 2 to 6 inches long, 1 to 2 inches wide, oblique at the base, dark green and scabrous above, gray-green and soft-pubescent beneath, sharply serrate; petioles ⅛- to ½-inch long. Drupes about ⅜-inch in diameter, dark purple, on peduncles ½-inch to 1 inch long.

Native to southern Europe, western Asia, and northern Africa. Occasionally cultivated.

### 2. Úlmus L. ELM

(The ancient Latin name of the elm.)

Deciduous or rarely evergreen trees.

Fig. 185. Celtis australis L.
Leaf and fruit, × ½.

Leaves simple, alternate, serrate, straight-veined, usually oblique at the base. Flowers inconspicuous, perfect or rarely polygamous apetalous, in axillary clusters or racemes, appearing before or rarely after the leaves. Fruit a slightly compressed orbicular or oval nutlet surrounded by a membranous wing.

About 16 species, native in the northern hemisphere, but not found west of the Rocky Mountains in North America. The elms are usually long-lived trees and are extensively planted in parks and along avenues and highways. Several horticultural varieties of the species given below are also in cultivation.

### KEY TO THE SPECIES

Leaves simply serrate, the two sides nearly equal at base, 1 to 3 inches long.
    Leaves leathery, glossy, mostly evergreen; fruit about ⅓-inch long
        1. *U. parvifolia.*
    Leaves not leathery or glossy, deciduous; fruit about ½-inch long
        2. *U. pumila.*
Leaves doubly serrate, the two sides unequal at base.
    Leaves glabrous and not rough to the touch above.
        Leaves 2 to 3½ inches long..................................................3. *U. foliacea.*
        Leaves 3 to 6 inches long....................................................4. *U. hollandica.*
    Leaves rough to the touch above.
        Leaves only slightly rough to the touch above; fruit densely ciliate on the margins.
            Leaves 1 to 2½ inches long; most of the branches with corky wings........................................................................................5. *U. alata.*
            Leaves 2½ to 5 inches long; branches not corky-winged
                6. *U. americana.*
        Leaves decidedly rough to the touch above; fruit not ciliate on the margins.
            Leaves 2 to 3½ inches long, with axillary tufts of hairs beneath; branches often with corky wings...................7. *U. campestris.*
            Leaves 3 to 8 inches long; branches never with corky wings.
                Leaves ciliate on the margins; fruit pubescent in the center
                    8. *U. fulva.*
                Leaves not ciliate on the margins; fruit glabrous..9. *U. glabra.*

### 1. Ulmus parvifòlia Jacq. CHINESE ELM. Fig. 186.

*U. chinensis* Pers.

A small or medium-sized evergreen tree, 20 to 60 feet high, with smooth bark and slender drooping branches forming a broad round-topped crown. Leaf-blades firm, elliptical to ovate, ¾-inch to 2½ inches long, ⅜- to ⅞-inch wide, only slightly unequal at the base, glabrous, dark green and glossy above, paler beneath, simply serrate; petioles ⅛- to ¼-inch long. Fruit elliptic-ovate, about ⅜-inch long, with the seed in the center.

Native to China, Korea, and Japan. Cultivated in parks, gardens, and along avenues for its graceful habit and glossy evergreen foliage.

2. **Ulmus pùmila** L. DWARF ASIATIC ELM. Fig. 187.

A small deciduous tree, 15 to 40 feet high, or rarely shrub-like.
Leaf-blades firm at maturity, elliptic to oblong-lanceolate, ¾-inch to
3 inches long, ½-inch to 1 inch wide,
nearly equal at the base, glabrous,
dark green and smooth above, paler
beneath, usually simply serrate;
petioles about ⅛-inch long. Fruit
obovate, about ½-inch long, deeply
notched at the apex.

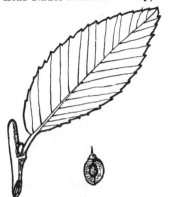

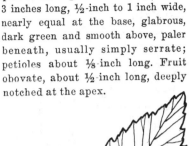

Fig. 186. Ulmus parvifolia Jacq.
Leaf, fruit, × 1.

Native in northern China, eastern Siberia,
and Turkestan. Occasionally cultivated in
gardens and along streets and highways.

Fig. 187. Ulmus pumila L.
Leaf, fruit, × 1.

3. **Ulmus foliàcea** Gilib. SMOOTH-
LEAF ELM. Fig. 188.

*U. nitens* Moench.

A deciduous tree, 40 to 70 feet high,
with deeply fissured bark and often
corky branches, commonly suckering.
Leaf-blades elliptic to ovate or obovate, 2 to 3½ inches long, very unequal at the base, dark green, glossy,

Fig. 188. Ulmus foliacea Gilib.
Fruit, leaf, × ½.

and nearly smooth above, paler and glabrous beneath except for axillary
tufts of hairs, doubly serrate; petioles ¼- to ½-inch long. Fruit obovate or elliptic, about ¾-inch long, wedge-shaped at the base, the seed
near the closed apical notch.

Native to Europe, western Asia, and northern Africa. A variable species, with many garden forms occasionally cultivated in parks and gardens.

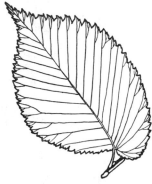

Fig. 189. Ulmus hollandica Mill. Leaf, × ½.

### 4. Ulmus hollándica Mill. DUTCH ELM. HOLLAND ELM. Fig. 189.

Usually a large broad-topped tree, with fissured bark, spreading branches, and pendulous branchlets, commonly suckering. Leaf-blades typically ovate or broadly elliptic, 3 to 6 inches long, 2 to 3 inches wide, glossy, dark green, glabrous, and nearly smooth above, paler and somewhat pubescent beneath, very unequal at the base, doubly serrate, petioles ¼- to ⅜-inch long. Fruit oval-obovate, ¾-inch to 1 inch long, with the seed near the apical notch.

According to Rehder (*Manual of Cultivated Trees and Shrubs*, p. 187) this species includes a number of elms which are supposedly of hybrid origin, the parents being *U. glabra* Huds. and *U. foliacea* Gilib. Some are more or less intermediate between the two parents. The various forms are occasionally cultivated in gardens and parks.

### 5. Ulmus alàta Michx. WAHOO ELM. WINGED ELM. Fig. 190.

A small deciduous tree, 25 to 40 feet high, with very corky-winged branches. Leaf-blades oblong or oblong-lanceolate, 1 to 2½ inches long, unequal at the base, only slightly rough above, pubescent beneath, doubly and sharply serrate; petioles ⅛-inch or less long. Fruit oblong, about ⅓-inch long, densely ciliate on the margins, pubescent on the surfaces.

Native from Virginia south to Florida, west to Illinois, Arkansas, and Texas. Occasionally cultivated in gardens and parks.

Fig. 190.
Ulmus alata Michx.
Portion of corky branchlet, leaf, × ½.

### 6. Ulmus americàna L. WHITE ELM. AMERICAN ELM. Fig. 191.

A large deciduous tree, 60 to 100 feet high, with light gray and fissured bark, spreading branches, and pendulous branchlets forming a broad-topped crown. Leaf-blades oval or obovate-oblong, 2½ to 5 inches long, 1 to 2¼ inches wide, very unequal at the base, glabrous, glossy, and slightly rough above, pubescent or nearly glabrous beneath, usually without axillary tufts of hairs, doubly serrate; petioles ¼- to ⅜-inch long. Fruit elliptic-ovate, ⅜- to ½-inch long, densely ciliate on the margins, glabrous on the surfaces, the seed at the base of the deep, almost closed, apical notch.

Native from the eastern United States and Canada west to the base of the Rocky Mountains. Extensively planted as a street and park tree.

Fig. 191. Ulmus americana L.
Fruit, × 1. Leaf, × ½.

7. **Ulmus campéstris** L. ENGLISH ELM. Fig. 192.

*U. procera* Salisb.

A tall deciduous tree, 40 to 120 feet high, with deeply fissured bark, spreading or ascending branches, and often corky-winged branchlets forming a round-topped or oval crown, commonly suckering. Leaf-blades broadly oval, 2 to 3½ inches long, 1 to 2¼ inches wide, very unequal at the base, dark green and rough above, pubescent and with axillary tufts of hairs beneath, doubly serrate; petioles ⅛- to ¼-inch long. Fruit

Fig. 192. Ulmus campestris L.
Fruit, leaf, × ½.

broadly oval to elliptic-oblong, ¾-inch to 1 inch long, glabrous on the margins and surfaces, the seed near the center and touching the deeply closed apical notch.

Native to England and western and southern Europe. Planted as a street, lawn, and park tree.

8. **Ulmus fúlva** Michx. SLIPPERY ELM. Fig. 193.

A small or medium-sized deciduous tree, 30 to 50 feet high. Bud-scales densely brown-tomentose. Leaf-blades obovate to oval-oblong, 4 to 8 inches long, 2 to 3 inches wide, very unequal at the base, dark green and very rough above, pubescent beneath, especially on the veins, doubly serrate, ciliate on the margins; petioles ¼- to ⅓-inch long. Fruit broadly elliptical to almost round, ⅜- to ¾-inch long, brownish

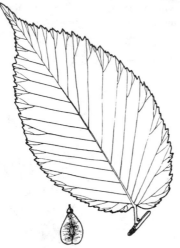

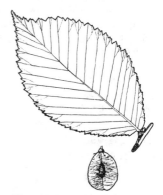

Fig. 193. Ulmus fulva Michx.
Fruit, leaf, × ½.

Fig. 194. Ulmus glabra Huds.
Fruit, leaf, × ½.

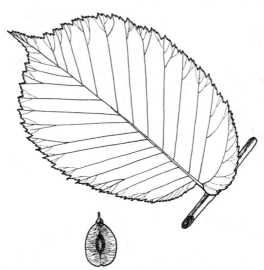

Fig. 195. Ulmus glabra var. camperdowni Rehd.
Fruit, leaf, × ½.

pubescent in the center, not ciliate on the margins, with the seed in the middle.

Native to the central and eastern United States and southeastern Canada. Occasionally cultivated.

9. **Ulmus glàbra** Huds. WYCH ELM. SCOTCH ELM. Fig. 194.

*U. scabra* Mill. *U. campestris* L., in part. *U. montana* With.

A wide-spreading deciduous tree, 50 to 100 feet high, not suckering. Leaf-blades broadly oval or broadly oblong elliptic to obovate, 3 to 6 inches long, 2 to 4 inches wide, very unequal at the base, dark green and very rough above, pubescent or nearly glabrous beneath except for small axillary tufts of hairs, doubly serrate; petioles about ⅛-inch long. Fruit broadly elliptic, ¾-inch to 1 inch long, glabrous, slightly notched at the apex, with the seed in the middle.

Native to north and central Europe and western Asia. A variable species with many forms in cultivation. The variety **cámperdòwni** Rehd., CAMPERDOWN ELM, fig. 195, is cultivated for its long drooping branches and branchlets.

### 3. **Zelkòva** Spach

(From Zelkoua, the Caucasian vernacular name of one species.)

Five species native in western and eastern Asia.

1. **Zelkova serràta** Mak. SAWLEAF ZELKOVA. Fig. 196.

*Z. acuminata* Planch.

A deciduous elm-like tree, 50 to 90 feet high, with a short trunk dividing into several ascending stems, spreading slender branches, and glabrous branchlets. Leaves simple, alternate; the blades ovate to oblong-ovate, 2 to 3 (or 6) inches

Fig. 196.
Zelkova serrata Mak.
Leaf, fruit, × ⅔.

long, ¾-inch to 1¼ inches wide, dark green and rough above, nearly glabrous beneath, sharply serrate; petioles about ⅛-inch long. Flowers inconspicuous, polygamous. Fruit an oblique drupe, about ⅛-inch wide.

Native in Japan. Occasionally cultivated for its broad round-topped crown of clean handsome foliage.

## **Moraceae.** Mulberry Family
### 1. **Fìcus** L. FIG

(The ancient Latin name.)

Trees, shrubs, and vines, with milky juice. Leaves simple, alternate, evergreen or deciduous. Flowers inconspicuous, monoecious or rarely dioecious, borne inside a closed receptacle which ripens into a fleshy fruit.

About 600 species, native in semitropical or tropical regions. Many species are in cultivation for their edible fruits, handsome foliage, and habits of growth. **Ficus benghalensis** L. is the Banyan Tree of India, which sends down aerial roots that grow as trunks and thus spread the tree over great areas.

KEY TO THE SPECIES

Leaves deeply lobed...................................................................1. *F. carica.*
Leaves not lobed.
    Leaves 4 to 12 inches long.
        Leaves pale and glabrous beneath; primary lateral veins more than
           15 pairs.....................................................................2. *F. elastica.*
        Leaves brownish beneath; primary lateral veins less than 15 pairs
           3. *F. macrophylla.*
    Leaves 2 to 4 inches long, densely rusty-tomentose beneath
           4. *F. rubiginosa.*

### 1. **Ficus càrica** L. Fig. 197.

A medium-sized deciduous tree, 20 to 30 feet high, with the trunk often branching from near the base, smooth gray bark, and spreading branches forming a broad round-topped crown. Leaf-blades 4 to 12 inches long, nearly as broad, palmately veined, 3- to 5-lobed, the lobes more or less toothed or wavy, harsh to the touch; petioles 3 to 5 inches long. Fruit 1 to 3 inches long, very variable.

Native to the Mediterranean region. Many horticultural varieties are grown for the edible fruits.

Fig. 197. Ficus carica L.
Leaf, $\times$ ¼.

### 2. **Ficus elástica** Roxbg. INDIAN RUBBER PLANT. INDIAN RUBBER TREE. Fig. 198.

An evergreen tree, 30 to 100 feet high in its native habitat but usually 6 to 15 feet high when grown as an ornamental pot plant. Leaf-blades thick and leathery, oblong to elliptic, 4 to 12 inches long, 2 to 4 inches wide, straight-veined from the midrib, dark green, glossy, and glabrous above, lighter and glabrous beneath, entire; petioles about 1 inch long. Fruit ½-inch long, rarely maturing on cultivated plants.

Native to the East Indies and southwestern Asia. Commonly grown as a house and conservatory plant.

3. **Ficus macrophylla** Desf. MORETON BAY FIG. Fig. 199.

A large evergreen tree, 40 to 60 feet high, with a broad rounded crown. Leaf-blades thick and leathery, elliptical or oval to oblong, 6 to

10 inches long, 3 to 4 inches wide, rounded at the base, dark green and glabrous above, brownish beneath, entire; petioles ¾-inch to 2 inches long. Fruit nearly globular, ¾-inch to 1 inch in diameter.

Fig. 198. Ficus elastica Roxbg.
Leaf, × ¼.

Native in Australia, where it is used as an avenue tree. Cultivated in southern California as a specimen tree. Not easily killed by frost.

4. **Ficus rubiginòsa** Desf. RUSTYLEAF FIG. Fig. 200.

*F. australis* Willd.

A large massive evergreen tree, 40 to 70 feet high, with widely spreading branches forming a broad round-topped crown. Leaf-blades thick and leathery, broadly ovate to ovate-elliptical, 2½ to 4 inches long, 1¼ to 2½ inches wide, rounded to subcordate at the base, dark green and puberulent to glabrous above, rusty-tomentose beneath, entire; petioles ½-inch to 1¼ inches long. Fruit globular, ⅓- to ½-inch in diameter.

Fig. 199.
Ficus macrophylla
Desf.
Leaf, × ¼.

Native to Australia, where it spreads by aerial roots like the Banyan Tree. Extensively cultivated in gardens and parks in southern California. Cannot stand heavy frosts.

### 2. **Maclùra** Nutt. OSAGE-ORANGE

(Named for Wm. Maclure, an American geologist.)
One species, native to the south central United States.

1. **Maclura pomífera** Schneid. OSAGE-ORANGE. Fig. 201.

*M. aurantiaca* Nutt.

A medium-sized deciduous tree, 25 to 60 feet high, with erect or spreading branches bearing stout axillary spines ½-inch to 2 inches long, and dark orange deeply furrowed bark. Leaves simple, alternate; the blades ovate to ovate-lanceolate, 2 to 5 inches long, 1½ to 2 inches wide, dark or light green and glossy above, paler and slightly pubescent but becoming glabrous be-

Fig. 200.
Ficus rubiginosa Desf.
Leaf, × ⅓.

neath, entire; petioles slender, ¾-inch to 2 inches long. Flowers inconspicuous, dioecious, axillary; the staminate pedicelled, in broad or narrow racemes ½-inch to 1 inch long; the pistillate sessile, in dense globose short-peduncled heads. Fruit globose, 4 to 6 inches in diameter, yellow when ripe, not edible, composed of numerous drupelets collected into a head, its juice very white and sticky.

Occasionally cultivated for hedge trees. Much used in the central and middle western parts of the United States for hedges before wire fences were introduced. The bark yields a yellow dye.

Fig. 201.
Maclura pomifera Schneid.
Leaf, × ½. Fruit, × ¼.

### 3. **Mòrus** L. MULBERRY

(The ancient Latin name of the mulberry.)

Deciduous trees or shrubs, often with milky juice. Leaves simple, alternate, dentate or lobed, 3-veined from the base. Flowers inconspicuous, monoecious or dioecious, in small drooping catkins, the pistillate catkins ripening into a fleshy aggregate blackberry-like fruit.

About 10 closely related species, restricted to the northern hemisphere, 3 of which are planted on the Pacific Coast.

Fig. 202. Morus alba L.
Fruit, leaf, × ½.

KEY TO THE SPECIES

Leaves mostly smooth and glabrous
above......................................1. *M. alba.*
Leaves rough above, variously pubescent beneath.
  Leaves usually broadly ovate, heart-shaped at base, coarsely serrate and usually not lobed
                2. *M. nigra.*
  Leaves usually ovate, rarely heart-shaped at base, coarsely toothed, and commonly lobed on the younger shoots..........3. *M. rubra.*

1. **Morus álba** L. WHITE MULBERRY. Fig. 202.

A small tree, 20 to 45 feet high, with a broad round-topped crown. Leaf-blades ovate to broadly ovate, 2½ to 6 inches long, 1½ to 3 inches wide, rather thin, smooth, light green, gla-

brous and shining on both surfaces or pubescent on the veins beneath, serrate or variously lobed; petioles ½-inch to 1 inch long. Fruit white, pinkish or purplish when mature, sweet.

Native to China. Cultivated and escaped in Europe and America. The variety **péndula** Dipp., WEEPING MULBERRY, has pendulous branches and is sometimes used as an ornamental. The variety **tatárica** Loud., RUSSIAN MULBERRY, with smaller leaves and dark red or white fruit, is a small very hardy bushy tree and is occasionally seen in cultivation.

Fig. 203. Morus nigra L.
Leaf, × ⅓.

2. **Morus nìgra** L. BLACK MULBERRY. Fig. 203.

A small tree, 20 to 30 feet high, with a short trunk and wide-spreading branches. Leaf-blades broadly ovate, 2½ to 5 (or 8) inches long,

2 to 3½ inches wide, heart-shaped at the base, dark green and rough-ish above, lighter and variously pubescent beneath, coarsely serrate, rarely lobed; petioles ½-inch to 1 inch long. Fruit oblong, ¾-inch to 1 inch long, dark red, purple, or black.

Native to western Asia. Occasionally cultivated for its fruit and dark green foliage.

3. **Morus rùbra** L. RED MULBERRY. Fig. 204.

A medium-sized tree, 20 to 60 feet high, with a short trunk and wide-spreading branches forming a broad round-topped crown. Leaf-

Fig. 204. Morus rubra L.
Portion of branchlet with leaves and pistillate catkin, × ½.

blades ovate to almost round, 3 to 5 (or 8) inches long, 1½ to 3½ inches wide, dark green and roughish above, densely soft-pubescent beneath, sharply serrate, or lobed on young shoots; petioles ¾-inch to 1¼ inches long. Fruit dark purple-red, about 1 inch long.

Native to the central and eastern United States. Occasionally cultivated for its fruits and as an ornamental.

## Proteaceae. Protea Family
### 1. Grevillea R. Br.

(Named for C. F. Greville, once vice-president of the Royal Society of England.)

About 200 species of trees and shrubs, mostly native to Australia.

**1. Grevillea robústa** Cunn. SILK-OAK. Fig. 205.

A tall much branched evergreen tree, 30 to 70 feet high (to 150 in Australia). Leaves twice-divided into lanceolate toothed or entire segments, fern-like, alternate, 6 to 12 inches long, the margins revolute. Flowers numerous, orange-colored, with long slender showy styles, in 1-sided racemes 2 to 4 inches long. Fruit a follicle with 1 or 2 flat winged seeds.

Native to Australia. Frequently planted in California parks and gardens. Blossoms from May to July, but the flowers usually do not set in cool foggy regions.

### 2. Hàkea Schrad.

(Named after Baron von Hake, a German friend of botany.)

Evergreen shrubs or trees. Leaves simple, alternate, entire or pinnately parted. Flowers showy, borne in pairs in the axils of deciduous bracts, crowded into close globose heads or short racemes. Fruit a hard woody capsule.

About 100 species, native to Australia. Very drought-resistant and planted chiefly in the South (U.S.).

KEY TO THE SPECIES

Leaves more than ⅜-inch wide, with
  3 to 7 nearly parallel veins
                1. *H. laurina.*
Leaves or their divisions less than
  ¼-inch wide, terete
                2. *H. suaveolens.*

**1. Hakea laurìna** R. Br. SEA-URCHIN HAKEA. Fig. 206.

A small tree or large shrub, 10 to 30 feet high. Leaf-blades elliptic to lanceolate, 3 to 6 inches long, ½-inch to 1 inch wide, 3- to 7-veined

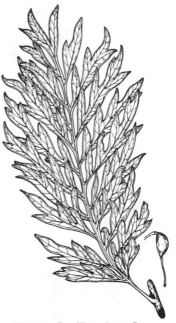

Fig. 205. Grevillea robusta Cunn.
Leaf, single fruit, × ⅓.

from the base, pale gray-green and glabrous on both surfaces, entire, tapering to a very short petiole. Flowers crimson, in globular involucrate heads 1 to 1½ inches across, the styles golden yellow and long-protruding in all directions. Capsule about 1 inch long.

Native to Australia. The evergreen Black Acacia-like foliage, large buds with white pubescent bracts, and beautiful clusters of crimson flowers make this a highly ornamental plant suitable for lawn and park planting. As yet rarely seen in cultivation.

**2. Hakea suavèolens** R. Br. Fig. 207.

*H. pectinata* Colla.

Usually a shrub, 6 to 12 feet high, but occasionally a small tree to 15 feet high. Leaves usually parted into 2 to 9 irregular sharp-pointed terete divisions, but sometimes unbranched and needle-like, 2 to 4 inches long, dark green and glabrous. Flowers small, white, fragrant, in dense racemes. Capsules about 1 inch long.

Native to western Australia. Much cultivated in California as a park or garden plant. Suitable for hedges and easily grown in very dry habitats.

**3. Leucadéndron** R. Br.

(From the Greek *leukos*, white, and *dendron*, tree, in reference to the silvery white foliage.)

About 60 species of trees and shrubs, native to southern Africa.

Fig. 206. Hakea laurina R. Br. Branchlet with leaves and bud, × ½.

**1. Leucadendron argénteum** R. Br. SILVER TREE. Fig. 208.

A small tree, 10 to 30 feet high, with few ascending branches densely covered with silvery white foliage. Leaves simple, alternate, evergreen, sessile, lanceolate, 3 to 6 inches long, ½-inch to 1¼ inches

Fig. 207. Hakea suaveolens R. Br. Leaves, fruit, × ½.

wide, densely silvery-hairy on both surfaces, entire. Flowers dioecious, numerous, in terminal sessile showy heads subtended by spreading involucral bracts, longer than the flowers. Fruit a small ellipsoidal nut with persistent calyx and style.

Native to the Table Mountain region of South Africa. Occasionally cultivated in parks and gardens of central and southern California.

#### 4. **Macadàmia** F. Muell.

(Named after John Macadam, secretary of the Philosophical Institute, Victoria, Australia.)

About 6 species of trees or shrubs, native to Australia.

1. **Macadamia ternifòlia** F. Muell. QUEENSLAND NUT. Fig. 209.

An evergreen tree, 20 to 50 feet high, with dense dark green foliage. Leaves simple, in whorls of 3 or 4; the blades thick and leathery, oblong or lanceolate, 5 to 12 inches long, 1 to 2 inches wide, dark green, glabrous, and shining above, lighter beneath, irregularly spinose-serrate or entire; petioles ¼-inch or less long, rarely absent. Flowers small, white, borne in pairs, forming simple elongated racemes 3 to 8 inches long. Fruit a globose drupe; the exocarp 2-valved and leathery; the endocarp smooth, shining, thick, and hard.

Fig. 208.
Leucadendron
argenteum
R. Br.
Leaf, × ½.

Native to Queensland and New South Wales. Occasionally cultivated in central and southern California.

## Magnoliaceae. Magnolia Family

### 1. **Liriodéndron** L. TULIP TREE

(From the Greek *lirion*, lily, and *dendron*, tree, in reference to the flowers, which resemble lilies.)

Two species, native in North America and China.

1. **Liriodendron tulipífera** L. TULIP TREE. Fig. 210.

A large beautiful deciduous tree, 60 to 150 feet high, with a tall columnar trunk devoid

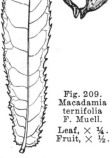

Fig. 209.
Macadamia
ternifolia
F. Muell.
Leaf, × ¼.
Fruit, × ½.

of branches for some height, and spreading branches forming a pyramidal crown. Leaves simple, alternate; the blades very broadly ovate to squarish in outline, 3 to 5 inches long and about as broad, 2- to 4-lobed, truncate or broadly notched at the apex and with a small lobe on either side of the notch, rounded to subcordate at the base, glabrous on both surfaces, paler beneath; petioles 2 to 4 inches long. Flowers large, 1½ to 2 inches high, showy, tulip-shaped, greenish white outside and orange-colored within; sepals 3, spreading, deciduous;

petals 6, erect; stamens numerous; pistils numerous, inserted with and above the stamens on an elongated cylindrical receptacle forming a cone-like brown fruit about 2½ inches long, each pistil 1- or 2-seeded and developing into a nutlet with a long narrow wing, falling from the receptacle when mature.

Native in woods from the northeastern United States southward to Florida. Occasionally cultivated as a park or lawn tree.

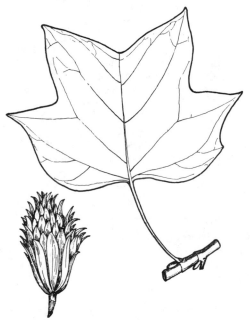

Fig. 210. Liriodendron tulipifera L.
Fruit, leaf, × ½.

## 2. **Magnòlia** L. MAGNOLIA

(Named after Pierre Magnol, director of the botanic garden at Montpelier.)

Evergreen or deciduous trees or shrubs with large buds enclosed in a single large bract. Leaves simple, alternate, entire, petioled. Flowers large, solitary, terminal, usually very showy; sepals 3, petal-like; petals 6 to 15; stamens numerous; pistils numerous, adnate with the stamens to an elongated cone-like receptacle. Fruit cone-like, com posed of numerous 1- or 2-seeded dehiscent carpels borne on a central axis. Seeds when ripe suspended from the cones by slender filamentous threads.

About 35 species, native to North and Central America and eastern Asia. The highly ornamental magnolias are popular trees because of their large showy, white, pink, or purplish flowers. One evergreen species, *M. grandiflora* L., and several deciduous species are cultivated in Pacific Coast parks and gardens. The deciduous species are more commonly used in Washington and Oregon than in California.

KEY TO THE SPECIES

Leaves 1 to 2½ feet long.
    Base of leaves tapering or rounded...................................1. *M. tripetala.*
    Base of leaves heart-shaped........................................2. *M. macrophylla.*
Leaves usually less than 12 inches long.
    Leaves 8 to 12 (rarely 14) inches long.........................3. *M. hypoleuca.*
    Leaves 8 inches or less long (or some leaves longer in *M. acuminata*).
        Leaves obovate or oval obovate; obtusely pointed.
            Sepals and petals alike.
                Leaves 3 to 6 inches long.....................................4. *M. conspicua.*
                Leaves 2 to 3½ inches long....................................5. *M. stellata.*
            Sepals and petals unlike.
                Flowers white...............................................................6. *M. kobus.*
                Flowers purplish outside.................................7. *M. soulangeana.*
        Leaves oval or broadly elliptical to oblong.
            Leaves thick and leathery, very glossy above, often with dense brown pubescence beneath, evergreen, 5 to 8 inches long
                                8. *M. grandiflora.*
            Leaves thin, soft-pubescent and light or gray-green below, deciduous, 5 to 10 inches long...........................9. *M. acuminata.*

### 1. Magnolia tripétala L. UMBRELLA MAGNOLIA.

A medium-sized deciduous tree, 15 to 40 feet high, with glabrous leaf-buds and spreading branches forming an open round-topped crown. Leaves clustered at the ends of the branches; the blades oblong-obovate, 1 to 2 feet long, 4 to 8 inches wide, acute or short-acuminate, tapering or rounded at the base, glabrous above, pale and pubescent beneath at least when young; petioles ½-inch to 2½ inches long. Flowers white, 7 to 10 inches across, with a strong odor. Fruit 3 to 6 inches long, rose-colored.

Native to the eastern United States. Occasionally cultivated for the very large leaves and flowers, which appear in May and June.

### 2. Magnolia macrophýlla Michx. BIGLEAF MAGNOLIA. LARGELEAF CUCUMBER TREE.

A medium-sized deciduous tree, 20 to 50 feet high, with stout spreading branches and tomentose young branchlets and buds. Leaf-blades oblong-obovate, 1 to 2½ feet long, 5 to 10 inches wide, blunt at the apex, somewhat heart-shaped and auriculate at the base, bright green and glabrous, or slightly pubescent above, paler and glaucescent beneath; petioles 2 to 4 inches long. Flowers creamy white, 10 to 12 inches across, fragrant. Fruit 2¼ to 5 inches long, rose-colored.

Native from Kentucky south to Florida and west to Arkansas and Louisiana. Cultivated for its very large flowers and leaves.

3. **Magnolia hypoleûca** Sieb. & Zucc. SILVER MAGNOLIA.

*M. obovata* Thunb.

A deciduous tree, 20 to 40 feet high, with a broad pyramidal crown and glabrous purplish branchlets. Leaf-blades obovate, 8 to 12 (or 14) inches long, 3 to 4½ inches wide, obtuse at the apex, rounded

Fig. 211.
Magnolia conspicua Salisb.
Leaf, × ⅓.

at the base, glabrous or slightly pubescent above, silvery white beneath; petioles 1 to 1½ inches long. Flowers white, 6 to 7 inches across, fragrant; the filaments purple. Fruit 5½ to 8 inches long, scarlet.

Native to Japan. Planted for its showy fragrant flowers and large leaves, silvery white beneath.

4. **Magnolia conspícua** Salisb. YULAN. Fig. 211.

*M. denudata* Desrouss.

A small deciduous tree, 10 to 20 feet high, with spreading branches and pubescent young branchlets and buds. Leaf-blades obovate or obovate-oblong, 4 to 6 inches long, 2½ to 4 inches wide, short-acuminate, tapering or rounded at the base, dark green and glabrous or nearly so above, lighter green and slightly pubescent beneath; petioles ¾-inch to 1½ inches long. Flowers white, 4½ to 6 inches across, appearing before the leaves; petals and sepals of nearly similar size. Fruit 3 to 4 inches long, brownish.

Native to China. A showy species, occasionally cultivated. The variety **purpuráscens** Maxim. has flowers which are rose-red outside and pale pink inside.

5. **Magnolia stellàta** Maxim. STAR MAGNOLIA.

*M. halleana* Hort.

A deciduous shrub or small tree, 6 to 15 feet high, with densely pubescent young buds and branchlets. Leaf-blades obovate or obovate-oblong, 1½ to 4 inches long, 1 to 2 inches wide, obtusely pointed, tapering at the base, dull green and glabrous above, light green and pubescent beneath or becoming glabrous in age; petioles ⅛- to ⅜-inch long. Flowers white, about 3 inches across, appearing before the leaves; sepals and petals alike. Fruit about 2 inches long, red.

Native to Japan. Planted for its numerous early-blooming fragrant flowers.

6. **Magnolia kòbus** Thunb. KOBUS MAGNOLIA.

A small deciduous tree, 15 to 30 feet high, or usually shrubby in cultivation, with slender glabrous branchlets and pubescent flower-buds. Leaf-blades obovate, 2½ to 4 inches long, 1½ to 2 inches wide, abruptly pointed, tapering to a wedge-shaped base, bright green and glabrous above, paler and pubescent at least on the veins beneath; petioles ⅜- to ⅝-inch long. Flowers white, about 4 inches across, appearing before the leaves; sepals 3, narrow, shorter than the petals. Fruit 4 to 5 inches long, brown.

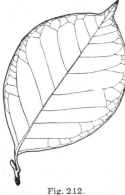

Fig. 212.
Magnolia soulangeana Soul.
Leaf, × ⅛.

Native to Japan. Cultivated in lawns and parks. The variety **boreàlis** Sarg. is a larger tree of pyramidal habit with larger leaves, flowers, and fruits.

7. **Magnolia soulangeàna** Soul. SAUCER MAGNOLIA. Fig. 212.

A deciduous large shrub or small tree, 8 to 20 feet high. Leaf-blades obovate, 4 to 7 inches long, glabrous or nearly so above, slightly pubescent beneath. Flowers white or purplish outside, appearing before the leaves; sepals variable, ½ as long to as long as the petals, usually colored.

A hybrid between *M. denudata* and *M. liliflora*. Several forms of this hybrid are extensively planted.

8. **Magnolia grandiflòra** L. SOUTHERN MAGNOLIA. BULL-BAY. Fig. 213.

*M. foetida* Sarg.

A medium-sized to tall evergreen tree, 25 to 60 (or 80) feet high, with stout ascending or spreading branches forming a round-topped or pyramidal crown, and rusty-tomentose branchlets and buds. Leaf-blades very thick and leathery,

Fig. 213. Magnolia grandiflora L.
Leaf, × ¼.

broadly elliptic or obovate-oblong, 4 to 10 inches long, 2 to 3½ inches wide, glossy, green, and glabrous above, lighter and often rusty-tomentose beneath especially when young; petioles ½- to ⅞-inch long. Flowers white, 6 to 10 inches across, fragrant. Fruit 3 to 4 inches long, rusty-tomentose.

Native from North Carolina south to Florida and west to Texas. Much cultivated for its large white flowers and glossy evergreen foliage.

### 9. Magnolia acumin&agrave;ta L. CUCUMBER TREE. Fig. 214.

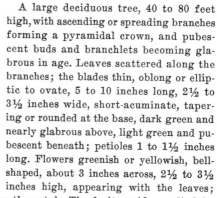

Fig. 214.
Magnolia acuminata L.
Leaf, × ⅓.

A large deciduous tree, 40 to 80 feet high, with ascending or spreading branches forming a pyramidal crown, and pubescent buds and branchlets becoming glabrous in age. Leaves scattered along the branches; the blades thin, oblong or elliptic to ovate, 5 to 10 inches long, 2½ to 3½ inches wide, short-acuminate, tapering or rounded at the base, dark green and nearly glabrous above, light green and pubescent beneath; petioles 1 to 1½ inches long. Flowers greenish or yellowish, bell-shaped, about 3 inches across, 2½ to 3½ inches high, appearing with the leaves; the sepals much smaller than the petals. The fruit ovoid or cylindric, 2 to 3½ inches long, and red or rose-colored.

This tree is native from New York south to Georgia and west to Arkansas. It is only occasionally cultivated on the Pacific Coast.

## Lauraceae. Laurel Family

### 1. Cinnam&ograve;mum Bl. CINNAMON

(From *kinnamomon*, the ancient Greek name.)

About 50 species of trees and shrubs, native in tropical and subtropical regions.

### 1. Cinnamomum camph&ograve;ra Nees. & Eberm. CAMPHOR TREE. Fig. 215.
*Camphora officinarum* Nees.

A handsome medium-sized evergreen tree, 20 to 40 feet high, with an enlarged base, spreading branches forming a round-topped crown, and dense glossy light green foliage. Leaves simple, alternate, with the odor of camphor when crushed; the blades rather thin but firm, ovate-elliptic, 2 to 5 inches long, 1 to 2 inches wide, acuminate, often 3-veined

from near the base, glabrous and shining above, pale and glaucous beneath, entire; petioles ½-inch to 1 inch long. Flowers small, yellowish, in branched axillary clusters. Fruit a small globose 1-seeded berry, about ⅜-inch in diameter.

Native to China and Japan. Grown as a street and lawn tree, especially in southern California.

### 2. Láurus L. LAUREL. SWEET-BAY

(The ancient classical name.)

Two species, native to the Mediterranean region and the Canary Islands.

#### 1. Laurus nóbilis L. GRECIAN LAUREL. SWEET-BAY. Fig. 216.

A small evergreen tree, 20 to 40 feet high, with ascending or somewhat spreading branches. Leaves simple, alternate; the blades stiff, lance-olate or oblong-elliptic, 2½ to 4 inches long, ¾-inch to 1¾ inches wide, dark or yellow-green

Fig. 215.
Cinnamomum
camphora
Nees. & Eberm.
Leaf, fruit, × ½.

and glabrous, entire, often slightly revolute and undulate; petioles about ¼-inch long. Flowers inconspicuous, yellowish, in axillary clusters. Fruit a small purple berry, ¼- to ⅜-inch in diameter.

Native to the Mediterranean region. Cultivated in parks and gardens and often grown as a tub plant.

### 3. Pèrsea Mill.

(Ancient Greek name of some oriental tree.)

Evergreen trees or shrubs. Leaves simple, alternate, leathery, entire. Flowers small, without petals, borne in compound clusters. Fruit a globose or pear-shaped berry or drupe.

About 50 species, native in tropical and

Fig. 216. Laurus nobilis L.
Leaf, × ½. Fruit, × 1.

subtropical America and one in the Canary Islands. The avocado of commerce belongs to this genus.

KEY TO THE SPECIES

Leaves 4 to 10 inches long; fruit fleshy, 2 to 8 inches long
1. *P. americana.*

Leaves 3 to 6 inches long; fruit scarcely fleshy, ¾-inch to 1 inch long
2. *P. indica.*

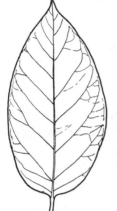

### 1. **Persea americàna** Mill. Avocado. Fig. 217.

*P. gratissima* Gaertn.

A much branched tree, 20 to 60 feet high. Leaf-blades elliptic-oblong to oval, 4 to 10 inches long, 1½ to 3 inches wide, green and glabrous above, paler and often glaucous beneath; petioles ½-inch to 1½ inches long. Fruit a large fleshy drupe, pear-shaped or ellipsoidal to ovoid, 2 to 8 inches long, green or purplish.

Native in tropical America. Extensively cultivated in southern California for the large rich buttery fleshy fruit.

Fig. 217.
Persea americana Mill.
Leaf, × ¼.

### 2. **Persea índica** Spreng. Madeira-Bay. Fig. 218.

A small handsome tree, 10 to 25 feet high. Leaf-blades oblong to lanceolate-oblong, 3 to 6 inches long, 1½ to 2½ inches wide, mostly acute at the apex, glabrous above, glabrous or slightly pubescent beneath; petioles ½-inch to 1 inch long. Fruit oblong or ovoid, ¾-inch to 1 inch long, scarcely fleshy.

Native in the Canary Islands. Occasionally planted as an ornamental tree.

Fig. 218. Persea indica Spreng.
Leaf, × ½.

### 4. **Sássafras** Nees. & Eberm.

(From the Spanish *Salsafras*, *Saxifraga*, in reference to medicinal properties supposedly similar to those of *Saxifraga*.)

Two species, one in northeastern America and one in China.

### 1. Sassafras variifòlium Kuntze. COMMON SASSAFRAS. Fig. 219.

*S. officinale* Nees.

A medium-sized to tall deciduous tree, 30 to 50 (or 80) feet high, with very rough ridged aromatic bark and bright green pubescent young

branches. Leaves simple, alternate; leaf-blades broadly oval and entire, or 1- to 3-lobed to near the middle, 3 to 5 inches long, 2 to 4 inches wide, light green and pubescent above when young, becoming glabrous, paler and pubescent beneath at least along the veins, turning various shades of yellow and orange in the fall; petioles ¾-inch to 1½ inches long. Flowers small, dioecious, yellowish, borne in several-flowered racemes. Fruit an oblong drupe, about ⅓-inch long, blue.

Native in eastern North America. Occasionally cultivated in gardens and parks, especially in Washington and Oregon, for its ornamental flowers, fruit, and fall coloration of the leaves.

Fig. 219.
Sassafras variifolium Kuntze.
Leaf, × ⅓.

### 5. Umbellulària Nutt.

(From the Latin *umbella,* a sunshade, referring to the arrangement of the flowers.)

One species, native in Oregon and California.

### 1. Umbellularia califórnica Nutt. CALIFORNIA-LAUREL. BAY TREE. Fig. 220.

An evergreen tree, 20 to 75 feet high, or shrub-like in open dry habitats, with aromatic dark green foliage and ascending branches spreading in old trees and thus forming a broad round-topped crown. Leaves simple, alternate; the blades thick and leathery, oblong-lanceolate or lanceolate, 3 to 5 inches long, ¾-inch to 1½ inches wide, dark green and glabrous, entire; petioles about ¼-inch long. Flowers small, yellowish, several in stalked umbels. Fruit an ovoid drupe, about

Fig. 220. Umbellularia
californica Nutt.
Fruit, leaf, × ½.

1 inch long, yellowish green, becoming purplish when mature.

Native on canyon slopes and along streams of the Coast Ranges and

the Sierra Nevada, south to the mountains of southern California, and north to the Umpqua River in Oregon, extending from sea level to about 3000 feet elevation. Known also as Pepperwood and Oregon-Myrtle. Occasionally cultivated for its dark green foliage which emits an odor of bay rum when crushed.

## Pittosporaceae. Pittosporum Family

### 1. Pittósporum Banks. PITTOSPORUM

(From the Greek *pitte*, to tar or pitch, and *sporos*, seed, in reference to the seeds embedded in a pitchy substance.)

Evergreen shrubs or trees. Leaves simple, alternate or apparently whorled in some species. Flowers in variable clusters or solitary. Fruit a globose or obovoid capsule, with leathery valves and few to numerous seeds embedded in a viscous pitchy substance.

About 70 species, native in Australia, New Zealand, and on the surrounding islands. The pittosporums are hardy evergreen trees or shrubs and are extensively used in California for hedges, street and avenue trees, and for general yard and lawn planting. *P. tobira* (a shrub) and *P. undulatum* have very fragrant flowers with an odor somewhat resembling that of orange blossoms.

KEY TO THE SPECIES

Leaves white-tomentose beneath.
    Leaves oblong-elliptic, 2 to 5 inches long, usually not tapering to the petioles, entire and flat along the margins..................1. *P. ralphi.*
    Leaves obovate to oblanceolate, 2 to 3 inches long, tapering to the petioles, recurved along the margins...................2. *P. crassifolium.*
Leaves glabrous beneath.
    Leaves ¼-inch or less wide, 2 to 3½ inches long; branches drooping
                                          3. *P. phillyraeoides.*
    Leaves ¾-inch to 2 inches wide.
        Leaves coarsely toothed.........................................4. *P. rhombifolium.*
        Leaves entire.
            Leaves 1 to 2 (rarely 3) inches long; flowers purplish to black
                                    5. *P. tenuifolium.*
            Leaves 2 to 5 inches long; flowers greenish yellow or white.
                Leaves light yellowish green, the midrib and petiole whitish; flowers greenish yellow...........................6. *P. eugenioides.*
                Leaves dark green above.
                    Leaves oblong-ovate or lanceolate, 3 to 5 inches long; flowers white.........................................7. *P. undulatum.*
                    Leaves obovate to oblanceolate, 2 to 3½ inches long, very obtuse at apex; flowers greenish yellow
                                        8. *P. viridiflorum.*

### 1. Pittosporum rálphi Kirk. Fig. 221.

An openly branched shrub or small tree, 10 to 20 feet high, with smooth gray bark. Leaf-blades thick, oblong-elliptic or oblong, 2 to 5 inches long, 1 to 1½ inches wide, scarcely tapering to the petiole, dark green and glabrous above, densely white-tomentose beneath, the margins entire and flat; petioles ½- to ¾-inch long. Flowers nearly black, about ½-inch long, in terminal clusters. Capsules broadly ovoid, ½- to nearly ¾-inch long, pubescent. Seeds less than ⅛-inch long.

Native to New Zealand. Occasionally cultivated in parks and gardens.

2. **Pittosporum crassifòlium** Soland. KARO. Fig. 222.

A tall shrub or small tree, 15 to 30 feet high, with dark brown bark, erect branches, and densely pubescent branchlets. Leaf-blades thick and leathery, obovate or oblanceolate, 2 to 3 inches long, ¾-inch to 1¼ inches wide, obtuse at the apex, tapering to the petioles, dark green and gla-brous above, densely white-tomentose beneath, the margins entire and slightly recurved. Flowers nearly black, about ½-inch long, in terminal clusters. Capsules subglobose, ¾-inch to 1¼ inches long, densely short-pubescent. Seeds ⅛-inch or more long.

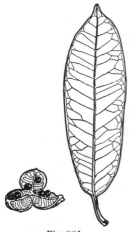

Fig. 221.
Pittosporum ralphi Kirk.
Fruit, leaf, × ½.

Native to New Zealand. Commonly cultivated for windbreaks, hedges, and ornamental groupings.

3. **Pittosporum phillyraeoìdes** DC. WILLOW PITTOSPORUM. NARROWLEAF PITTOSPORUM. Fig. 223.

A graceful small tree, 15 to 30 feet high, with drooping willow-like branches and branchlets. Leaf-blades linear to narrow-oblong, 2 to 3 inches long, about ¼-inch wide, acuminately tipped at the recurved apex, glabrous, light green, entire; petioles about ¼-inch long. Flowers yellow, about ⅜-inch long, usually solitary and pedicelled in the leaf-axils. Cap-

Fig. 222.
Pittosporum crassifolium Soland.
Fruit, leaf, × ½.

sules ovoid, approximately ½-inch long, yellow and somewhat granular.

Native to Australia. Occasionally used in landscaping.

4. **Pittosporum rhombifòlium** A. Cunn. QUEENSLAND PITTOSPORUM. Fig. 224.

A large tree, 30 to 50 feet high (or grown as a pot plant), with a pyramidal habit. Leaf-blades rhomboid, 2½ to 4 inches long, 1 to 2

Fig. 223.
Pittosporum
phillyraeoides DC.
Fruit, leaf, × ½.

Fig. 224. Pittosporum
rhombifolium A. Cunn.
Seed, × ½. Capsule, × 1. Fruit-
cluster, × ½. Leaf, × ½.

inches wide, glabrous, coarsely and irregularly toothed; petioles ½- to ¾-inch long. Flowers white, about ¼-inch long, in terminal clusters. Capsules globose, about ¼-inch in diameter, glabrous, varying from green through yellow to orange when maturing. Seeds 2 or 3, black.

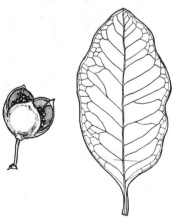

Fig. 225.
Pittosporum tenuifolium Gaertn.
Fruit, leaf, × 1.

Native to Australia. Grown occasionally in southern California as a street or lawn tree for its clean foliage and bright-colored capsules, which persist through the winter.

5. **Pittosporum tenuifòlium** Gaertn. TAWHIWHI. Fig. 225.

*P. nigricans* Hort.

A large shrub or small tree, 6 to 25 feet high, with black bark and numerous ascending branches forming a compact crown. Leaf-blades thin, oblong-ovate to somewhat obovate, 1 to 3 inches long, ½-inch to 1 inch wide, dark green, glabrous, and shining above, paler beneath, the margins entire but slightly undulate; petioles ⅛- to ¼-inch long. Flowers dull purple or black (rarely yellow), ¼- to ½-inch long, solitary or rarely fascicled in the leaf-axils. Capsules globose to ovoid, about ½-inch long, glabrous and minutely roughened when mature.

Native to New Zealand. Extensively cultivated for hedges, mass planting, and as a single tree. **Pittosporum buchanani** Hook. is a closely related species with larger leaves.

6. **Pittosporum eugenioìdes** A. Cunn. TARATA. Fig. 226.

Fig. 226.
Pittosporum eugenioides
A. Cunn.
Fruit-cluster, leaf, × ½.

A small tree, 15 to 30 feet high (or sometimes shrub-like), with ascending or spreading stems and branches, and yellowish green foliage. Leaves often crowded toward the ends of the branchlets; the blades elliptic-oblong or oblong-ovate, 2½ to 4 inches long, ¾-inch to 1¼ inches wide, glabrous, the midrib and petioles often whitish, entire but usually conspicuously wavy-margined; petioles ¼- to ½-inch long. Flowers greenish yellow, about ¼-inch long, numerous, in terminal clusters. Capsules ovoid, about ⅜-inch long, glabrous when mature.

Native to New Zealand. The most extensively cultivated pittosporum in California. Suitable for hedges when pruned or as a lawn tree when permitted to attain normal growth.

7. **Pittosporum undulàtum** Vent. ORANGE PITTOSPORUM. VICTORIAN-BOX. Fig. 227.

A handsome tree, 20 to 40 (or 60) feet high, with spreading branches forming a round-topped crown, or often trimmed to a shrub. Leaves often crowded to-

Fig. 227.
Pittosporum undulatum Vent.
Fruit, × 1. Leaf, × ½.

ward the ends of the branchlets; the blades oblong-ovate or lanceolate, 3 to 5 inches long, 1 to 1¾ inches wide, glabrous, dark green and glossy above, paler beneath, the margins entire, undulate or flat;

petioles ⅜- to ¾-inch long. Flowers white, about ½-inch long, in terminal clusters, very fragrant. Capsules globose, about ½-inch in diameter, smooth, brown, many-seeded.

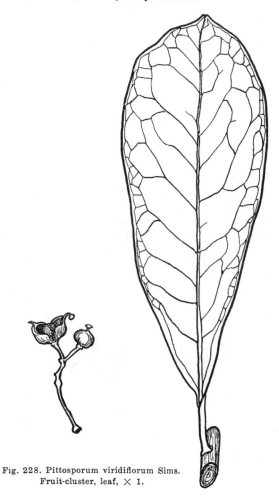

Fig. 228. Pittosporum viridiflorum Sims.
Fruit-cluster, leaf, × 1.

Native to Australia. Commonly cultivated in gardens and parks for the glossy dark green foliage and fragrant orangeblossom-like flowers. Suitable for large hedges.

**8. Pittosporum viridiflòrum** Sims. CAPE PITTOSPORUM. Fig. 228.

A large shrub or small tree, 10 to 25 feet high, with dark green foliage. Leaf-blades obovate to oblanceolate, 2 to 3½ inches long, ¾-inch to 1½ inches wide, obtuse or abruptly acute at the apex, gradually narrowed to the petiole, glabrous, dark green above, paler beneath, entire, often slightly revolute; petioles ¼- to ⅜-inch long. Flowers greenish yellow, in dense terminal clusters. Capsules subglobose, about ¼-inch long, glabrous. Seeds 3 or 4, reddish brown.

Native to southern Africa. Occasionally cultivated in southern California for its mass of dark green foliage and numerous fruits exposing the reddish brown seeds when ripe.

**2. Hymenòsporum** R. Br.

(From the Greek *hymen*, membrane, and *sporos*, seed, in reference to the winged seeds.)

One species, native to Australia. Closely allied to *Pittosporum*, but differs in having flat winged seeds not embedded in a sticky substance.

**1. Hymenosporum flàvum** F. Muell. Fig. 229.

An evergreen shrub or tree, 10 to 40 feet high. Leaves simple, alternate,

Fig. 229.
Hymenosporum flavum F. Muell.
Seeds, capsule, leaf, × ½.

crowded toward the ends of the branchlets; the blades elliptical or obovate, 2 to 6 inches long, 1 to 2 inches wide, glabrous, abruptly acute or acuminate, cuneate at the base, entire, sometimes undulate; petioles ¼- to ½-inch long. Flowers yellowish, about 1 inch broad, very fragrant, in loose terminal clusters. Fruit a stalked capsule, about 1 inch long, with numerous compressed winged seeds.

Native to Australia. Cultivated for its glossy dark green foliage and large clusters of fragrant flowers.

## Hamamelidaceae. Witch-Hazel Family

### 1. Liquidámbar L.

(From the Latin *liquidus*, fluid, and Arabic *ambar*, amber, in reference to the fragrant resin from the bark of *L. orientalis*.)

Four species, native in North and Central America and in western and eastern Asia.

**1. Liquidambar styraciflua** L. SWEET GUM. Fig. 230.

A deciduous tree, 60 to 125 feet high, with a straight trunk and deeply furrowed bark. Leaves simple, alternate; the blades round to broadly ovate in outline, deeply palmately 5- or 7-lobed, 4 to 6 inches long and as broad, the lobes acuminate and finely serrate, subcordate or trun-

cate at the base, dark green and glabrous above, paler beneath and glabrous except for tufts of hairs in the axils of the veins; petioles 5 to 6 inches long. Flowers small, without petals, and monoecious; the staminate in small heads forming terminal racemes; the pistillate in globose heads with slender peduncles. Fruit a capsule with 1 or 2 winged seeds, borne in head-like clusters about 1 inch in diameter, armed with the persistent styles.

Native in the eastern part of the United States, westward through Ohio, Indiana, and Illinois, and south-ward to Missouri, Arkansas, Texas, Mexico, and Central America. Cul-

Fig. 230. Liquidambar styraciflua L. Fruit-cluster, leaf, × ⅓.

tivated in parks and gardens for its handsome foliage, which turns bright crimson in the fall.

## Platanaceae. Plane Tree Family

### 1. Plátanus L. PLANE TREE. SYCAMORE

(From the Greek *platys*, broad, in reference to the broad leaves.)

Deciduous trees, with pale or whitish bark exfoliating in thin plates, and ovoid sharp-pointed buds concealed by the hollow bases of the petioles. Leaves large, simple, alternate, palmately veined and lobed, long-petioled. Flowers small, monoecious, greenish, in dense globose heads. Fruit a small 1-seeded nutlet surrounded by long hairs, borne in compact clusters on a globose receptacle and forming a globular fruiting head disintegrating when mature.

Five or 6 species, native in the eastern and western United States, Mexico, Central America, southeastern Europe, and southwestern Asia. One is native on the Pacific Coast.

#### KEY TO THE SPECIES

Leaves usually lobed to or below the middle, the lobes entire or with callous-tipped teeth terminating the main veins; stipules large and persistent..........................................................................1. *P. racemosa.*

Leaves lobed to about ⅓ the length of the blade, the lobes broadly tri-angular, the middle one longer than broad; and the stipules early deciduous........................................................................2. *P. acerifolia.*

1. **Platanus racemòsa** Nutt. WESTERN SYCAMORE. CALIFORNIA PLANE TREE. Fig. 231.

A large tree, 40 to 90 feet high, with a main trunk often dividing into ascending or almost prostrate secondary stems, and with thick stout contorted branches forming an irregular round-topped crown. Leaf-blades thick and firm, truncate or cordate, 5 to 10 (or 18) inches broad, 4 to 9 (or 12) inches long, 3- or 5-lobed to or below the middle, the lobes longer than wide, acuminate or acute, entire or with callous-

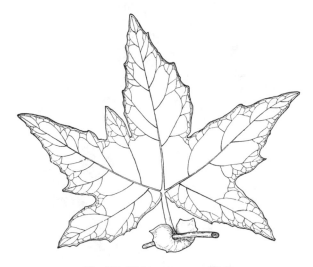

Fig. 231. Platanus racemosa Nutt.
Leaf, × ¼.

tipped teeth terminating the main veins, light green and glabrous above, paler and pubescent and becoming rusty-hairy beneath; petioles ½-inch to 3 inches long; stipules conspicuous, often sheathing the stem and remaining after the fall of the leaves. Fruiting heads 3 or more on a zigzag axis, globose, ¾-inch to 1¼ inches in diameter, the nutlets falling away in winter.

Native on stream banks and adjacent flats in the more arid parts of California, but not on the deserts. Western Sycamore occurs abundantly in the South Coast Ranges, and extends southward to southern California and Mexico, northward in the Great Valley and Sierra Nevada foothills to Tehama County. Occasionally cultivated.

**2. Platanus acerifòlia** Willd. LONDON PLANE TREE. Fig. 232.

A medium-sized to large tree, 30 to 70 feet high. Leaf-blades truncate or cordate, 5 to 8 (or 12) inches wide, usually broader than long, 3- to 5-lobed to about ⅓ the length of the blade, the lobes broadly triangular and coarsely toothed to almost entire, the middle lobe as long or longer than broad, glabrous or nearly so; petioles 1½ to 4 inches long. Fruiting heads 2 (rarely 1 or 3), about 1 inch in diameter, bristly, on a pendulous stalk.

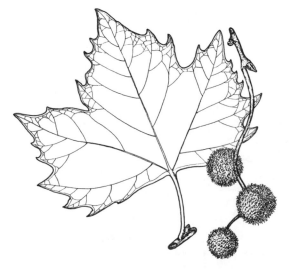

Fig. 232. Platanus acerifolia Willd.
Leaf, fruiting heads, × ⅓.

Probably a hybrid between **P. orientalis** L. and **P. occidentalis** L., with characters of both, these usually intermediate between the two. The London Plane Tree is the species of *Platanus* most commonly cultivated on the Pacific Coast. Most nurseries sell this tree under the name of Oriental Plane Tree, *P. orientalis*, which is not grown here.

## Rosaceae. Rose Family

### 1. Cercocárpus Kunth. MOUNTAIN-MAHOGANY

(From the Greek *kerkis*, a shuttle, and *karpos*, fruit, in reference to the type of fruit.)

Deciduous or partly evergreen shrubs or small trees with short stout branchlets. Leaves simple, alternate and often apparently fascicled. Flowers small, solitary or in small clusters on short lateral branchlets; sepals greenish white, united into a cylindrical elongated calyx-tube abruptly expanded at the summit into 5 deciduous small divisions; petals absent. Fruit an achene, partly or wholly enclosed in the persistent calyx-tube, with an elongated twisted plumose style.

About 5 species, with many closely related varieties or forms, native in western North America from Washington to Mexico.

### KEY TO THE SPECIES

Leaves narrowly lanceolate, entire.........................................1. *C. ledifolius.*
Leaves elliptic or ovate to obovate, toothed at least near apex.
   Leaves white-tomentose beneath........................................2. *C. traskiae.*
   Leaves pale to whitish pubescent beneath or nearly glabrous
                                       3. *C. betuloides.*

### 1. Cercocarpus ledifòlius Nutt. MOUNTAIN-MAHOGANY. Fig. 233.

A shrub or small tree, 6 to 25 feet high. Leaf-blades thick and leathery, narrowly lanceolate, ½-inch to 1 inch long, about ¼-inch wide,

dark green and glabrous above at least in age, pale and pubescent beneath, the margins entire and revolute; petioles ⅛-inch or less long. Flowers solitary or clustered.

Native on arid mountain slopes, usually on those bordering the desert areas, from Washington and Montana south through the Great Basin region to California and Arizona. In California it occurs on the eastern slopes of the Sierra Nevada from Modoc County south to the mountains of southern California and west into Siskiyou County.

### 2. Cercocarpus tráskiae Eastw. TRASK-MAHOGANY. Fig. 234.

Fig. 233. Cercocarpus ledifolius Nutt. Cluster of fruits, leaves, × 1.

A large shrub or small tree, 6 to 20 feet high. Leaf-blades elliptical or broadly ovate, ¾-inch to 2 inches long, ½-inch to 1½ inches wide, densely white-tomentose beneath. Flowers usually 4 or 5 (or 1 to 9) in a cluster.

Native on the southern side of Santa Catalina Island.

3. **Cercocarpus betuloìdes** Nutt. HARD TACK. Fig. 235.

*C. parvifolius* var. *betuloides* Sarg.

A shrub or small tree, 5 to 20 feet high. Leaf-blades obovate, ½-inch to 1½ inches long, ½-inch to 1 inch wide, serrate above the middle, entire along the wedge-shaped base, dark green and glabrous above, paler and often pubescent and conspicuously straight-veined beneath; petioles ¼-inch or less long. Flowers 2 or 3 in a cluster.

Native on the dry chaparral and adjacent areas of the Coast Ranges and the Sierra Nevada of California, extending north to Oregon, east to the Rocky Mountain states, and south to Mexico. Hard Tack is a variable plant in

Fig. 234. Cercocarpus traskiae Eastw.
Cluster of fruits, leaf, × ½.

Fig. 235. Cercocarpus betuloides Nutt.
Cluster of fruits, leaf, × 1.

leaf characters and many variants have been given varietal or species-names.

## 2. Crataègus L. HAWTHORN

(From the Greek *kratos*, strength, in reference to the hardness of the wood.)

Usually spiny shrubs or small trees. Leaves simple, alternate, deciduous or evergreen, often lobed. Flowers showy, white or red, in corymbs, sepals and petals usually 5, or the petals numerous in some horticultural varieties. Fruit pome-like, with 1 to 5 bony 1-seeded nutlets.

About 1000 species have been described from the temperate regions of the northern hemisphere, one of which is native on the Pacific Coast. Several species are cultivated for their ornamental flowers and fruits.

KEY TO THE SPECIES

Leaves distinctly lobed; veins extending to the tips of the lobes and to the sinuses.
    Spines stout, 1 inch or less long; fruit ⁵⁄₁₆- to ½-inch in diameter.
        Leaves 3- to 5-lobed, the lobes serrulate................1. *C. oxyacantha.*
        Leaves deeply 3- to 7-lobed, the lobes entire or only sparingly toothed....................................................................2. *C. monogyna.*
    Spines slender, to 3 inches long; fruit about ¼-inch in diameter
                                                                                3. *C. cordata.*
Leaves not lobed or only slightly so; veins extending to the tips of the lobes but not to the sinuses.
    Leaves soft-pubescent beneath at least on the veins, 2½ to 4 inches long, sharply and doubly serrate.................................4. *C. mollis.*
    Leaves glabrous, ¾-inch to 3 inches long, sharply serrate above the entire base.
        Thorns 3 to 6 inches long; fruit red..............................5. *C. crusgalli.*
        Thorns 1 to 2 inches long; fruit black..........................6. *C. douglasi.*

1. **Crataegus oxyacántha** L. ENGLISH HAWTHORN. Fig. 236.

A small deciduous tree, 10 to 18 feet high, with stout spines about 1 inch long. Leaf-blades broadly ovate, ¾-inch to 2 inches long, ½-inch

to 1¼ inches wide, glabrous, 3- to 5-lobed, the lobes serrulate; petioles ½-inch to 1 inch long. Flowers red or white, about ½-inch broad, in clusters of 5 to 12. Fruit ½-inch long, bright red, with 2 stones.

Native to Europe and northern Africa. Many varieties of this species are in cultivation. The variety **páuli** Rehd. is one of the most showy. It has double bright scarlet flowers blooming in April and May. Some forms have single or double white flowers and red or yellow fruits.

2. **Crataegus monógyna** Jacq. ENGLISH HAWTHORN. Fig. 237.

Similar to the preceding species,

Fig. 236. Crataegus oxyacantha L. Leaf and thorn, × 1.

but the leaf-blades are paler beneath and more deeply 3- to 7-lobed. The lobes have only a few teeth or are entire. The fruit has only one stone.

Native to Europe, northern Africa, and western Asia.

3. **Crataegus cordàta** Ait. WASHINGTON HAWTHORN. Fig. 238.

    *C. phaenopyrum* Medikus.

A small deciduous tree, 15 to 30 feet high, with slender spines 1½ to 3 inches long. Leaf-blades triangular-ovate, 1 to 2½ inches long and

Fig. 237. Crataegus monogyna Jacq.
Portion of branchlet with leaf and thorn,
cluster of fruits, × 1.

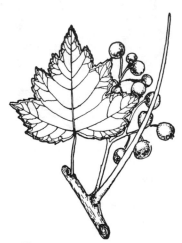

Fig. 238. Crataegus cordata Ait.
Branchlet with leaf, thorn, and
cluster of fruits, × ⅔.

Fig. 239. Crataegus mollis Scheele.
Leaf, × ½.

about as wide, truncate to slightly cordate at the base, bright green and glabrous above, paler and glabrous beneath, 3- or 5-lobed, the lobes sharply serrate; petioles ½-inch to 1¼ inches long, slender. Flowers white, about ½-inch broad, in many-flowered glabrous corymbs. Fruit depressed-globose, about ¼-inch in diameter, shining, bright red, with 3 to 5 stones, the calyx deciduous.

Native to the eastern United States from western North Carolina to Georgia, westward to Illinois and Missouri. Planted in Washington, Oregon, and northern California as a street and park tree. The bright fall coloration of the foliage, the large clusters of bright red fruits which remain on the branches until January, and its relative freedom from disease make this hawthorn one of the best for planting in small gardens and in parkways.

### 4. Crataegus móllis Scheele. DOWNY HAWTHORN. Fig. 239.

A small tree, 15 to 30 feet high, with stout spines 1 to 2 inches long. Leaf-blades broadly ovate, 2½ to 4 inches long, 2 to 3 inches wide, densely pubescent beneath but becoming almost glabrous except on the veins, sharply and doubly serrate or with several short acute lobes; petioles 1 to 1½ inches long. Flowers large, white, about 1 inch broad, in many-flowered tomentose clusters. Fruit short-ellipsoidal or pear-shaped, ¾-inch to 1 inch long, bright red, more or less pubescent, with 4 or 5 nutlets.

Native in the north central United States. A fine species, cultivated in gardens and parks for its bright green foliage, large flowers, and colored fruits.

### 5. Crataegus crusgálli L. COCKSPUR THORN. Fig. 240.

A small tree, 10 to 25 feet high, with slender straight or slightly curved thorns 3 to 4 inches long, becoming compound and 6 to 8 inches long on the old branches and trunk. Leaf-blades thick and somewhat leathery, obovate to oblong-ovate, 1 to 4 inches long, ½-inch to 1 inch wide, wedge-shaped at the base, dark green and shining above, paler beneath, sharply serrate along upper half of the margins; petioles ½- to ¾-inch long.

Fig. 240.
Crataegus crusgalli L.
Portion of branchlet with
leaf and thorn, × ½.

Flowers white, about ¾-inch broad, in many-flowered glabrous clusters. Fruit subglobose, ½-inch long, bright or dull red, covered with whitish bloom, with 2 nutlets, persisting on the branches during the winter.

Native in the northeastern United States and southeastern Canada. A handsome small tree, cultivated in gardens for the dark green foliage, which turns to orange and red in the fall, and for the numerous decorative red fruits.

#### 6. Crataegus doúglasi Lindl. WESTERN BLACK HAWTHORN. Fig. 241.

A deciduous shrub or small tree, 15 to 25 feet high, with reddish branchlets and short stout spines. Leaf-blades ovate to obovate,

Fig. 241.
Crataegus douglasi Lindl.
Portion of branchlet with
thorn and leaf, × ⅔.

1 to 3 inches long, ½-inch to 1½ inches wide, doubly serrate above the entire wedge-shaped base and often incised near the apex; petioles ½- to ¾-inch long. Flowers white, about ½-inch wide, in clusters. Fruit subglobose, about ½-inch long, black and shining, usually with 5 nutlets.

Native on the Pacific Coast from British Columbia south to northern California and east through the northern Rocky Mountains.

#### 3. Cydònia Mill. QUINCE

(From the native place, Cydon, now Canea, in Crete.)

#### 1. Cydonia oblónga Mill. QUINCE. Fig. 242.
*C. vulgaris* Pers.

A deciduous shrub or small tree, 10 to 25 feet high, with thornless branches. Leaves simple, alternate; the blades oblong-ovate, 2 to 4 inches long, 1 to 2 inches wide, rounded or heart-shaped at the base, dull green above, paler and densely pubescent beneath, entire; petioles ⅜- to ¾-inch long. Flowers large, white, 1½ to 2 inches across, solitary and terminal at the ends of leafy shoots. Fruit a many-seeded fragrant pear-shaped villous yellow pome.

Native in central Asia. Cultivated for the fruits and sometimes for the showy flowers.

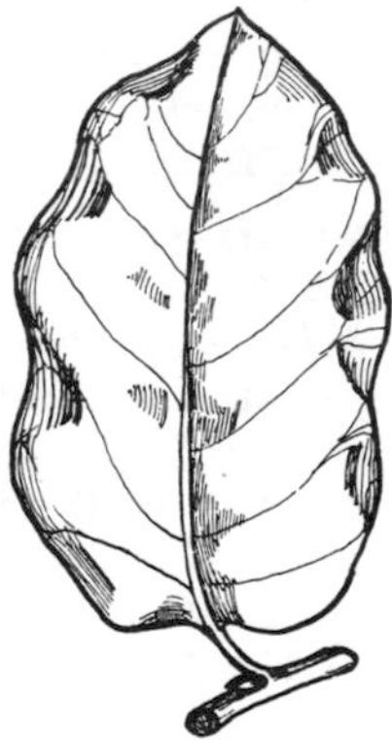

#### 4. Eriobótrya Lindl. LOQUAT

(From the Greek *erion*, wool, and *botrys*, cluster, in reference to the downy flower-clusters.)

About 10 species of trees or shrubs, native to China, Japan, and southern Asia.

Fig. 242.
Cydonia oblonga Mill.
Leaf, × ½.

1. **Eriobotrya japónica** Lindl. LOQUAT. Fig. 243.

*Photinia japonica* Gray.

A small evergreen tree, 10 to 20 feet high, with a dense round-topped crown. Leaves simple, alternate and often apparently clustered at the ends of the branches; the blades elliptic-oblong to obovate, 6 to 10 inches long, bright green and glossy above, paler and usually rusty-tomentose beneath, entire or remotely toothed along upper half of the margin; petioles about ¼-inch long, pubescent. Flowers white, about ⅜-inch across, fragrant, in rusty-tomentose panicles 3 to 6 inches long. Fruit a yellow pear-shaped pome with 1 to 4 large seeds, edible.

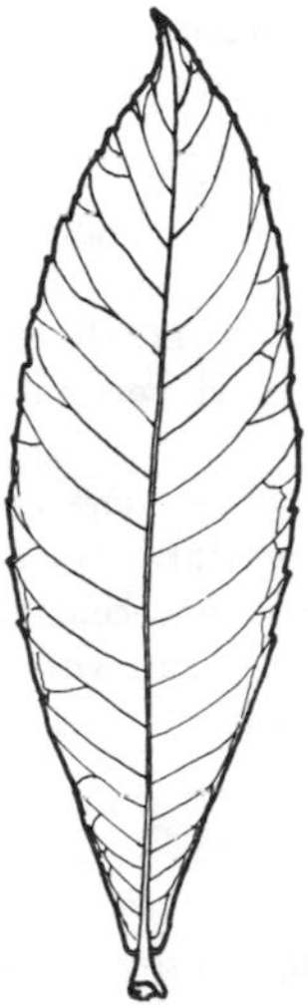

Native to Japan and China. Cultivated in gardens and parks for the large handsome leaves, clusters of fragrant flowers, and edible fruits.

### 5. **Lyonothámnus** Gray

(Named for W. S. Lyon, the discoverer of the tree, and from the Greek *thamnos*, shrub.)

One species.

1. **Lyonothamnus floribúnda** Gray. CATALINA IRONWOOD. Fig. 244.

*L. floribundus* var. *asplenifolius* Greene.

A slender evergreen tree, 25 to 50 feet high, with a single straight trunk and thin bark

Fig. 243.
Eriobotrya
japonica Lindl.
Leaf, × ¼.

peeling off in long strips, or often shrub-like with several stems from the base. Leaves opposite, dimorphic, fern-like, dark green and glabrous above, paler beneath. Leaf-blades of simple leaves lanceolate, 3 to 6 inches long, about ½-inch wide, nearly entire or pinnately cut into wedge-shaped lobes; petioles ½-inch to 1 inch long. The pinnately compound leaves 3 to 6 inches long, with 3 to 7 leaflets similar to the simple

Fig. 244.
Lyonothamnus floribunda Gray.
Leaf, × ⅓.

leaves; petioles 1 to 2 inches long, often winged. Flowers small, about ¼-inch across, white, in broad compound terminal clusters 4 to 8 inches broad. Fruit consisting of 2 woody glandular carpels, splitting along the ventral side and partially along the dorsal side, about ¼-inch long.

Native on Santa Catalina, San Clemente, Santa Rosa, and Santa Cruz islands off the coast of southern California. The simple leaf form is confined to Santa Catalina Island. Occasionally grown in parks and gardens for its dark green glossy fern-like foliage. This handsome tree should be more extensively cultivated. It cannot stand prolonged freezing weather.

### 6. **Màlus** Mill. APPLE

(Ancient Latin name for the apple tree.)

Deciduous trees or shrubs. Leaves simple, alternate, serrate or incisely lobed, petioled. Flowers white or pink, in short terminal clusters. Fruit a pome.

About 25 species, native in the temperate regions of North America, Europe, and Asia. The apple trees of orchards are horticultural varieties of **Malus pumila** Mill., native to southeastern Europe and central Asia. Several varieties are cultivated for their showy flowers and attractive fruits.

Fig. 245. Malus fusca Schneid. Flower-cluster, leaf, fruit-cluster, × ½.

1. **Malus fúsca** Schneid. OREGON CRAB APPLE. Fig. 245.

*Pyrus rivularis* Dougl. *Malus rivularis* Roem.

A deciduous shrub or small tree, 10 to 30 feet high. Leaf-blades ovate to elliptic, 1 to 4 inches long, ¾-inch to 1½ inches wide, dark green and glabrous or sparingly pubescent above, pale and pubescent beneath, sharply serrate with glandular teeth, or some slightly 3-lobed or coarsely toothed; petioles 1 to 1½ inches long. Flowers white, about ¾-inch across, in short 4- to 10-flowered clusters. Fruit oblong-ovoid, ½- to ¾-inch long, yellow-green, becoming purplish in age.

Native on the Pacific Coast from Alaska to northern California, where it occurs in the outer and middle Coast Ranges from Humboldt County south to Sonoma County. Oregon Crab Apple attains its greatest development in Oregon and Washington.

### 7. **Photínia** Lindl.

(From the Greek *photeinos*, shining, in reference to the glossy foliage.)

Evergreen or deciduous shrubs or trees. Leaves simple, alternate, usually serrulate, short-petioled. Flowers small, white, in terminal clusters. Fruit a small 1- to 4-seeded pome.

About 25 species, native in Europe and southern Asia, one in California and two in Mexico.

KEY TO THE SPECIES

Leaves 2 to 4 inches long, acute at apex

1. *P. arbutifolia.*

Leaves 4 to 7 inches long, acuminate

2. *P. serrulata.*

1. **Photinia arbutifòlia** Lindl. TOYON. CHRISTMAS BERRY. Fig. 246.

*Heteromeles arbutifolia* Roem.

A large evergreen shrub or small tree, 10 to 20 feet high. Leaf-blades thick and leathery, oblong or elliptical, 2 to 4 inches long, 1 to 1¾ inches wide, glabrous, dark green and glossy above, paler beneath, serrate; petioles ½-inch to 1 inch long. Flower-clusters 2 to 4 inches long and almost as broad. Fruit ovoid,

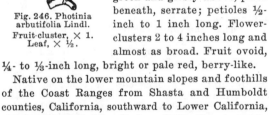

Fig. 246. Photinia arbutifolia Lindl. Fruit-cluster, × 1. Leaf, × ½.

¼- to ⅓-inch long, bright or pale red, berry-like.

Native on the lower mountain slopes and foothills of the Coast Ranges from Shasta and Humboldt counties, California, southward to Lower California, and in the Sierra Nevada from Butte County southward to Mariposa County. Extensively cultivated in gardens and parks for its handsome evergreen foliage and clusters of red berry-like fruits, which ripen from October to December.

2. **Photinia serrulàta** Lindl. LOW PHOTINIA. CHINESE PHOTINIA. Fig. 247.

A tall evergreen shrub or small tree, 10 to 30 feet high. Leaf-blades thick and leathery, oblong or oblong-ovate, 4 to 7 inches long, 1½ to 2¼ inches wide, glabrous, dark green and glossy above, very light green beneath, finely serrate; petioles 1 to 1½ inches long. Flower-clusters 4 to 7 inches broad. Fruit globose, about ¼-inch in diameter, red.

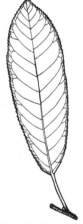

Fig. 247. Photinia serrulata Lindl. Leaf, × ⅓.

Native to China. Cultivated in gardens and parks for the large clusters of white flowers and handsome glossy foliage. Not all but some of the leaves on a single plant turn a brilliant red in the fall and remain on the plant for many months.

### 8. **Prùnus** L. PLUM

(The ancient Latin name.)

Deciduous or evergreen shrubs or trees, usually with bitter juice. Leaves simple, alternate, usually serrate. Flowers white, pink, or red, in clusters or sometimes solitary. Fruit a drupe, usually edible.

A large genus of 150 to 200 species, native in the north temperate zone, extending into tropical America and southern Asia. Five tree-like species are native on the Pacific Coast. Many species and horticultural varieties are cultivated for their edible fruits and showy flowers, which may be single or double and varicolored. The cultivated plums, peaches, apricots, almonds, nectarines, and cherries belong to this genus.

KEY TO THE SPECIES

Leaves purple-red, finely serrate.................1. *P. cerasifera* var. *pissardi*.
Leaves green.
  Leaves thick and leathery, evergreen.
    Leaves 1 to 2 inches long, usually spinose-toothed; petioles ½-inch
      or less long; drupe ½- to ¾-inch in diameter......2. *P. ilicifolia*.
    Leaves 2 to 6 inches long.
      Leaves entire or some finely denticulate or bristly toothed; peti-
        oles ⅛- to ½-inch long.
        Leaves broadly ovate or oblong-ovate, often folded inward
          along the midrib; drupe ¾-inch to 1¼ inches in diameter
                                       3. *P. lyoni*.
      Leaves usually elliptical, or oblong-elliptical to oblong-obovate,
        not folded inward.
        Leaves 1¼ to 2¼ inches wide, usually distantly small-
          toothed....................................................4. *P. laurocerasus*.
        Leaves ¾-inch to 1¼ inches wide.................5. *P. caroliniana*.
      Leaves regularly and finely serrate; petioles ½-inch to 1 inch
        long; drupe about ¼-inch long........................6. *P. lusitanica*.
  Leaves thin and not leathery, deciduous, finely serrulate.
    Leaves folded lengthwise in the bud; drupe globular.
      Petioles about ½-inch long, with a few glands near base of
        blades; flowers in racemes....................................7. *P. demissa*.
      Petioles ¼-inch or less long, without glands; flowers in corymbs
                                    8. *P. emarginata*.
    Leaves rolled inward from edge to edge in bud; drupe oblong;
      flowers in umbels................................................9. *P. subcordata*.

### 1. **Prunus cerasífera** var. **pissárdi** Koehne. PURPLELEAF PLUM. Fig. 248.

A small slender deciduous tree, 15 to 25 feet high, with ascending or spreading branches forming a round-topped crown. Leaf-blades purple-red, elliptic, ovate, or obovate, 2 to 4 inches long, 1 to 1½ inches wide, finely serrate. Flowers usually solitary, white to pale pink, ¾-inch to 1 inch broad, on pedicels ½- to ¾-inch long. The drupe is subglobose, ½- to 1 inch in length, and red- or yellow-colored.

An horticultural variety, much planted as a lawn or street tree.

**2. Prunus ilicifòlia** Walp. HOLLYLEAF CHERRY. Fig. 249.

An evergreen shrub or a small tree, 10 to 25 feet high. Leaf-blades rather thick and leathery, ovate or elliptic, 1 to 2 inches long, 1 to 1½ inches wide, dark green and glossy above, paler and yellowish beneath, spinosely toothed

Fig. 248. Prunus cerasifera var. pissardi Koehne. Portion of branchlet with leaves and fruit, × ½.

or rarely almost entire; petioles ⅛- to ½-inch long. Flowers white, about ⅛-inch broad, in short peduncled racemes. Drupe subglobose, ½- to ¾-inch in diameter, red or dark purple to black, the flesh sweet but very thin, ripe in October and November.

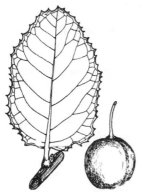

Fig. 249. Prunus ilicifolia Walp. Leaf, fruit, × ⅔.

Native in the Coast Ranges of California from Napa and Solano counties southward to the Tehachapi Mountains and the lower slopes of the mountains of southern and Lower California. Also known as Islay and Evergreen Cherry. Commonly cultivated as an ornamental tree or used for hedges.

**3. Prunus lỳoni** Sarg. CATALINA CHERRY. Fig. 250.

*P. ilicifolia* var. *integrifolia* Sudw. *P. integrifolia* Sarg.

An evergreen tree, 15 to 35 feet high. Leaf-blades thick and leathery, broadly ovate or oblong-ovate to lanceolate, 2½ to 5 inches long, 1 to 2½ inches wide, dark green, glossy, and glabrous above, paler beneath, entire or nearly so; petioles ⅛- to ½-inch long. Flowers white, about ¼-inch broad, in dense racemes that are 3 to 5 inches long. The drupe is sub-globose, approximately 1 inch in diameter, dark purple or black, and sweetish. It becomes ripe in October and November.

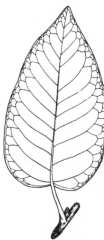

Fig. 250.
Prunus lyoni Sarg.
Leaf, × ½.

Fig. 252. Prunus
caroliniana Ait.
Leaf, × ½.

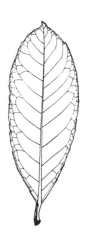

Fig. 251. Prunus
laurocerasus L.
Leaf, × ⅓.

Native on the islands off the coast of southern California and Lower California. Very closely related to *P. ilicifolia*, and may be only a variety of that species. Often cultivated as an ornamental tree.

4. **Prunus laurocérasus** L. CHERRY-LAUREL. Fig. 251.

*Laurocerasus officinalis* Roem.

An evergreen shrub or small tree, 10 to 20 feet high. Leaf-blades thick and leathery, elliptical or oblong to oblong-obovate, 2½ to 6 inches long, 1¼ to 2½ inches wide, glabrous, dark green and glossy above, paler beneath, entire or finely serrulate; petioles ⅛- to ⅜-inch long. Flowers white, about ¼-inch broad, in racemes 2½ to 5 inches long. Drupe ovoid, about ¼-inch long, black, ripe in August and September.

Native in southeastern Europe and Asia Minor. Frequently cultivated for the dark green foliage and fragrant white flowers.

5. **Prunus caroliniàna** Ait. CAROLINA CHERRY-LAUREL. MOCK-ORANGE. Fig. 252.

*Laurocerasus caroliniana* Roem.

A medium-sized evergreen tree, 20 to 40 feet high, with slender horizontal branches and dense dark green foliage forming a broad crown. Leaf-blades leathery, oblong-lanceolate or elliptical, 2 to 4 inches long, ¾-inch to 1¼ inches wide, tapering or rounded at the base, acuminate, glabrous, dark green and shining above, paler beneath, usually entire but sometimes remotely and irregularly spinulose-serrate, somewhat revolute; petioles about ¼-inch long. Flowers cream-colored, in dense racemes shorter than the leaves. Drupe subglobose to ellipsoidal, about ½-inch long, black and shining, ripening in the autumn and remaining on the tree until after the following flowering period in the spring.

Native in the southeastern United States. Cultivated as a street and park tree because of its glossy evergreen foliage, habit of growth, and clusters of flowers and fruits.

Fig. 253.
Prunus lusitanica L.
Leaf, × ½.

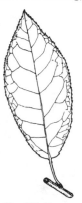

Fig. 254. Prunus
demissa Walp.
Leaf, × ½.

### 6. Prunus lusitánica L.
PORTUGAL-LAUREL. Fig. 253.
*Laurocerasus lusitanica* Roem.

An evergreen shrub or small tree, 15 to 35 feet high. Leaf-blades thick and leathery, ovate or oblong-ovate, 2½ to 4 inches long, 1½ to 2¼ inches wide, glabrous, dark green and glossy above, paler beneath, finely serrate; petioles ½-inch to 1 inch long. Flowers white, ¼- to ⅜-inch broad, in racemes 5 to 10 inches long. Drupe ovoid, about ¼-inch long, dark purple, ripe in August.

This tree is native to Spain and Portugal. It is cultivated for its evergreen and glossy foliage.

### 7. Prunus demíssa Walp. WESTERN CHOKE CHERRY. Fig. 254.
*P. virginiana* var. *demissa* Torr.
*Cerasus demissa* Nutt.

A deciduous shrub or small tree, 10 to 20 feet high. Leaf-blades oblong or oblong-obovate to rarely oval, 1½ to 4 inches long, ¾-inch to 1¾ inches wide, folded lengthwise in the bud, dark green and glossy above, paler and often slightly pubescent beneath, finely serrate; petioles about ½-inch long, with a few glands

Fig. 255.
Prunus emarginata Walp.
Leaf, × 1.

near the junction with the blades. Flowers white, ⅓- to ½-inch across, in racemes 3 to 6 inches long. Drupe ovoid to subglobose, about ⅓-inch long, red or purple, ripe in August and September.

Native in western North America from British Columbia south to California and eastward to the Rocky Mountains.

8. **Prunus emargināta** Walp. BIT-TER CHERRY. Fig. 255.

*Cerasus emarginata* Dougl.

A deciduous shrub or small tree, 10 to 20 feet high, with smooth and reddish bark. Leaf-blades broadly elliptic or oblong-obovate, ¾-inch to 2½ inches long, ½-inch to 1 inch wide, with 1 to 4 large glands near the base, folded lengthwise in the bud, emarginate or obtuse at the apex, dark green and glabrous above, paler beneath, finely serrulate; petioles

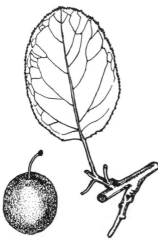

Fig. 256. Prunus subcordata Benth. Fruit, leaf, × ¾.

¼-inch or less long. Flowers white, about ½-inch broad, in corymbs. Drupe globose, ⅓- to ½-inch in diameter, red to purple-black, ripe in August.

Native in western North America from British Columbia south to southern California and eastward to Arizona, Nevada, Idaho, and Montana.

9. **Prunus subcordàta** Benth. SIERRA PLUM. Fig. 256.

A deciduous shrub or small tree, 10 to 20 feet high, with more or less spinescent branchlets. Leaf-blades elliptical or broadly ovate to almost round, 1 to 2½ inches long, ½-inch to 1½ inches wide, obtuse or truncate to subcordate at the base, dark green and glabrous above, paler and often pubescent beneath, rolled inward from edge to edge in the bud, sharply serrate; petioles ¼-inch or less long. Flowers white, about ½-inch wide, in 2- to 4-flowered umbels. Drupe

Fig. 257. Pyrus communis L. Leaf, × ½.

ovoid or short-oblong, ¾-inch to 1 inch long, red or yellowish, ripe in August and September.

Native on the lower slopes of the Sierra Nevada from Tulare County to Modoc County, California, in the Coast Ranges from the Santa Cruz Mountains north to Humboldt County, eastward in Siskiyou County, and north to southern Oregon. Several local varieties differing in fruit and leaf characters have been described. The plants in the Coast Ranges rarely set good fruits.

#### 9. Pýrus L. PEAR

(From *Pirus,* ancient Latin name of the pear tree.)

Deciduous trees or shrubs. Leaves simple, alternate, serrate or lobed, involute in the bud. Flowers white or pink, in corymbs. Fruit an ovoid or pear-shaped pome.

About 20 species, native in Europe, eastern Asia, and northern Africa. Extensively cultivated for their fruits and as ornamental trees. The common pear is **P. communis L.,** Fig. 257.

#### 10. Sórbus L. MOUNTAIN-ASH

(The ancient Latin name.)

Deciduous trees or shrubs. Leaves simple or odd-pinnate, alternate. Flowers white or

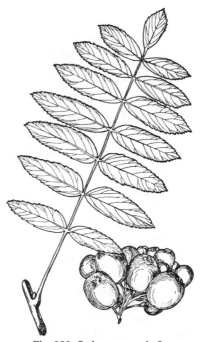

Fig. 258. Sorbus aucuparia L. Leaf, × ½. Fruit-cluster, × 1.

rarely pink, in terminal compound clusters. Fruit a small globose berry-like pome.

About 80 species, native in the northern part of the northern hemisphere, 3 or 4 of which occur in North America.

KEY TO THE SPECIES

Leaflets ¾-inch to 2 inches long; inflorescence pubescent; flowers about ⅓-inch across................................................................................1. *S. aucuparia.*
Leaflets 1¾ to 4½ inches long; inflorescence glabrous; flowers ⅛- to ¼-inch across................................................................................2. *S. americana.*

**1. Sorbus aucupària** L. ROWAN TREE. EUROPEAN MOUNTAIN-ASH. Fig. 258.

*Pyrus aucuparia* Gaertn.

A small deciduous tree, 15 to 30 feet high. Leaves 5 to 7 inches long, with 9 to 15 leaflets; the leaflets oblong to oblong-lanceolate, ¾-inch to 2 inches long, ½- to ¾-inch wide, dull green above, pale and pubescent beneath, serrate on the upper ⅔ of the margin, sessile. Flowers about ⅓-inch across, in pubescent clusters 4 to 6 inches broad. Fruit globose, about ⅜-inch in diameter, bright red.

Native in Europe, extending eastward to western Asia and Siberia. Rowan Tree; with several varieties, is frequently cultivated in gardens and parks.

**2. Sorbus americàna** Marsh. AMERICAN MOUNTAIN-ASH.

*Pyrus americana* DC.

A small deciduous tree, 15 to 30 feet high, or sometimes shrubby. Leaves 6 to 8 inches long, with 11 to 17 leaflets; the leaflets lanceolate, 1¾ to 4½ inches long, ½-inch to 1 inch wide, light green above, paler and glabrous beneath or slightly pubescent when young, serrate on upper ⅔ of the margin, sessile, or the terminal leaflet petiolulate. Flowers about ¼-inch across, in dense glabrous clusters 3 to 5½ inches broad. Fruit globose, about ¼-inch in diameter, bright red.

Native from Newfoundland west to Manitoba, south to Michigan and North Carolina. Occasionally cultivated in gardens and parks.

## Leguminosae. Pea Family
### 1. Acàcia Willd.

(From the Greek *akakie,* derived from *ake,* a point, in reference to the prickles.)

Evergreen or rarely deciduous trees or shrubs. Leaves bipinnate or reduced to phyllodia (flattened petioles and rachises resembling simple entire leaves), alternate or whorled, usually with one or more glands on the margins or rachises. Flowers small, commonly yellow, in globular heads or spikes; the stamens numerous, distinct, long-exserted. Fruit a 2-valved pod (legume), rarely indehiscent.

A large genus of about 500 species, native in subtropical regions and a few in the warmer parts of the temperate zone. One is native in the southwestern United States.

KEY TO THE SPECIES

I. Plants usually with spiny or thorny branches

Leaves bipinnately compound, deciduous.................................1. *A. greggi.*
Leaves reduced to phyllodia, evergreen, with 2 spines at the base of each phyllodium........................................................................................2. *A. armata.*

II. Plants not spiny or thorny, evergreen

*A. Leaves reduced to phyllodia*

1. Phyllodia 1 inch or less long

Phyllodia less than ⅛-inch wide, linear, sharp-pointed, whorled
3. *A. verticillata.*

Phyllodia more than ⅛-inch wide, alternate, gray or gray-green.

Phyllodia ½-inch to 1 inch long.

Phyllodia obliquely ovate to elliptical; somewhat sickle-shaped; midvein eccentric............................................4. *A. cultriformis.*

Phyllodia broadly elliptical...............................6. *A. podalyriaefolia.*

Phyllodia ¼- to ½-inch long, somewhat wedge-shaped
5. *A. pravissima.*

2. Phyllodia 1 inch or more long

*a. Phyllodia grayish or bluish green*

Phyllodia 1 to 1½ inches long.

Phyllodia broadly elliptical...............................6. *A. podalyriaefolia.*

Phyllodia obliquely ovate to elliptical, somewhat sickle-shaped
4. *A. cultriformis.*

Phyllodia 2 to 8 (rarely to 12) inches long.

All phyllodia less than ¾-inch wide; funicle folded around the seed
7. *A. retinodes.*

Some phyllodia more than ¾-inch wide.....................11. *A. cyanophylla.*

*b. Phyllodia dark green or yellowish green*

(1) Phyllodia ¼-inch or less wide

Phyllodia linear, recurved or hooked at apex; flowers in heads
8. *A. calamifolia.*

Phyllodia linear-lanceolate, not hooked at apex; flowers in cylindrical spikes...............................13a. *A. longifolia* var. *floribunda.*

(2) Phyllodia more than ¼-inch wide

Phyllodia with 1 main vein from base.

Phyllodia falcate-lanceolate, the marginal gland about ½-inch from base...............................................9. *A. pycnantha.*

Phyllodia lanceolate or linear-lanceolate.

Some phyllodia 1 inch or more wide, the marginal gland near base or absent...............................................10. *A. saligna.*

All phyllodia less than 1 inch wide, the marginal gland ¼- to ½-inch from base...............................................7. *A. retinodes.*

Phyllodia with 2 or more main veins from base.

Phyllodia lanceolate to oblanceolate, somewhat sickle-shaped; flowers in heads...............................................12. *A. melanoxylon.*

Phyllodia oblong-lanceolate, both margins curved outward; flowers in spikes.

Phyllodia ⅜- to ⅝-inch wide...............................13. *A. longifolia.*

Phyllodia 1 to 2 inches wide...............................14. *A. latifolia.*

*B. Leaves bipinnately compound*

Leaves with 2 to 10 pairs of gray pinnae...............................15. *A. baileyana.*

Leaves with 8 to 25 pairs of pinnae.

Leaflets grayish green; glands on rachis only at bases of pinnae
16a. *A. decurrens* var. *dealbata.*

Leaflets dark green; glands at bases of pinnae and additional ones between...............................................16b. *A. decurrens* var. *mollis.*

**1. Acacia gréggi** Gray. CATCLAW. TEXAS-MIMOSA. Fig. 259.

Usually a shrub but occasionally a tree, 12 to 18 feet high, with armed branches and deciduous gray-green foliage. Spines short, about ¼-inch long, curved, scattered along the branches. Leaves twice-pinnately compound, 1 to 2 inches long, with 1 to 3 pairs of pinnae, each pinna ½- to ¾-inch long and composed of 4 to 6 pairs of leaflets ⅛- to ¼-inch long. Flowers yellow, in cylindrical spikes 1 to 2½ inches long. Legumes

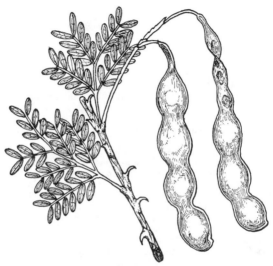

Fig. 259. Acacia greggi Gray.
Portion of branchlet with leaves, thorns, and
fruit-pod, × ¾.

flattened, 1 to 6 inches long, about ¾-inch wide, much constricted between the 1 to 10 seeds.

Native in desert regions of the southwestern United States. Flowers in early spring.

**2. Acacia armàta** R. Br. KANGAROO-THORN. Fig. 260.

Usually a tall spreading shrub or rarely tree-like, 8 to 20 feet high, with many slender spiny branches covered with dark evergreen foliage. Phyllodia obliquely oblong-ovate, ½-inch to 1 inch long, ⅛- to ⅜-inch wide, the outer margin curved and wavy, the inner almost straight, the midrib eccentric and terminated by a sharp point. Flowers golden yellow, in solitary pedunculate heads. Legumes flat, 1½ to 2 inches long, ³⁄₁₆-inch wide, sometimes curved, tomentose.

Native to Australia. An excellent ornamental plant for hedges and screens. It thrives well in either moist or very dry conditions and is very satisfactory for covering banks and tolerant enough for planting near eucalypti. Flowers in early spring.

Fig. 260. Acacia armata R. Br. Portion of branchlet with leaves, thorns, and fruit-pod, × 1.

sometimes planted singly. Flowers in the spring.

### 4. Acacia cultrifórmis Cunn. KNIFE ACACIA. Fig. 262.

Usually a tall shrub but sometimes a small tree, 10 to 20 feet high, with spreading willowy branches and gray foliage thickly covering the branches. Phyllodia obliquely ovate to elliptical, somewhat sickle-shaped, ¾-inch to 1¼ inches long, ¼- to ½-inch wide, gray on both surfaces, the midrib eccentric, the marginal gland about ⅓ of the distance from the base on the margin turned toward the

### 3. Acacia verticillàta Willd. WHORL-LEAF ACACIA. STAR ACACIA. Fig. 261.

A small tree, 10 to 20 feet high, or more often a spreading shrub, with very dark green foliage. Phyllodia whorled, or often spirally arranged on leading shoots, linear, ½-inch to 1 inch long, less than ⅛-inch wide, pungently pointed, sessile. Flowers light yellow, in spikes 1 inch or less long. Legumes flat, 2 to 3 inches long, ⅛-inch wide.

Native to Australia. Used for group planting and for hedges, or

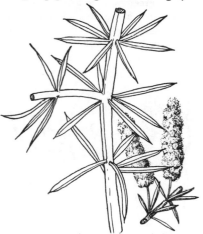

Fig. 261. Acacia verticillata Willd. Portion of branch with leaves, × 1½. Flowering spikes, × ½.

stem. Flowers bright golden yellow, in heads arranged in axillary racemes near the ends of the branches. Legumes 1½ to 3 inches long, ¼-inch wide, sometimes constricted between the seeds.

Native to Australia. Used as an ornamental plant because of its gray foliage and clusters of bright yellow flowers appearing in the spring.

5. **Acacia pravíssima** F. v. M. SCREWPOD ACACIA. Fig. 263.

A small tree, 15 to 20 feet high, with pendulous slender branches and branchlets covered with short grayish foliage. Phyllodia triangular-ovate or wedge-shaped, ¼- to ½- (1) inch long, ¼- to ½-inch wide,

Fig. 262. Acacia cultriformis Cunn.
Leaves, × 1.

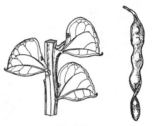

Fig. 263.
Acacia pravissima F. v. M.
Leaves, × 1. Fruit, × ½.

grayish green, with a large gland ¼ to ½ the distance from the base on margin facing stem, indistinctly 2-veined, sessile. Flowers yellow, in heads ⅛-inch in diameter and arranged in racemes longer than the phyllodia. Legumes 1½ to 3 inches long, about ¼-inch wide, twisted.

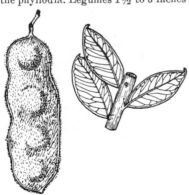

Native in Victoria, Australia. A showy species planted for its fast-growing habit, long drooping branches, and clusters of yellow flowers in early spring.

6. **Acacia podalyriaefòlia** Cunn. PEARL ACACIA. Fig. 264.

A shrub or small tree, 8 to 20 feet high, with slender branches, gray-pubescent branchlets, and grayish foliage. Phyllodia ovate to oblong, 1 to 1½ inches long, ½-inch to 1¼ inches wide, pubescent, the midrib eccentric

Fig. 264. Acacia podalyriaefolia Cunn.
Fruit, leaves, × ⅔.

and ciliate, the margins ciliate, with a gland at about the middle of the margin nearest the midrib. Flowers yellow, in heads borne in simple racemes. Legumes flat, gray, 1 to 3 inches long, about ¾-inch wide.

Native in Queensland, Australia. Cultivated for its gray foliage and abundant yellow flower-clusters appearing in early spring.

7. **Acacia retinòdes** Schlecht. WATER WATTLE. Fig. 265.

A small tree, 15 to 25 feet high, or sometimes shrubby, with spreading branches drooping at the ends, and blue or yellow-green foliage. Phyllodia lanceolate, broadest above the middle, $3\frac{1}{2}$ (rarely 2) to 6 inches long, $\frac{1}{4}$- to $\frac{3}{4}$-inch wide, the marginal gland $\frac{1}{4}$- to $\frac{1}{2}$-inch from the base. Flowers yellow, in globose heads arranged in lateral racemes. Legumes 3 to 4 inches long, about $\frac{1}{4}$-inch wide. The funicle folded around the seed.

Native in Australia. Commonly cultivated in lawns and along streets. Flowers almost the year round. Often sold by nurserymen, especially in southern California, as *A. floribunda*.

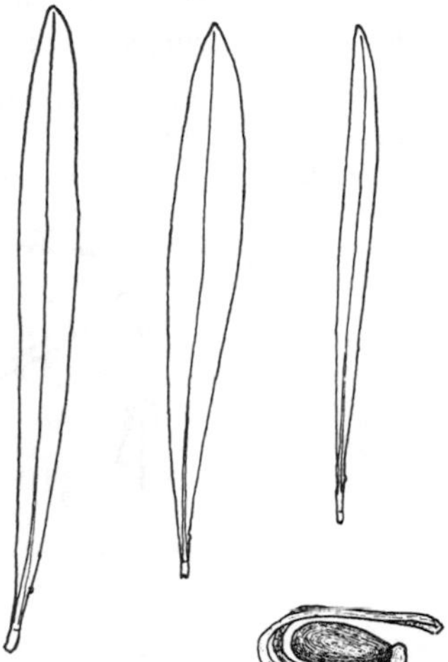

8. **Acacia calamifòlia** Sweet. BROOM WATTLE. Fig. 266.

Fig. 265. Acacia retinodes Schlecht. Leaves, $\times$ $\frac{1}{2}$. Seed, $\times$ 2.

A small tree, 8 to 15 feet high, or sometimes shrubby, with spreading slender branches and pendulous reddish branchlets. Phyllodia linear, 2 to 4 inches long, about $\frac{1}{16}$-

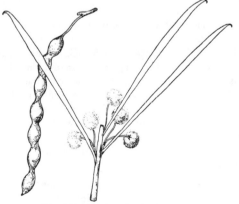

Fig. 266. Acacia calamifolia Sweet.
Fruit, $\times$ $\frac{1}{2}$. Leaves and flowering heads, $\times$ 1.

inch wide, short-hooked at the apex, dark green. Flowers yellow, in globular solitary heads or 2 or 3 in short racemes. Legumes 1¼ to 2 inches long, about ¼-inch wide, somewhat constricted between the seeds.

Native in Australia. Occasionally cultivated as a specimen tree. Flowers in the spring and early summer.

### 9. Acacia pycnántha Benth. GOLDEN WATTLE. Fig. 267.

A small or medium-sized openly branched tree, 15 to 35 feet high, with crooked brash branches and glaucous branchlets. Phyllodia fal-

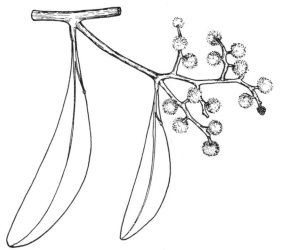

Fig. 267. Acacia pycnantha Benth.
Portion of branchlet with leaves and flowering heads, × ½.

cate-lanceolate or oblong-ovate, 3 to 6 inches long, ½-inch to 1½ inches wide, somewhat sickle-shaped, 1-veined from the base, dull green on both surfaces, the midrib often eccentric, the marginal gland about ½-inch from the base. Flowers bright golden yellow, in globular heads arranged in long axillary racemes. Legumes 2 to 5 inches long, ¼-inch wide, slightly constricted between the seeds.

Native in Australia. Cultivated for its rapid growth and showy flowers appearing in early spring. Short-lived and soon becoming a very unsymmetrical tree.

### 10. Acacia salígna Wendl. WILLOW ACACIA. GOLDENWREATH. Fig. 268.

A tall shrub or low tree, 10 to 20 feet high, with drooping branches. Phyllodia lanceolate or linear-lanceolate to oblanceolate, 3 to 8 inches long (or the lower phyllodia 12 inches long), ¼-inch to 1½ inches wide, green, 1-veined from the base, the marginal gland near the base or absent. Flowers golden yellow, in globular heads (about ½-inch in diameter) borne in large terminal racemes or scattered in groups of 3 to 5 or solitary. Legumes 3 to 5 inches long, about ¼-inch wide, constricted between the seeds. Funicle short, not surrounding the seed.

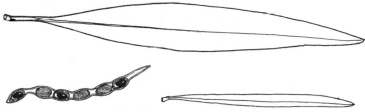

Fig. 268. Acacia saligna Wendl.
Leaves, fruit, × ½.

Native in western Australia. A variable species in size and color of the leaves and in arrangement of the flower-heads. Cultivated for its "weeping" habit of growth and handsome golden yellow flower-clusters, which appear in the spring and somewhat throughout the year.

Fig. 269. Acacia melanoxylon R. Br.
Portion of branchlet with leaves and flowering heads, × ⅓.

### 11. **Acacia cyanophýlla** Lindl. BLUELEAF WATTLE.

A tall shrub or small tree, 10 to 20 feet high, with drooping branches. Very close to *A. saligna*, and perhaps only a blue-leaved form of that species.

### 12. **Acacia melanóxylon** R. Br. BLACKWOOD ACACIA. Fig. 269.

A medium-sized to large tree, 25 to 70 feet high, with spreading branches and dense dull dark green foliage forming a narrow or broadly pyramidal crown. Phyllodia lanceolate to oblanceolate or oblong-ovate, somewhat sickle-shaped, 2½ to 5 inches long, ½-inch to 1 inch wide, with 3 to 5 veins from the base, the marginal gland very

close to the base. The leaves on young trees or new shoots often bipinnate. Flowers creamy white, in globular heads arranged in short racemes.
Legumes 3 to 5 inches long, about ⅜-inch wide, twisted or bent.

Native in Australia. Abundantly planted throughout the coastal
regions and sometimes in the interior of California. Flowers in late
spring.

13. **Acacia longifòlia** Willd. SYDNEY GOLDEN WATTLE. Fig. 270.

A small tree, 15 to 30 feet high, or often shrub-like, with light green
foliage. Phyllodia oblong-lanceolate, 2 to 6 inches long, ¼- to ⅝-inch

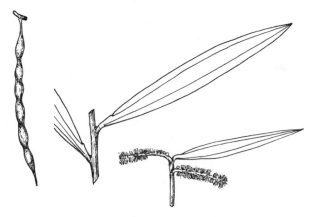

Fig. 270. Acacia longifolia Willd.
Fruit, leaf, flowering spikes, × ½.

wide, with both margins convex, 2- or 4-veined from the base, with
a gland very near the base. Flowers bright yellow, in spikes ¾-inch
to 2¼ inches long. Legumes rounded, 1½ to 4 inches long.

Native in Australia. Frequently planted as a park and avenue tree
or as a ground cover. Flowers in the spring.

13*a*. Var. **floribúnda** F. v. M. GOSSAMER WATTLE.

Similar to the species, except that the phyllodia are narrower,
linear-lanceolate, ⅛- to ¼-inch wide, mostly at the ends of the
branches, thus giving the plant a very compact crown. The flowers
whitish yellow.

14. **Acacia latifòlia** Benth. BROADLEAF ACACIA.

Similar to *A. longifolia,* except that the phyllodia are 1½ to 2 inches
wide. Usually shrub-like.

15. **Acacia baileyàna** F. v. M. BAILEY ACACIA. COOTAMUNDRA WATTLE. Fig. 271.

A small tree, 15 to 30 feet high, or often shrub-like, with many spreading glaucous branches and striking gray foliage. Leaves twice-pinnately compound, 1 to 3 inches long, with 2 to 5 pairs of pinnae, usually with a gland at the base of each pair, each pinna about 1 inch long and composed of 20 to 30 pairs of leaflets about ¼-inch long.

Fig. 271. Acacia baileyana F. v. M.
Leaf, × 1. Fruit, × ½.

Flowers golden yellow, in globular heads borne in long axillary or terminal racemes. Legumes 2 to 4 inches long, ½-inch or less wide, sometimes constricted between the seeds.

Native in Australia. Extensively cultivated as a lawn tree for its feathery gray foliage and the abundant golden yellow flower-clusters, which appear in early spring.

16a. **Acacia decúrrens** var. **dealbàta** F. v. M. SILVER WATTLE. Fig. 272.

A large tree, 30 to 60 feet high, with wide-spreading or ascending branches, glaucous branchlets, and dense gray-green feathery foliage. Leaves twice-pinnately compound, 4 to 6 inches long, with 8 to 25 pairs of pinnae 1 to 2 inches long, with a single gland at the base of each pair, each pinna composed of 30 to 60 pairs of leaflets about 3/16-inch long.

Flowers yellow, in globular heads borne in axillary and often compound racemes. Legumes 2 to 4 inches long, ¼-inch or more wide.

Native in Australia. Extensively cultivated in parks and gardens. Flowers in early spring.

**16b. Acacia decúrrens** var. **móllis** Lindl. BLACK WATTLE. Fig. 273.

*A. mollissima* Willd.

Similar to the preceding, except for the dark green and larger leaves, lighter yellow flowers which bloom most abundantly from April to June, and additional glands on the internodes of the leaf-rachises.

Native in Australia. Not so frequently planted as the variety *dealbata*.

Fig. 272. Acacia decurrens var. dealbata F. v. M. Portion of compound leaf with glands on rachis, × 1½.

OTHER TREE-SPECIES RARELY CULTIVATED ON THE PACIFIC COAST

**17. Acacia eláta** Cunn. CEDAR WATTLE.

Leaves twice-compound, 12 inches long, 8 to 10 inches wide. Flowers in heads borne in compound racemes.

Native in Australia.

**18. Acacia kòa** Gray. KOA.

Phyllodia sickle-shaped, 4 to 5 inches long, about ½-inch wide, with 3 to 5 prominent veins from the base. Flowers in heads borne solitary or in short racemes.

Native in the Hawaiian Islands.

Fig. 273. Acacia decurrens var. mollis Lindl. Portion of compound leaf with glands on rachis, × 1½.

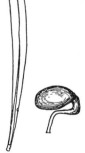

Fig. 274. Acacia neriifolia Cunn. Leaf, × ½. Seed, × 2.

**19. Acacia neriifòlia** Cunn. BALD ACACIA. Fig. 274.

Phyllodia oblanceolate, 1½ to 5 inches long, ¼- to ⅜-inch wide, with 1 main vein from the base. Flowers in heads borne in simple or compound racemes.

Native in Australia. Often confused with *A. retinodes*, but may be distinguished from that species by the short funicle not folded around the seed.

**20. Acacia pruinòsa** Cunn. BRONZE ACACIA.

Leaves twice-compound, with 2 to 4 pairs of pinnae, each pinna 2½ to 4 inches long and composed of 20 to 30 leaflets ¼- to ¾-inch long. Flowers in heads borne in racemes. Native in Australia.

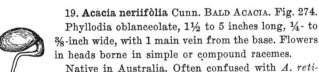

## 2. **Albízzia** Durazz.

(Named in honor of a noble Italian family, Albizzi, one member of which was the first to introduce plants of this genus into Italy.)

Small trees or shrubs resembling acacias. Leaves twice-pinnately compound, usually deciduous but evergreen in one species grown on the Pacific Coast. Flowers small and numerous, yellow, white, or pink, in globose heads or cylindrical spikes; stamens numerous, usually joined at the base, long-exserted. Fruit a 2-valved flat pod.

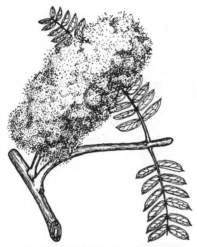

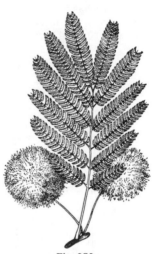

Fig. 275. Albizzia lophantha Benth.
Portion of branchlet with basal part of compound leaf and flowering spike, × ¾.

Fig. 276.
Albizzia julibrissin Durazz.
Leaf and flowering heads, × ⅓.

About 25 species, native in tropical and subtropical regions of Asia, Africa, Australia, and one in Mexico.

### KEY TO THE SPECIES

Leaves evergreen; flowers yellow, in cylindrical spikes....1. *A. lophantha.*
Leaves deciduous; flowers pink, in globular heads..........2. *A. julibrissin.*

### 1. **Albizzia lophántha** Benth. PLUME ALBIZZIA. Fig. 275.

*Acacia lophantha* Willd.

An evergreen shrub or small tree, 10 to 20 feet high, with glabrous or rusty-tomentose young branchlets. Leaves 5 to 8 inches long, 4 to 6 inches wide, with 14 to 24 pinnae, each pinna with 40 to 60 leaflets ¼- to ⅜-inch long, the midrib nearer the upper margin, the petiole with a gland ½-inch to 1 inch from the base. Flowers greenish yellow, distinctly pedicelled, in axillary paired spikes 1 to 2 inches long. Pods about 3 inches long.

Native in Australia. Cultivated in California as a tub or lawn tree. Flowers from February to June. It cannot withstand as severe a climate as can *Albizzia julibrissin*.

### 2. **Albizzia julibríssin** Durazz. SILK TREE. Fig. 276.

*Acacia julibrissin* Willd.

A deciduous tree, 20 to 30 feet high, with light green feathery foliage. Leaves 5 to 8 inches long, 4 to 5 inches wide, with 8 to 26 pinnae, each with 30 to 40 leaflets ¼- to ⅜-inch long, the midrib close to the upper margin, the petiole with a gland ½-inch to 1 inch from the base. Flowers pink, in clustered heads on the upper ends of the branches. Pods 4 to 6 inches long.

Native in Persia, extending eastward to Japan. Occasionally cultivated for its mass of pink flowers blooming in late spring and summer.

### 3. **Bauhínia** L. ORCHID TREE

(Named for John and Caspar Bauhin, herbalists of the sixteenth century, the paired leaflets suggesting two brothers.)

About 150 species of shrubs, trees, and vines, native in the tropics of both hemispheres.

### 1. **Bauhinia purpùrea** L. ORCHID TREE. PURPLE BAUHINIA. Fig. 277.

A small evergreen tree, 15 to 25 feet high. Leaves simple, alternate; the blades leathery, somewhat heart-shaped, 2 to 4 inches long and about as wide, bilobed, 7- to 11-veined from the base,

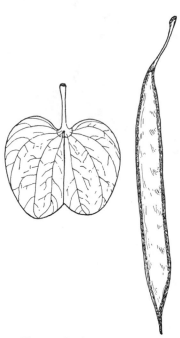

Fig. 277. Bauhinia purpurea L.
Leaf, fruit, × ½.

glabrous, entire; petioles about 1 inch long. Flowers red, purple, or white, fragrant, 3 to 4 inches wide, few in axillary or terminal clusters, suggesting orchid flowers. Fruit-pods flat, 4 to 10 inches long.

Native to India, Burma, and China. A very handsome and striking ornamental small tree, cultivated from Berkeley, California, southward to San Diego for its showy orchid-like flowers blooming in winter and early spring.

#### 4. Cássia L. SENNA

(From the Greek name of a plant, *kasian*, of the Bible.)

About 400 species of herbs, shrubs, or trees, native in the tropics and warmer parts of the temperate zones.

1. **Cassia tomentòsa** L. WOOLLY SENNA. Fig. 278.

An evergreen shrub or small tree, 10 to 20 feet high, with tomentose branches and foliage. Leaves even- and once-pinnately compound, alternate, 2½ to 4 inches long; leaflets 8 to 16, opposite, elliptical, ¾-inch to 1½ inches long, ¼- to ½-inch wide, dark green and short-tomentose above, lighter and densely tomentose beneath, almost sessile. Flowers deep yellow, about 1 inch broad, nearly regular, borne in great profusion in terminal clusters. Fruit a 2-valved tomentose pod, 3 to 6 inches long.

Native in Mexico. Frequently cultivated for its large deep yellow flowers, which bloom during the winter and somewhat throughout the year.

Fig. 278. Cassia tomentosa L. Leaf, flower, × ½.

#### 5. Ceratònia L. CAROB

(From the Greek *keratonia*, horn, in reference to the large fruit-pods; the Carob Tree.)

A single species.

1. **Ceratonia síliqua** L. ALGAROBA. ST. JOHN'S BREAD. CAROB. Fig. 279.

A small to medium-sized evergreen tree, 20 to 45 feet high. Leaves even- and once-pinnately compound, alternate, 6 to 9 inches long; leaflets 6 to 10, leathery, broadly elliptical to rounded, 1 to 3 inches long, ¾-inch to 1½ inches wide, dark green, glossy, and glabrous above, paler beneath, short-stalked. Flowers small, red, in short single or clustered lateral racemes. Fruit-pods 4 to 10 inches long.

Native in the eastern Mediterranean region, where it is extensively cultivated for its very palatable fruit. Cultivated in California as a street, park, and garden tree.

### 6. **Cercídium** Tul. PALO VERDE

(From the Greek *kerkidion,* a weaver's shuttle, in reference to the fruit.)

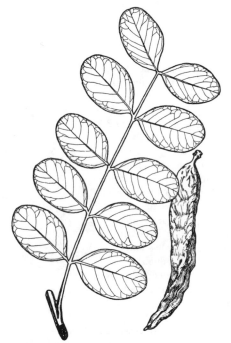

Fig. 279. Ceratonia siliqua L.
Leaf, fruit, × ⅓.

About 11 species of trees and shrubs, native in the southwestern United States, Mexico, and northwestern South America.

### 1. **Cercidium torreyànum** Sarg. PALO VERDE. Fig. 280.
*Parkinsonia torreyanum* Wats.

A small tree or large shrub, 15 to 35 feet high, with slender, glabrous, yellowish green, armed branches and gray foliage, or leafless for most of the year. Leaves twice-pinnately compound, alternate, with 2 (rarely

4 or 6) pinnae—each about 1 inch long, with 2 to 4 pairs of oblong leaflets ⅛- to ⅜-inch long—from a short petiole. Flowers yellow, almost
regular, ½- to ¾-inch broad, on jointed pedicels, in axillary racemes
2 to 4½ inches long. Fruit-pods oblong, 2 to 4 inches long, pointed
at each end, 1- to 8-seeded, sometimes constricted between the flat seeds.

Native on the Colorado Desert of California, east to Arizona, and
south to Lower California and Mexico. Grown in southern California.

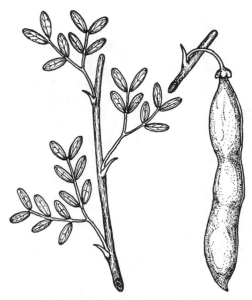

Fig. 280. Cercidium torreyanum Sarg.
Portion of branchlet with leaves, thorns,
and fruit, × ¾.

Palo Verde is the Spanish-Mexican name, and it refers to the smooth
green bark, which stands out in contrast to the gray of the desert.
The pods, abundant, and ripening in July, furnish food for stock.

### 7. Cércis L. Judas Tree. Redbud

(From *Kerkis*, the ancient Greek name of the oriental Judas Tree.)

Deciduous trees or shrubs. Leaves simple, alternate, palmately veined,
entire, petioled. Flowers somewhat irregular, reddish purple, borne
in umbel-like clusters from the old wood. Fruit an oblong flat legume.

Eight species, native in North America, southeastern Europe, Asia, and Japan. One species is native on the Pacific Coast.

Leaves rounded or notched at apex.
  Flowers ½-inch long; fruit-pods 1½ to 3 inches long
                                         1. *C. occidentalis.*
  Flowers ¾-inch long; fruit-pods 3 to 4 inches long....2. *C. siliquastrum.*
Leaves abruptly short-acuminate..........................................3. *C. canadensis.*

### 1. Cercis occidentàlis Torr. WESTERN REDBUD. Fig. 281.

A tall deciduous shrub or small tree, 8 to 20 feet high. Leaf-blades round, 2 to 3½ inches across, heart-shaped at the base, round or notched at the apex, somewhat palmately veined, glabrous, entire; petioles ½-inch to 1 inch long.

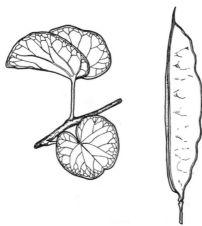

Flowers about ½-inch long, red-purple, appearing before the leaves in simple umbellate clusters at the alternate nodes. Legumes 1½ to 3 inches long, ½- to ⅝-inch wide, dull red when mature.

Native to California's inner North Coast Ranges, the foothills of the Sierra Nevada, and sparingly in the mountains of southern California. Western Redbud is one of the most attractive native flowering trees or shrubs and is rather extensively planted throughout the Pacific Coast in gardens and parks.

Fig. 281. Cercis occidentalis Torr.
Leaves, fruit, × ½.

### 2. Cercis siliquástrum L. EUROPEAN REDBUD.

A small deciduous tree, 15 to 30 feet high. Leaf-blades round, 3 to 5 inches wide, deeply cordate, rounded at the apex, glabrous; petioles ¾-inch to 1¼ inches long. Flowers about ¾-inch long, purplish rose. Legumes 3 to 4 inches long, dull brownish red when mature.

This tree is native to southeastern Europe and western Asia. Occasionally it is cultivated in gardens and parks. It is easily confused with *C. occidentalis* Torr., from which it usually differs in having larger flowers, leaves, and fruits.

### 3. Cercis canadénsis L. AMERICAN REDBUD. Fig. 282.

A small to medium-sized tree, 20 to 35 feet high, with a straight trunk usually branching 8 to 10 feet above the ground. Leaf-blades broadly ovate or almost round, 3 to 5 inches long and almost as wide, truncate to cordate at the base, short-acute or abruptly acuminate, glabrous except for axillary tufts of whitish hairs beneath, entire; petioles 2 to 4 inches long. Flowers rose-colored, rarely white, borne in clusters of 4 to 8, appearing before the leaves. Legumes 2½ to 3½ inches long, rose-colored.

Native in the eastern United States. Cultivated as an ornamental tree in parks and gardens. More commonly planted in Washington and Oregon than in California.

### 8. Cladrástis Raf.

(From the Greek *klados,* branch, and *thraustos,* fragile, in reference to the brittle branches.)

Four species of trees, one in the southeastern United States, two in western China, and one in Japan.

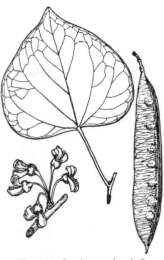

Fig. 282. Cercis canadensis L.
Flower-cluster, leaf, fruit, × ½.

### 1. Cladrastis lùtea Koch. YELLOW-WOOD. Fig. 283.
#### *C. tinctoria* Raf.

A medium-sized deciduous tree, 30 to 50 feet high, with a trunk 1 to 2 feet in diameter, usually dividing 6 to 10 feet above the ground into 2 or 3 secondary stems, with yellow wood, and wide-spreading more or less pendulous brittle branches. Leaves odd-pinnately compound, alternate, 8 to 12 inches long; leaflets alternate, 7 or 9, broadly oval, 2 to 4 inches long, 1½ to 2 inches wide, thin, glabrous, entire, short-petiolulate, the terminal leaflet rhombic-ovate, 2 to 3½ inches wide. Flowers papilionaceous, white, more than 1 inch long, fragrant, borne in drooping panicles 10 to 18 inches long. Fruit a narrow-oblong flattened legume, 2 to 4 inches long, 2- to 6-seeded.

Native in the southeastern United States. Occasionally cultivated in parks and on lawns for its handsome foliage and large panicles of white flowers blooming in May and June.

### 9. **Erythrina** L. Coral Tree

(From the Greek *erythros*, red, in reference to the color of the flowers.)

Trees, shrubs, or herbs, mostly spiny. Leaves alternate, pinnately compound, with 3 broad entire leaflets. Flowers large and showy,

Fig. 283. Cladrastis lutea Koch.
Leaf, × ⅛.

reddish, nearly papilionaceous, borne in loose racemes. Fruit a long legume, constricted between the usually highly colored seeds.

About 50 species, native in warm-temperate and tropical regions. Two or 3 species are occasionally cultivated in the warmer parts of the Pacific Coast. **E. cristagálli** L., Common or Cockspur Coral Tree, fig. 284, is a shrub or small tree, with oblong-lanceolate or broadly elliptical leaflets, 2 to 4 inches long, and spiny petioles and midribs. Native to Brazil. **E. corallodéndron** L., Coral Tree, is a small tree, 15 to 20 feet high, usually spiny, but sometimes unarmed, with rhombic-ovate leaflets 2 to 5 inches long, and usually unarmed petioles. The bright scarlet flowers appearing before the leaves are very ornamental. Native in tropical America.

### 10. Gledítsia L. Honey-Locust

(Named in honor of J. T. Gleditsh, a German botanist.)

About 10 species of trees, native in eastern North America, Asia, and western tropical Africa.

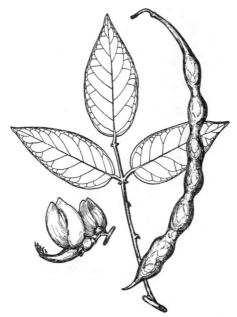

Fig. 284. Erythrina cristagalli L.
Flower, leaf, fruit, × ⅓.

### 1. Gleditsia triacánthos L. Honey-Locust. Fig. 285.

A deciduous tree, 25 to 70 feet high, with stout, simple or branched, glabrous, brown spines 2 to 4 inches long. Leaves once- or twice-pinnately compound, alternate, 6 to 10 inches long; the once pinnate leaves with 20 to 30 leaflets, the twice pinnate with 8 to 14 pinnae; leaflets oblong-lanceolate or elliptical, ¾-inch to 1½ inches long, ¼- to ½-inch wide, entire or remotely and finely toothed, short-petiolulate.

Flowers small, greenish, in narrow racemes 1½ to 3 inches long. Fruit-pods flat, 10 to 16 inches long, many-seeded, dark brown.

Native in the eastern United States. Sometimes planted as a highway and specimen tree.

### 11. Labúrnum Griseb. GOLDENCHAIN

(The ancient Latin name.)

Three species of deciduous trees or shrubs, native in southern Europe and western Asia.

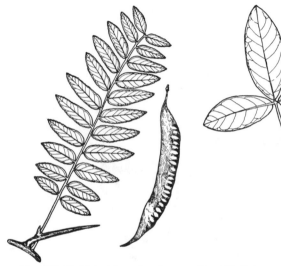

Fig. 285. Gleditsia triacanthos L.   Fig. 286. Laburnum vulgare Griseb.
Leaf and thorn, fruit, × ⅛.   Leaf, × ½.

### 1. Laburnum vulgàre Griseb. GOLDENCHAIN. Fig. 286.

*L. anagryoides* Medikus.

A deciduous shrub or small tree, 15 to 25 feet high. Leaves pinnately trifoliolate, alternate, or apparently clustered on the ends of short branchlets; long-petioled; leaflets elliptic-ovate, 1 to 2 inches long, light green above, paler and slightly pubescent beneath, entire. Flowers bright yellow, pea-like, in pendulous racemes about 6 inches long. Fruit a 2-valved pod about 2 inches long.

Native to central and southern Europe. This species, with several garden varieties, is often cultivated in gardens for its large clusters of yellow flowers appearing in late spring and early summer.

## 12. Ólneya Gray

(Named for Stephen T. Olney, a botanist of Rhode Island.)
A single species.

**1. Olneya tesòta** Gray. DESERT IRONWOOD. Fig. 287.

A small spreading spiny tree, 15 to 25 feet high, with grayish foliage. Leaves once-pinnately compound, alternate; leaflets 10 to 16, cuneate-obovate, ¼- to ½-inch long, entire, sessile. Spines stout, slightly curved, ⅛- to ⅓-inch long, in pairs at the bases of the leaves. Flowers violet purple, pea-like, about ⅜-inch long, in axillary racemes ½-inch to 1½ inches long. Fruit-pods ½-inch to 3 inches long, 1- to 8-seeded, constricted between the seeds.

Native on the Colorado Desert of California, east to Arizona, and south to Mexico. Rarely cultivated.

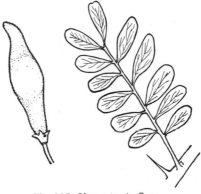

Fig. 287. Olneya tesota Gray.
Fruit, leaf, × ½.

## 13. Parkinsònia L.

(Named for James Parkinson, author of botanical treatises and herbalist to King James I.)

Often leafless shrubs or small trees. Leaves twice-pinnately compound, alternate or apparently fascicled. Flowers yellow, borne in racemes. Fruit a 2-valved pod.

Three species, two native in subtropical and tropical America and one in southern Africa.

### KEY TO THE SPECIES

Leaves 6 to 12 inches long; branches with spines, the branchlets not spinose-tipped........................................................1. *P. aculeata.*
Leaves about 1 inch long; branches without spines, the branchlets spinose-tipped............................................................2. *P. microphylla.*

**1. Parkinsonia aculeàta** L. PARKINSONIA. HORSE-BEAN. JERUSALEM-THORN. Fig. 288.

A small tree, 15 to 30 feet high, with sparse gray-green foliage and spines ½-inch to 1 inch long at the bases of the leaf-clusters. Leaves clustered on short branchlets, with 1 or 2 pairs of pinnae from a very short common petiole, or almost sessile and thus appearing as 2 once pinnate leaves, each pinna 6 to 12 inches long; leaflets 40 to 80, from flattened gray-green leaf-rachises, ovate or obovate, about ⅛-inch long, early deciduous. Flowers yellow, almost regular, ½- to ¾-inch wide, in slender erect racemes 3 to 6 inches long. Fruit-pods linear-cylindric, 2 to 6 inches long, much constricted between the seeds.

Native in the Bahamas, West Indies, Mexico, Panama, Arizona, and from Texas to Florida. Planted as a park and street tree in the desert towns of southern California.

**2. Parkinsonia microphýlla** Torr. MALE PALO VERDE. Fig. 289.

A large shrub or small tree, 5 to 25 feet high, with stiff spinose-tipped branchlets. Leaves twice-pinnately compound, with 1 or rarely 2 pairs of

Fig. 288. Parkinsonia aculeata L. Leaf, fruit, × ⅓.

pinnae from a very short petiole, or apparently sessile, appearing as once pinnate leaves; each pinna ½-inch to 1¼ inches long, with 8 to 16 elliptic entire leaflets about 1/16-inch long. Flowers pale yellow, almost regular, in loose racemes about 1 inch long, nearly sessile. Fruit-pods linear-cylindric, 1 to 3 inches long, 1- to 3-seeded, constricted between the seeds.

Native to Arizona, Mexico, and the Whipple Mountains of southeastern San Bernardino County, California.

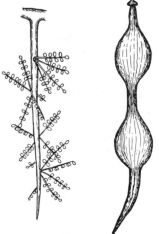

Fig. 289. Parkinsonia microphylla Torr. Portion of branchlet with leaves, × ½. Fruit, × 1.

**14. Parosèla** Cav. PEABUSH. DALEA

*Dalea* L.

(The Latin name. *Parosela* is an anagram of *Psoralea*, the two genera being very closely related.)

About 150 species of trees and shrubs, chiefly native to Mexico, but extending into South America and the southern part of the United States.

1. **Parosela spinòsa** Heller. SMOKE TREE. Fig. 290.

*Dalea spinosa* Gray.

An intricately branched nearly leafless shrub or small tree, 6 to 30 feet high, with yellowish green or ashy gray spinose branchlets sparsely dotted with glands. Leaves simple, alternate, early deciduous, linear-oblong to spatulate, ¼-inch to 1 inch long, $\frac{1}{16}$- to ⅛-inch wide, entire to irregularly dentate, almost sessile. Flowers dark blue or violet-purple, ⅓- to ½-inch long, in short spike-like racemes ½-inch to 1 inch long. Fruit-pods canescent and glandular-dotted, ¼- to ½-inch long, with 1 or 2 seeds.

Fig. 290. Parosela spinosa Heller. Branchlet with flowers, × 1.

Native on the Colorado Desert of California, east to Arizona, and south to Mexico. Rarely cultivated.

### 15. **Prosòpis** L. MESQUITE

(From the Greek *Prosopis*, the ancient name of the Burdock.)

Shrubs or trees, with spiny or thorny branches. Leaves twice-pinnately compound, alternate, deciduous. Flowers small, numerous, regular, sessile, in axillary pedunculate cylindrical spikes. Fruit an indehiscent many-seeded pod.

About 25 species, native in Africa and the desert regions of southwestern North America.

KEY TO THE SPECIES

Leaves 2 inches or more long; pinnae with 20 to 36 leaflets; pods straight or curved but not spirally coiled
1. *P. juliflora* var. *glandulosa*.
Leaves less than 2 inches long; pinnae with 10 to 22 leaflets; pods spirally coiled................................................2. *P. pubescens*.

1. **Prosopis juliflòra** var. **glandulòsa** Ckll. HONEY MESQUITE. Fig. 291.

Usually a much branched shrub or small tree, 6 to 20 feet high, with a short trunk dividing into many crooked branches. Thorns usually 1 or 2 in the axils of the leaves, ¼-inch to 1¼ inches long. Leaves usually with 2 pinnae 2½ to 4 inches long, each pinna with 20 to 36 linear entire leaflets ½-inch to 1 inch long; petioles slender, ¾-inch to 1½ inches long, enlarged and glandular at the base. Flowers greenish yellow, in slender cylindrical spikes 2 to 3½ inches long. Fruit-pods linear, 3 to 8 inches long, curved, flat or becoming thickened, irregularly constricted between the seeds.

Fig. 291. Prosopis juliflora
var. glandulosa Ckll.

Portion of branchlet with
leaves and fruit, × ½.

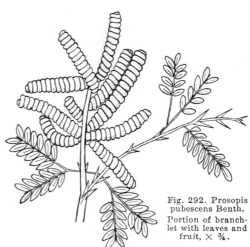

Fig. 292. Prosopis
pubescens Benth.

Portion of branch-
let with leaves and
fruit, × ¾.

Native on the Colorado and Mohave deserts of California, east to eastern Texas and southern Kansas, south to Lower California and Mexico. Also reported from Peru and Chile. The large fruit-pods are very nutritious and are used by the Indians and Mexicans for food. The flowers, blooming from April to June, are a source of an excellent honey, hence the name, Honey Mesquite.

2. **Prosopis pubéscens** Benth. SCREW-BEAN MESQUITE. Fig. 292.

*Strombocarpus pubescens* Gray.

A shrub or small tree, 10 to 30 feet high, with stout whitish stipular spines ¼- to ½-inch long. Leaves 1½ to 3 inches long, usually with 2 pinnae from a slender petiole about ½-inch long, each pinna 1 to 2 inches long, with 10 to 22 oblong leaflets ⅛- to ⁷⁄₁₆-inch long. Flowers yellowish, in slender cylindrical spikes 2 to 3 inches long. Fruit-pods coiled into a narrow cylindrical body 1 to 1½ inches long, borne in clusters of 2 to 15.

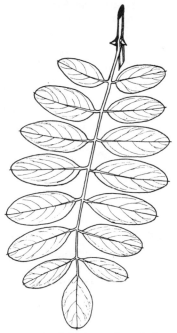

Native on the Colorado and Mohave deserts of southern California, east to Nevada, Arizona, New Mexico, and Texas, and south to Lower California and Mexico. The pods and early spring growth furnish food for livestock. The Indians and Mexicans grind the seeds into a meal used in baking.

### 16. Robínia L. Locust

(Named in honor of John and Vespasian Robin, who first cultivated the locust tree in Europe.)

Deciduous trees or shrubs, often with stipular spines. Leaves odd-pinnately compound, alternate, with stalked entire leaflets. Flowers pea-like, in long axillary drooping racemes. Fruit a flat several-seeded 2-valved dehiscent pod.

Fig. 293. Robinia pseudoacacia L. Leaf, × ⅓.

About 8 species, native in the United States and Mexico, 3 of which are arborescent. Scions from the pink-flowering shrubby species are grafted on stocks of the Black Locust.

Leaves, branchlets, and fruit glabrous.........................1. *R. pseudoacacia.*
Leaves, branchlets, and fruit pubescent or glandular-hispid.
    Peduncles and branchlets glandular-hispid; leaflets ¾-inch to 1¼
      inches long...................................................2. *R. neomexicana.*
    Peduncles and branches hispid, not glandular; leaflets 1 to 1¾ inches
      long.................................................3. *R. hispida* var. *macrophylla.*

**1. Robinia pseudoacàcia** L. BLACK LOCUST. Fig. 293.

A tall tree, 40 to 80 feet high, with erect or spreading usually prickly glabrous branches. Leaves 4 to 12 inches long, with 7 to 19 broadly elliptical glabrous leaflets 1 to 2 inches long. Stipular spines ½-inch to 1 inch long, persistent. Flowers white, very fragrant, about ¾-inch long, in drooping racemes 4 to 6 inches long. Fruit-pods 1 to 4 inches long, about ½-inch wide, flat, glabrous, 1- to 7-seeded.

Native in the central and eastern United States. Frequently cultivated as shade and avenue trees. Withstands poor soil and semidrought conditions. Flowers in May and June. The variety **bessoniàna** Nichols has slender unarmed branches. The variety **decaisneàna** Carr. has light rose-colored flowers.

Fig. 294.
Robinia neomexicana Gray.
Leaf, × ⅛.

**2. Robinia neomexicàna** Gray. HAIRY LOCUST. Fig. 294.

A large shrub or small tree, 15 to 25 feet high, with pubescent or glandular-hispid branchlets, foliage, and inflorescences. Leaves 5 to 10 inches long, with 11 to 21 elliptic-oblong pubescent leaflets ¾-inch to 1¼ inches long. Stipular spines ¼- to ½-inch long. Flowers rose-pink, about 1 inch long, in short compact glandular-hispid racemes 2 to 4 inches long. Fruit-pods 2 to 4 inches long, about ⅓-inch wide, glandular-hispid.

Native in Colorado, New Mexico, Arizona, and Nevada. Occasionally cultivated in parks and gardens for the rose-pink flowers which bloom in May and June. A closely related species, **R. viscosa** Vent., occurs in the southeastern United States.

**3. Robinia híspida** var. **macrophýlla** DC. ROSE-ACACIA. Fig. 295.

This variety is a large-leaflet form of the species **R. hispida** L. which grows as a shrub in the southeastern United States. In cultivation on the Pacific Coast, it is grown as a top-grafted specimen. The branches are "semiweeping" and bear large clusters of rose-colored flowers in the spring.

### 17. Sóphora L.

(An Arabian name for a tree with pea-shaped flowers.)

About 25 species of woody or rarely herbaceous plants, native in the temperate and subtropical regions in Asia and North America.

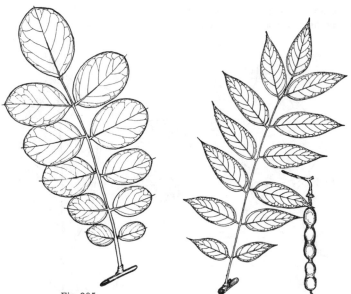

Fig. 295.
Robinia hispida var. macrophylla DC.
Leaf, × ⅓.

Fig. 296. Sophora japonica L.
Leaf, fruit, × ⅓.

1. **Sophora japónica** L. JAPANESE PAGODA TREE. CHINESE SCHOLAR TREE. Fig. 296.

A deciduous tree, 20 to 40 feet high, with spreading green branches and dark green foliage. Leaves odd-pinnately compound, alternate, 6 to 10 inches long; leaflets 7 to 17, ovate-lanceolate, 1 to 2 inches long, dark green above, paler and pubescent beneath, entire, short-stalked. Flowers yellowish white, pea-like, about ½-inch long, borne in loose compound clusters 8 to 12 inches long. Fruit-pods linear, 2 to 3 inches long, almost round in cross section, 1- to 6-seeded.

Native to China. Occasionally cultivated for its dark green foliage and large clusters of flowers blooming in the summer. A form with long pendulous branches, variety **péndula** Loud., is sometimes grown as specimen tree.

## Rutaceae. Rue Family
### 1. Cítrus L.

(An ancient name of a fragrant African wood, but later transferred to the citron.)

Small or medium-sized evergreen trees, with spreading sometimes spiny branches and dense dark green foliage. Leaves unifoliolately

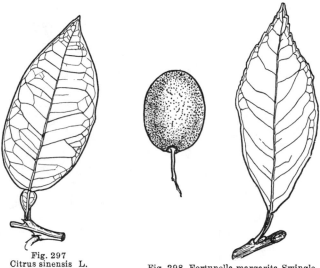

Fig. 297
Citrus sinensis L.
Leaf, × ½.

Fig. 298. Fortunella margarita Swingle.
Leaf, fruit, × ¾.

compound but apparently simple, alternate, glandular-dotted, with the petioles usually narrowly winged and jointed at the apex. Flowers white or pinkish, borne in clusters or rarely solitary, fragrant. Fruit a hesperidium of several segments filled with juicy pulp.

About 6 species, native to tropical and subtropical Asia and the Malayan Archipelago. Many varieties and hybrids are cultivated in California for their edible fruits and evergreen foliage. These include *C. aurantifolia,* Lime, *C. aurantium,* Sour Orange, *C. limonia,* Lemon, *C. nobilis* var. *deliciosa,* Tangerine Orange, *C. grandis,* Grapefruit, and *C. sinensis,* Sweet Orange, fig. 297.

## 2. **Fortunélla** Swingle. Kumquat

(Named in honor of Robert Fortune, who introduced the first kumquat into Europe in 1846.)

About 4 species, native to eastern Asia.

Evergreen shrubs or small trees. Leaves unifoliolately compound but apparently simple, alternate, glandular-dotted beneath. Flowers white, solitary or in few-flowered clusters. Fruit a small hesperidium with juicy pulp.

This genus can be distinguished from *Citrus* by its smaller fruits and fewer cells in the ovary. Two species are cultivated in California for their ornamental fruits, which are sometimes eaten fresh or preserved. **F. margaríta** Swingle, Oval Kumquat, fig. 298, is a shrub or small tree, with ovoid or ellipsoidal yellow-orange fruit 1 to 1½ inches long. **F. japónica** Swingle, Round Kumquat, is a much branched shrub, with globular bright orange-colored fruit about 1 inch in diameter.

## 3. **Ptèlea** L. Hop Tree

(The Greek name of the elm, in reference to the similar fruit.)

About 8 closely related species, native to North America.

### 1. **Ptelea báldwini** var. **crenulàta** Jepson. Hop Tree. Fig. 299.

A small straggling deciduous tree or shrub, 10 to 20 feet high, with aromatic foliage. Leaves 3-foliolate, alternate, long-petioled; leaflets ovate to oblong-elliptic, 1½ to 3 inches long, entire or crenulate-serrate, dark green above, paler beneath, glandular-dotted, sessile. Flowers small, greenish white, in axillary clusters. Fruit a flattened 2-seeded samara, orbicular, ½- to ¾-inch in diameter.

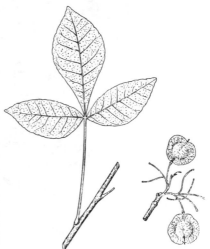

Fig. 299.
Ptelea baldwini var. crenulata Jepson.
Leaf, fruit, × ½.

Native to California in the canyons of the lower foothills of the inner North Coast Range from Mount Diablo north to Shasta County, south in the Sierra Nevada foothills to Fresno County.

**Ptelea trifoliàta** L., Hop Tree. Leaflets 3 to 5 inches long. Fruit about 1 inch long. Native in the eastern United States. Occasionally cultivated.

## Simarubaceae. Quassia Family

### 1. Ailánthus Desf.

(From the Chinese name *Ailanto*, the Tree of Heaven.)

Six to 9 species, native in central and southern Asia, northern Australia, and the East Indies.

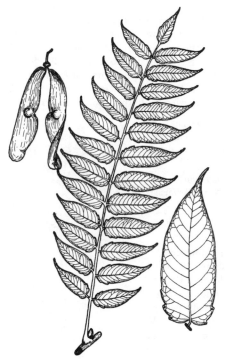

Fig. 300. Ailanthus glandulosa Desf.
Fruit, × ⅔. Leaf, × ⅙. Leaflet, × ⅓.

### 1. Ailanthus glandulòsa Desf. Tree of Heaven. Fig. 300.

*A. altissima* Swingle.

A tall slender sparingly branched deciduous tree, 30 to 70 feet high, with light green foliage having a very disagreeable odor. Leaves odd-pinnately compound, alternate, 1 to 3 feet long, long-petioled; leaflets

11 to 31, ovate or ovate-lanceolate, 3 to 5 inches long, 1 to 2 inches wide, entire except for 1 to 4 blunt glandular teeth near the base. Flowers small, greenish white, in large terminal clusters. Fruit a samara about 1½ inches long, with the seed in the middle.

Native to China. Much planted by early settlers, especially in the Sierra Nevada foothills, where it has become naturalized along many water courses. The Tree of Heaven should not be planted as an ornamental because of the disagreeable odor of its foliage and its habit of rapidly becoming a tree weed. Flowers in the summer.

## Meliaceae. Mahogany Family
### 1. Cedrèla L.

(From *kedros*, in reference to the resemblance of the wood to the cedar.)

About 18 species, native in tropical America, southeastern Asia, and Australia.

1. **Cedrela sinénsis** Juss. CHINESE CEDRELA. Fig. 301.

A deciduous tree, 15 to 40 feet high. Leaves even-pinnately compound, alternate, 10 to 20 inches long; leaflets 10 to 22, oblong to oblong-lanceolate, 3½ to 7 inches long, 1¼ to 2 inches wide, finely serrate to entire. Flowers small, white, fragrant, in drooping panicles 8 to 12 inches long. Fruit a woody capsule about 1 inch long; seeds winged.

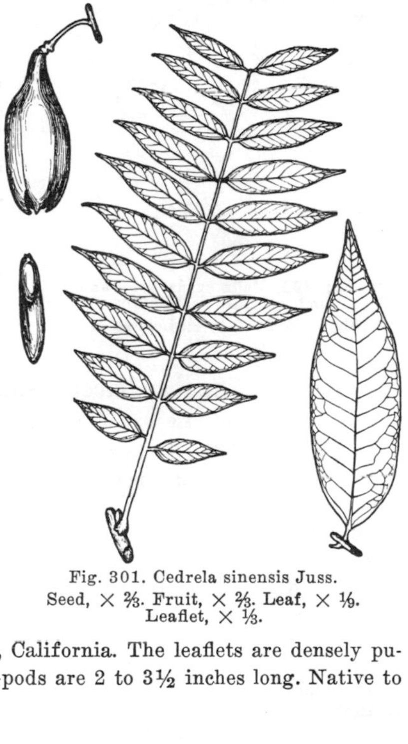

Native to China. Occasionally cultivated in parks and gardens for its feathery foliage and large fragrant flower-clusters. Flowers in June. A few fine specimens of **Cedrela físsilis** Vell. are cultivated as

Fig. 301. Cedrela sinensis Juss. Seed, × ⅔. Fruit, × ⅔. Leaf, × ⅙. Leaflet, × ⅓.

streot trees at Santa Barbara, California. The leaflets are densely pubescent beneath and the fruit-pods are 2 to 3½ inches long. Native to Brazil and Paraguay.

## 2. Mèlia L. Bead Tree

(From an ancient Greek name.)

About 10 species, native to Australia and southeastern Asia.

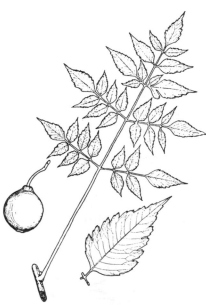

Fig. 302. Melia azedarach var. umbraculiformis Berckm. Fruit, × ⅔. Leaf, × ⅙. Leaflet, × ⅔.

**1. Melia azédarach var. umbraculifórmis** Berckm. Texas Umbrella Tree. Fig. 302.

A deciduous tree, 25 to 40 feet high, with ascending branches forming a flattened crown resembling an umbrella. Leaves twice-pinnately compound, alternate, 1 to 3 feet long; leaflets ovate to elliptic, 1 to 2½ inches long, ½-inch to 1 inch wide, sharply toothed or rarely lobed. Flowers purplish, about ¾-inch broad, borne in loose clusters 4 to 8 inches long. Fruit a subglobose yellowish drupe about ½-inch wide.

This variety, according to M. B. Coulston in Bailey's *Standard Cyclopedia of Horticulture*, originated probably in Texas. It is extensively cultivated as a street and shade tree for its compact umbrella-shaped crown.

## Anacardiaceae. Sumac Family
### 1. Schìnus L.

(From *Schinos*, the Greek name for the Mastic Tree, in reference to the mastic-like juices of some species.)

Evergreen trees. Leaves simple or compound, alternate. Flowers small, whitish, dioecious, in drooping panicles or racemes. Fruit a globose drupe.

About 15 species, chiefly native to South America, one to the Hawaiian Islands, one in Jamaica, and one on St. Helena Island.

KEY TO THE SPECIES

Leaflets 20 to 60................................................................1. *S. molle.*
Leaflets 7................................................................2. *S. terebinthifolius.*

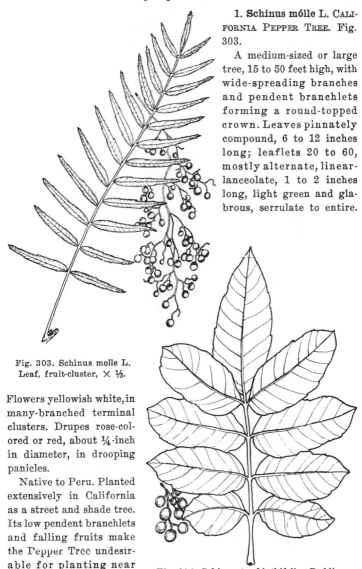

**1. Schinus mólle** L. CALI-
FORNIA PEPPER TREE. Fig.
303.

A medium-sized or large
tree, 15 to 50 feet high, with
wide-spreading branches
and pendent branchlets
forming a round-topped
crown. Leaves pinnately
compound, 6 to 12 inches
long; leaflets 20 to 60,
mostly alternate, linear-
lanceolate, 1 to 2 inches
long, light green and glabrous, serrulate to entire.

Fig. 303. Schinus molle L.
Leaf, fruit-cluster, × ⅛.

Flowers yellowish white, in
many-branched terminal
clusters. Drupes rose-colored or red, about ¼-inch
in diameter, in drooping
panicles.

Native to Peru. Planted
extensively in California
as a street and shade tree.
Its low pendent branchlets
and falling fruits make
the Pepper Tree undesir-
able for planting near
sidewalks.

Fig. 304. Schinus terebinthifolius Raddi.
Fruit-cluster, leaf, × ½.

**2. Schinus terebinthifòlius** Raddi. Brazilian Pepper Tree. Fig. 304.

An evergreen tree, 15 to 30 feet high, with a more rigid habit of growth than that of *S. molle*. Leaves once-pinnately compound, 6 to 8 inches long; leaflets 7, oblong, about 1 inch long, dark green above, paler beneath, serrate to almost entire. Flowers whitish, in racemes. Drupes bright red, about ⅛-inch in diameter.

Native to Brazil. Occasionally planted as a shade tree from San Diego to San Francisco, California.

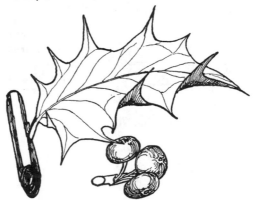

Fig. 305. Ilex aquifolium L.
Leaf, fruit, × 1.

## Aquifoliaceae. Holly Family

### 1. Ìlex L. Holly

(Ancient Latin name of *Quercus ilex*, in reference to the similar leaves.)

About 300 species of shrubs and trees, native in temperate and tropical regions.

**1. Ilex aquifòlium** L. English Holly. Fig. 305.

A small evergreen tree, 8 to 25 feet high in cultivation but to 70 feet in its native habitat. Leaves simple, alternate; the blades very thick and leathery, ovate to oblong-elliptic, 1½ to 3 inches long, 1 to 1½ inches wide, glabrous, dark green and glossy above, lighter beneath, the margins wavy and with large triangular spiny teeth or almost entire on older branches; petioles about ¼-inch long. Flowers white, fragrant, about ¼-inch wide, in axillary clusters or rarely solitary. Fruit globose, about ⅓-inch in diameter, berry-like, with 2 to 8 bony nutlets.

Native in southern and central Europe, western Asia, and China. Cultivated as a specimen tree for its dark green glossy foliage and clusters of dark red fruits persisting through the winter. Several garden forms with variations in the margins and color of leaves are in cultivation.

## Celastraceae. Staff Tree Family
### 1. Maytènus Feuill.

(A Chilean name.)

About 70 species of trees and shrubs, native in tropical and temperate South America and the West Indies.

**1. Maytenus boària** Molina. MAYTEN. Fig. 306.

Fig. 306.
Maytenus boaria Molina.
Fruit-cluster, leaf, × 1.

A small evergreen tree, 20 to 30 feet high (to 100 feet in Chile), with long pendulous branchlets. Leaves simple, alternate; the blades thin, lanceolate to ovate-lanceolate, 1 to 2 inches long, ¼- to ½-inch wide, glabrous, glandular-serrate; petioles ⅛-inch or less long. Flowers small, greenish white, in axillary clusters. Fruit a small 1- (rarely 2-) seeded capsule.

Native to Chile. Cultivated in parks and sometimes as a street tree because of its long pendulous branchlets.

## Aceraceae. Maple Family
### 1. Acer L. MAPLE

(Ancient Latin name of the maple.)

Deciduous trees or shrubs. Leaves opposite, usually simple and palmately veined, or pinnately 3- to 5-foliolate. Flowers small, dioecious or polygamous, in drooping clusters. Fruit a double samara with terminal wings.

About 120 species, native in North America, Asia, Europe, and northern Africa.

#### KEY TO THE SPECIES

Leaves compound; petals wanting..........................................1. *A. negundo.*
Leaves simple; petals present except in *A. saccharum* and *A. dasycarpum.*
    Under surface of leaves grayish or white-glaucescent.
        Leaves deeply 5-lobed to 5-cleft, 3 to 6 inches wide, the lobes
        deeply and irregularly toothed; petals wanting
                2. *A. dasycarpum.*

Leaves 3- or 5-lobed, the lobes coarsely crenate-serrate or serrate to entire.

Leaves 1½ to 2¾ inches wide, 3-lobed, the lobes entire or seldom irregularly serrate..............................3. *A. buergerianum.*

Leaves 3 to 7 inches wide.

Lobes crenate or serrate; petals distinct.

Lobes crenate-serrate, broadly acute or rounded at apex; leaves 3½ to 7 inches wide..............4. *A. pseudoplatanus.*

Lobes sharply serrate, acute to short-acuminate; leaves 2 to 4½ inches wide..........................................5. *A. rubrum.*

Lobes wavy-margined to sparsely toothed, acute to long-acuminate; petals wanting...............................6. *A. saccharum.*

Under surface of leaves green, often paler than above.

Leaves not lobed....................................................................7. *A. tataricum.*

Leaves lobed.

Leaves 5- to 11-lobed or parted, the lobes doubly serrate.

Leaves deeply 5- to 9-lobed or parted, 2 to 4 inches wide

8. *A. palmatum.*

Leaves 7- to 11-lobed.

Samara-wings in nearly a straight line; native species

9. *A. circinatum.*

Samara-wings spreading at an obtuse angle or nearly parallel; introduced species.............................10. *A. japonicum.*

Leaves 3- to 7-lobed, the lobes coarsely toothed or again lobed, or almost entire.

Leaves 1 to 3 (rarely 4) inches wide.

Leaves shallowly 3-lobed near apex, the lobes pointed, the margins entire along the lower half..3. *A. buergerianum.*

Leaves 3- to 5-lobed.

Lobes rounded, not serrate.........................11. *A. campestre.*

Lobes acute or acuminate, doubly serrate....12. *A. ginnala.*

Leaves 3½ to 12 (or 18) inches wide.

Leaves 6 to 12 (or 18) inches wide, deeply 3- to 5-lobed or cleft.................................................13. *A. macrophyllum.*

Leaves 3½ to 8 inches wide.

Leaves bright green and shining beneath, 5- to 7-lobed; juice from petiole-base milky; petals present

14. *A. platanoides.*

Leaves pale beneath, 3- to 5-lobed; juice from petiole-base not milky; petals wanting...................6. *A. saccharum.*

## 1. Acer negúndo L. Box-Elder.

A deciduous tree, 50 to 70 feet high, with pale green glabrous branchlets. Leaves pinnately compound; leaflets 3 or 5 (rarely 7 or 9), ovate to elliptic or obovate, 2 to 4 inches long, 1½ to 2½ inches wide, acuminate, rounded and often unequal at the base, bright green and nearly glabrous above, paler and slightly pubescent beneath, often with axillary tufts of hairs, coarsely and irregularly serrate above the middle, or almost entire, or the terminal one 3-lobed; petioles

2 to 3 inches long, glabrous. Flowers dioecious, apetalous, appearing before the leaves; the staminate fascicled, on slender pedicels; the pistillate in slender drooping racemes. Samaras 1½ to 2 inches long, glabrous, in drooping racemes 6 to 8 inches long, the wings diverging at an acute angle and usually incurved.

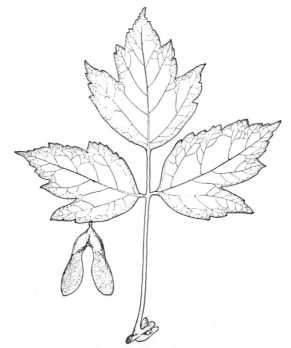

Fig. 307. Acer negundo var. californicum Sarg.
Fruit, × ⅔. Leaf, × ⅓.

Native in the eastern and midwestern United States. Occasionally planted as a shade tree in parks and along streets.

1*a*. Var. **califórnicum** Sarg. CALIFORNIA BOX-ELDER. Fig. 307.

A small or medium-sized tree, 20 to 40 feet high, with hoary-tomentose branchlets. Leaflets usually 3, the lateral pair sometimes divided, oblong-ovate, 2 to 4 inches long, glabrous on the upper surface except along the midrib and veins, closely and densely pubescent beneath, coarsely toothed above the middle or nearly entire. Samaras finely pubescent or rarely glabrous, the wings very slightly diverging.

Native to California, only occasional along stream banks and valley bottoms in the Coast Ranges, the Sierra Nevada foothills, the lower Sacramento Valley, the San Joaquin Valley, and in elevated canyons on the southern slopes of the San Bernardino Mountains. Extensively cultivated as a street tree and in parks and gardens throughout the Pacific Coast.

Fig. 308. Acẹr dasycarpum Ehrh.
Leaf, × ½.

1*b*. Var. **variegàtum** Carr. Variegated Box-Elder.

*A. negundo* var. *argentea-variegatum* Bonamy.

Leaflets with broad white margins and splotches.

Commonly cultivated in the cities of Oregon and Washington as street and specimen trees.

2. **Acer dasycárpum** Ehrh. Silver Maple. Fig. 308.

*A. saccharinum* L.

A large wide-spreading tree, 60 to 100 feet high. Leaf-blades 6 to 7 inches long and nearly as wide, cordate, bright green above, silvery white beneath, deeply 5-lobed, the lobes acuminate and irregularly double-serrate, the middle lobe often 3-lobed; petioles 3 to 4¾ inches long. Flowers greenish, apetalous, on short pedicels in axillary clusters, appearing before the leaves. Samaras 1½ to 2½ inches long, glabrous, the wings very divergent and falcate.

Native in the central and eastern parts of the United States. A rapid-growing tree often with pendulous branches. Commonly cultivated as a park and street tree. Several garden forms varying in habit of growth and lobing of the leaves are cultivated.

3. **Acer buergeriànum** Miq. TRIDENT MAPLE. Fig. 309.

*A. trifidum* H. & A.

A small tree, 10 to 25 feet high, with glabrous branchlets. Leaf-blades 1¾ to 3 inches long and as wide, rounded or widely tapering at the base, dark green and glabrous on the upper surface, pale green and glaucescent beneath, 3-lobed near the apex, the lobes triangular, acute, pointing forward, entire on the lower half, entire to finely and irregularly serrate on the upper half; petioles 1½ to 2½ inches long. Flowers yellowish, in broad pubescent erect panicles. Samaras 1¾ to 2 inches long, glabrous, the wings parallel.

Fig. 309. Acer buergerianum Miq. Leaf, fruit, × ½.

Native to Japan and China. Occasionally cultivated in parks and lawns.

4. **Acer pseudoplátanus** L. SYCAMORE MAPLE. Fig. 310.

A rather large spreading tree, 30 to 90 feet high. Leaf-blades 3 to 7 inches long and about as wide, cordate, dark green and glabrous above, glaucescent or puberulent beneath, 5-lobed, the lobes ovate, coarsely and crenately serrate; petioles 3 to 4 inches long. Flowers yellowish green, in long pendulous panicles 3 to 6 inches long, appearing after the leaves. Samaras 1¼ to 2 inches long, glabrous, the wings spreading at an acute to a right angle.

Native in Europe and western Asia. Planted as a park or street tree. Many horticultural forms, varying in pubescence, color, and lobing of the leaves, are sold by nurserymen.

**5. Acer rùbrum** L. RED MAPLE. SCARLET MAPLE. Fig. 311.

A tree, 30 to 100 feet high, with reddish branches. Leaf-blades 2 to 4 inches long and about as wide, subcordate, dark green and shining above, glaucescent or more or less pubescent beneath, especially along the veins, 3- or 5-lobed, the lobes triangular-ovate and unequally

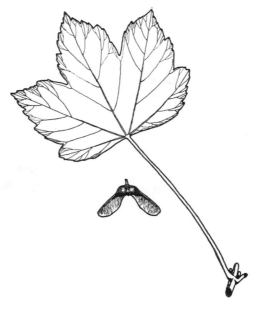

Fig. 310. Acer pseudoplatanus L.
Fruit, leaf, × ⅛.

toothed; petioles 2 to 4 inches long, often red. Flowers bright scarlet varying to yellowish red or yellow, in close clusters along the branches, appearing before the leaves. Samaras 2¼ to 4 inches long, glabrous, usually bright red when young, the wings diverging at an acute angle, or at maturity at nearly a right angle.

One of the commonest and most generally distributed trees native in eastern North America. Planted along streets and in parks. Several varieties occur in nature and are sometimes cultivated with the species.

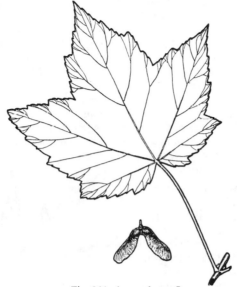

Fig. 311. Acer rubrum L.
Fruit, leaf, × ½.

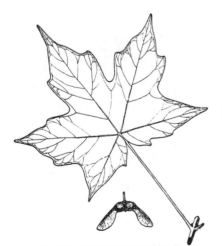

Fig. 312. Acer saccharum Marsh.
Fruit, leaf, × ⅓.

Fig. 313. Acer tataricum L.
Fruit, leaf, × ½.

6. **Acer sáccharum** Marsh. SUGAR MAPLE. ROCK MAPLE. Fig. 312.

*A. saccharinum* Wang., not L.

A large tree, 60 to 100 feet high. Leaf-blades 3 to 6 inches long and as wide, truncate or cordate at the base, dark green and glabrous above, pale or light green beneath, 3- or 5-lobed, the lobes acuminate and sparingly toothed; petioles 1½ to 3 inches long. Flowers greenish yellow, apetalous, on hairy pedicels in nearly sessile umbel-like corymbs, appearing with the leaves. Samaras ¾-inch to 1¼ inches long, glabrous, the wings widely diverging.

Native in southeastern Canada and the north central and eastern United States. Cultivated as a street, park, and lawn tree. A variety **rùgeli** Rehd., with 3 entire lobes, is sometimes cultivated.

7. **Acer tatáricum** L. TATARIAN MAPLE. Fig. 313.

A large shrub or small tree, 10 to 20 feet high. Leaf-blades broadly ovate to oblong-ovate, 1½ to 3½ inches long, 1¼ to 2 inches wide, cordate at the base, nearly glabrous, dark green above, slightly paler beneath, doubly serrate, sometimes slightly lobed; petioles ½-inch to 1¼ inches long. Flowers white or greenish yellow, in long-peduncled panicles, appearing after the leaves. Samaras about 1 inch long, glabrous, the wings nearly parallel, bright red in summer.

Native in southeastern Europe and western Asia. Occasionally cultivated in parks and gardens.

Fig. 314. Acer palmatum Thunb.
Fruit, leaf, × ¾.

8. **Acer palmàtum** Thunb. JAPANESE MAPLE. Fig. 314.

*A. polymorphum* Sieb. & Zucc.

A small tree, 10 to 25 feet high, with slender branches and glabrous branchlets, petioles, and peduncles. Leaf-blades 2 to 4 inches long and as wide, subcordate, green and glabrous on both surfaces, deeply 5- to 9-lobed or parted, the lobes lanceolate to lance-ovate and doubly serrate or incised; petioles ⅝-inch to 1¾ inches long. The flowers purple, in glabrous erect corymbs that appear after the leaves. The samaras about ¾-inch long, glabrous, the wings widely diverging.

Native to Japan and Korea. Numerous forms varying in color and in leaf-dissection are in cultivation in gardens and parks. All are small trees of graceful habit. In early spring the foliage is usually a reddish bronze, gradually changing to green and tinged with bronze as the season progresses. The variety *disséctum* of nurseries has deeply dissected fern-like foliage which remains reddish bronze during most of the spring and summer.

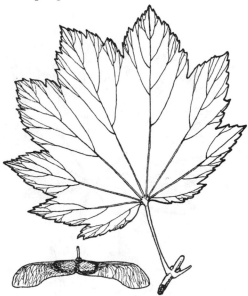

Fig. 315. Acer circinatum Pursh.
Fruit, leaf, × ¾.

**9. Acer circinàtum** Pursh. VINE MAPLE. Fig. 315.

A shrub or small tree, 5 to 35 feet high, or often vine-like or nearly prostrate. Leaf-blades thin, 2 to 6 inches long and about as broad, cordate, glabrous, light or dark green above, paler beneath, 7- to 11-lobed, the lobes triangular-ovate, acute and sharply serrate, the basal pair of lobes smaller; petioles 1 to 2 inches long, grooved. Flowers reddish purple, in peduncled umbel-like corymbs, appearing with the leaves. Samaras glabrous, the wings ½- to ¾-inch long, spreading in nearly a straight line, reddish when ripe.

Native from the coast of British Columbia southward in western Washington and Oregon to northern California. Vine Maple is a very handsome tree or shrub because of its delicate light green foliage which turns orange and red in the autumn, reddish flowers, and orange or reddish fruits. Rarely cultivated.

10. **Acer japónicum** Thunb. FULLMOON MAPLE. Fig. 316.

A small tree or shrub, 6 to 20 feet high. Leaf-blades 2 to 5 inches long and as wide, cordate, light green and glabrous except along the

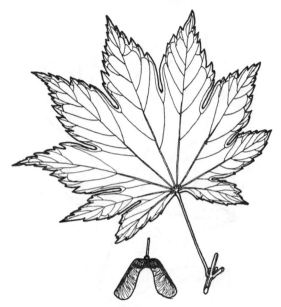

Fig. 316. Acer japonicum Thunb.
Fruit, leaf, × ¾.

veins beneath, 7- to 11-lobed, the lobes acuminate and doubly serrate; petioles ¾-inch to 1¾ inches long. Flowers purple, in drooping corymbs, appearing after the leaves. Samaras 1¾ to 2 inches long, glabrous at maturity, the wings widely diverging.

Native to Japan. Cultivated in parks and lawns. The variety **áureum** Schwerin has yellow leaves. The variety **pársonsi** Veitch has leaves divided nearly to the base into 9 to 11 pinnately dissected lobes.

11. **Acer campéstre** L. HEDGE MAPLE. ENGLISH CORKBARK MAPLE. Fig. 317.

A small round-headed tree, 15 to 25 feet high, with somewhat corky fissured bark. Leaf-blades 1 to 4 inches long and almost as wide, dull green and glabrous above, pubescent or rarely glabrous beneath, deeply 3- to 5-lobed, the lobes obtuse and entire or the middle one slightly 3-lobed; petioles 1 to 3 inches long. Flowers greenish, in loose erect pubescent corymbs. Samaras 1 to 1¼ inches long, pubescent, the wings spreading horizontally in nearly a straight line.

Native to Europe and western Asia. Occasionally cultivated in lawns and parks.

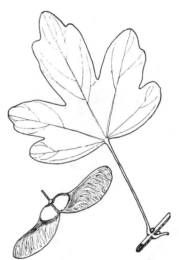

Fig. 317. Acer campestre L.
Fruit, leaf, × ½.

Fig. 318.
Acer ginnala Maxim.
Fruit, leaf, × ½.

12. **Acer ginnàla** Maxim. AMUR MAPLE. Fig. 318.

*A. tataricum* var. *ginnala* Maxim.

A shrub or small tree, 6 to 20 feet high. Leaf blades 1½ to 3½ inches long, 1¼ to 2½ inches wide, subcordate at the base, glabrous, dark green and shining above, light green and shining beneath, 3-lobed, the lobes doubly serrate, the terminal broadest and longest; petioles 1 to 1½ inches long. Flowers yellowish, in long-peduncled clusters. The samaras approximately 1 inch long, the wings spreading at an acute angle or very nearly parallel.

Native to Manchuria, north China, and Japan. Cultivated in gardens and parks.

13. **Acer macrophýllum** Pursh. BIGLEAF MAPLE. OREGON MAPLE. Fig. 319.

A medium-sized or large tree, 30 to 100 feet high, with spreading branches forming a broad crown. Leaf-blades 8 to 14 (or 18) inches long and almost as wide, cordate, dark green and slightly pubescent

Fig. 319. Acer macrophyllum Pursh.
Leaf, fruit, × ⅛.

above, pale green beneath, deeply 3- to 5-lobed, the lobes 2- to 4-toothed or almost entire; petioles 6 to 12 inches long. Flowers yellowish green, in drooping racemes 4 to 6 inches long, the staminate and perfect flowers in the same raceme, appearing after the leaves. Samaras 1¾ to 2 inches long, the body covered with dense short yellow or brown hairs, the wings nearly glabrous and slightly divergent.

Native on the Pacific Coast from Alaska to California. Planted as a street tree and in parks.

**14. Acer platanoides** L. NORWAY MAPLE. Fig. 320.

A medium-sized round-headed tree, 25 to 60 (or 80) feet high. Leaf-blades thin, 3 to 6 inches long and as wide, cordate, light green and glabrous on both surfaces, 5- or rarely 7-lobed, the lobes acute, with few sharp coarse teeth; petioles 2½ to 3½ inches long, with milky juice.

Flowers greenish white, in erect corymbs, appearing a little after the leaves. Samaras 1½ to 2 inches long, glabrous, the wings diverging in nearly a straight line.

Native to Europe and western Asia. Frequently cultivated as a street and lawn tree. The leaves resemble somewhat those of the sycamore.

Fig. 320. Acer platanoides L.
Fruit, leaf, × ⅓.

## Hippocastanaceae

Horse-Chestnut Family

**1. Aésculus** L. HORSE-CHESTNUT. BUCKEYE

(A name given by Pliny to an oak having edible acorns.)

Deciduous trees or shrubs. Leaves palmately compound, opposite. Flowers large, irregular, in terminal panicles. Fruit a large 3-valved capsule with 1 or 2 polished seeds.

About 15 species, native to America, southeastern Europe, and Asia.

KEY TO THE SPECIES

Leaflets with stalks ⅜-inch to 1 inch long.........................1. *A. californica.*
Leaflets sessile or with stalks ¼-inch or less long.
  Leaflets 5 to 7; flowers white...................................2. *A. hippocastanum.*
  Leaflets mostly 5; flowers pink or scarlet............................3. *A. carnea.*

**1. Aesculus califórnica** Nutt. CALIFORNIA BUCKEYE. Fig. 321.

A tree, 15 to 40 feet high, with smooth gray bark and a broad open crown, or sometimes shrub-like. Leaves long-petioled, early deciduous; leaflets usually 5, oblong-lanceolate, 3 to 6 inches long, 1½ to 2 inches wide, dark green above, paler beneath, nearly glabrous, serrate, on stalks ⅜-inch to 1 inch long. Flowers pinkish white, about ½-inch long, in large cylindrical erect clusters 6 to 10 inches long. Capsules somewhat pear-shaped, 1½ to 2½ inches long, with 1 or 2 large glossy brown seeds.

Native on canyon slopes and low dry hills of the Sierra Nevada and the Coast Ranges of California.

### 2. Aesculus hippocástanum L. COMMON HORSE-CHESTNUT.

A medium-sized to large tree, 25 to 60 feet high, with wide-spreading branches forming a rounded crown. Leaves long-petioled; leaflets 5 to 7, oblong-obovate, 4 to 10 inches long, 2 to 3½ inches wide, dark green and glabrous above, rusty-tomentose near the base on the lower

Fig. 321. Aesculus californica Nutt.
Leaf, × ¼. Fruit, × ½.

surface when young, irregularly and bluntly serrate, sessile. Flowers creamy white, spotted with yellow and red, about ¾-inch long, in large showy upright panicles 8 to 12 inches long. Capsules globose, about 2½ inches in diameter, prickly, with 1 or 2 large chestnut-colored seeds.

Native to Asia, but escaped from cultivation in the eastern part of the United States. Cultivated as a street and park tree.

### 3. Aesculus cárnea Hayne. RED HORSE-CHESTNUT. Fig. 322.

Leaflets usually 5, almost sessile. Flowers light to dark red, in panicles 5 to 8 inches long. Fruit subglobose, 1 to 1½ inches long, slightly prickly.

A hybrid between *A. hippocastanum* L. and *A. pavia* L. Cultivated in parks and as a street tree.

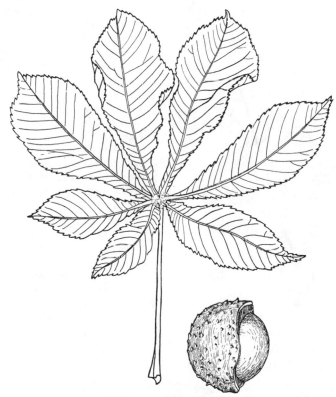

Fig. 322. Aesculus carnea Hayne.
Leaf, × ¼. Fruit, × ½.

## Sapindaceae. Soapberry Family
### 1. Koelreutèria Laxm.

(Named for Joseph G. Koelreuter, Professor of Natural History at Karlsruhe.)

About 5 species of trees, native to China and Japan.

1. **Koelreutèria paniculàta** Laxm. GOLDENRAIN TREE. Fig. 323.

A small deciduous tree, 10 to 30 feet high. Leaves odd-pinnately or sometimes bipinnately compound, alternate, 8 to 14 inches long;

Fig. 323.
Koelreuteria paniculata Laxm.
Leaf, × ⅛.

leaflets 7 to 15, ovate to oblong-ovate, 1¼ to 3 inches long, ¾-inch to 1¾ inches wide, dark green and glabrous above, paler and puberulent beneath, irregularly crenate-serrate to lobed. Flowers yellow, about ½-inch long, in broad loose clusters 8 to 14 inches long. Fruit an inflated papery-walled capsule, oblong-ovoid, 1 to 2 inches long, with 2 or 3 globular black seeds.

Native to China, Korea, and Japan. Cultivated in gardens and parks for its large clusters of yellow flowers blooming in late summer.

### 2. **Sapíndus** L. SOAPBERRY

(From the Latin *sapo*, soap, and *Indus*, in reference to the soapy nature of the berries and the West Indian habitat of the first species that was known to the Europeans.

About 15 species of trees or shrubs, native in subtropical and tropical regions.

### 1. **Sapindus saponària** L. SOUTHERN SOAPBERRY. Fig. 324.

A small evergreen tree, 15 to 30 feet high. Leaves pinnately compound, alternate, 6 to 7 inches long; leaflets 4 to 9, oblong-lanceolate, 3 to 4 inches long, 1 to 1½ inches wide, glabrous

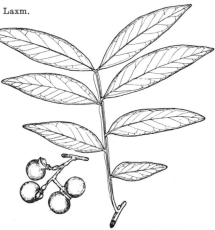

Fig. 324. Sapindus saponaria L.
Fruit, leaf, × ⅓.

above, tomentulose beneath, entire, sessile or short-stalked, the rachis usually narrowly winged. Flowers small, white, in large terminal panicles 6 to 10 inches long. Fruit berry-like, globose, about ¾-inch in diameter, orange-brown.

Native to Florida, the West Indies, and South America. Occasionally cultivated in southern California.

## Rhamnaceae. Buckthorn Family

### 1. Rhámnus L. BUCKTHORN

(From *Rhamnos*, the ancient Greek name of the buckthorn.)

About 100 species, native in the temperate and warm regions of the northern hemisphere. Five species with several varieties are native on the Pacific Coast. One of them is often a small tree.

1. **Rhamnus purshiàna** DC. CASCARA BUCKTHORN. CASCARA SAGRADA. Fig. 325.

A tall shrub or small tree, 10 to 35 feet high, with a trunk 4 to 12 inches in diameter. Leaves simple, alternate, deciduous; the blades oblong-elliptic, 2 to 6 (or 8) inches long, ¾-inch to 2½ inches wide, dark green and glabrous above, paler beneath, brownish tomentulose on the midrib and veins, entire to finely serrulate; petioles ⅜- to ¾-inch long, brownish tomentulose. Flowers small, greenish white, borne in umbellate axillary clusters. Fruit globose, ¼- to ½-inch in diameter, black, berry-like, with 3 or rarely 2 nutlets.

Fig. 325.
Rhamnus purshiana DC.
Leaf, fruit, × ½.

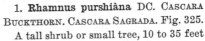

Fig. 326.
Zizyphus jujuba Mill.
Leaf, fruit, × ⅓.

Native from British Columbia south to Washington, Oregon, and northern California, east to Idaho and Montana. Cultivated as an ornamental tree, especially in Oregon and Washington. The bark is extensively used in drug manufacture, and its collection in Oregon and Washington constitutes a local industry.

## 2. Zízyphus Mill.

(From *Zizyphon*, ancient Greek name of *Z. jujuba*.)

About 40 species of shrubs and trees, native in tropical and sub-tropical regions.

**1. Zizyphus jujùba** Mill. COMMON JUJUBE. Fig. 326.

*Z. sativa* Gaertn.

A deciduous shrub or small tree, 10 to 30 feet high. Leaves simple, alternate, with stipular spines; the blades oblong-ovate to ovate, ¾-inch to 2½ inches long, 3-veined from the base, glabrous, entire to crenate-serrate; petioles about ³⁄₁₆-inch long. Flowers small, yellow, in axillary clusters. Fruit a subglobose drupe, ½- to ¾-inch long, dark red to almost black.

Native in southern Asia, Africa, and Australia. Grown as a specimen tree in southern California. Flowers from March to June.

## Tiliaceae. Linden Family

### 1. Tília L. BASSWOOD. LINDEN

(The ancient Latin name.)

Deciduous trees. Leaves simple, alternate, petioled; blades usually heart-shaped and toothed. Flowers small, yellowish, fragrant, in clusters on long leaf-like bracts. Fruit nut-like, subglobose, 1- or 2-seeded.

About 25 species, native in the north temperate zone and one in the mountains of Mexico. There is much confusion in the naming of species because of their great variability and the number of intermediate forms which have arisen as natural or garden hybrids.

KEY TO THE SPECIES

Leaves glaucous or whitish tomentose beneath.
    Leaves glaucous and glabrous beneath except for axillary tufts of brownish hairs.................................................1. *T. cordata.*
    Leaves whitish tomentose beneath, without axillary tufts of hairs.
      Petioles less than ½ the length of the blade; branches ascending
                                      2. *T. tomentosa.*
      Petioles more than ½ the length of the blade; branches pendulous
                                      3. *T. petiolaris.*
Leaves not glaucous or whitish tomentose beneath.
    Leaves glabrous beneath except for axillary tufts of hairs.
      Leaves 4½ to 10 inches long.........................................4. *T. americana.*
      Leaves 2½ to 4½ inches long..........................................5. *T. vulgaris.*
    Leaves pubescent beneath at least along the veins, with axillary tufts of hairs.................................................6. *T. platyphyllos.*

**1. Tília cordàta** Mill. SMALL-LEAVED LINDEN. LITTLELEAF EUROPEAN LINDEN. Fig. 327.

*T. ulmifolia* Scop.

A medium-sized to tall tree, 30 to 70 feet high, with spreading branches forming a rounded crown. Leaf-blades broadly heart-shaped to almost round, 1½ to 3 inches long and about as broad or often broader than long, dark green and glabrous above, glaucous and glabrous beneath except for axillary tufts of brownish hairs, regularly serrate; petioles ¾-inch to 1¼ inches long. Flowers 5 to 7, pedicelled, in erect cymes. Fruit globose, about ¼-inch in diameter, tomentose, slightly ribbed, thin-shelled.

Native to Europe. Occasionally cultivated as a park and street tree.

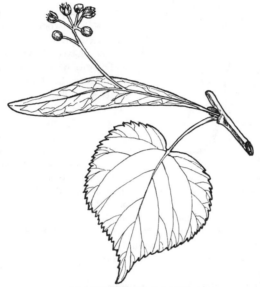

Fig. 327. Tilia cordata Mill.
Portion of branchlet with leaf and flower-cluster, × ¾.

2. **Tilia tomentòsa** Moench. WHITE LINDEN. SILVER LINDEN. Fig. 328.

A medium-sized to large deciduous tree, 30 to 60 (or 100) feet high, with upright or spreading branches and tomentose young branchlets. Leaf-blades nearly orbicular, 2 to 4 inches long and about as broad, obliquely truncate to heart-shaped at the base, abruptly acuminate, dark green and slightly pubescent above, becoming glabrous, white-tomentose beneath, sharply and irregularly double-serrate; petioles less than ½ the length of the blade. Flowers 7 to 10, in tomentose cymes. Fruit ovoid, ¼- to ⅜-inch long, slightly 5-angled, tomentose, hard-shelled.

Native in southeastern Europe and western Asia. Occasionally cultivated as a park and street tree.

3. **Tilia petiolàris** Hook. f. WEEPING WHITE LINDEN. Fig. 329.

　*T. tomentosa* var. *petiolaris* Kirchn.

Closely related to *T. tomentosa*, but can be distinguished by its "weeping" habit, longer petioles, finer serrations of the leaf-margins, and 5-grooved fruits.

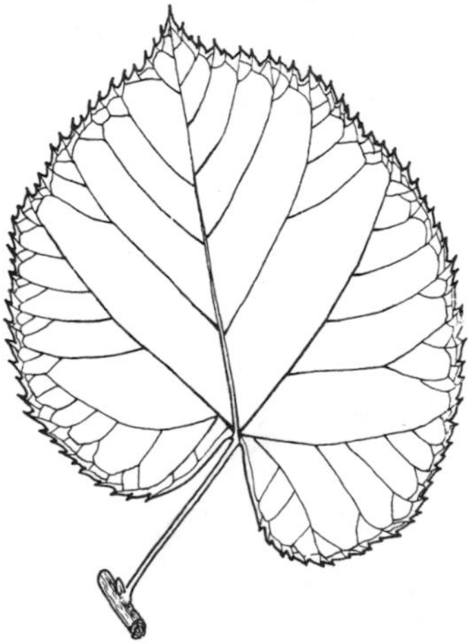

Native probably in southeastern Europe or western Asia. Cultivated as a park and street tree.

4. **Tilia americàna** L. AMERICAN LINDEN. BASSWOOD. Fig. 330.

　*T. glabra* Vent.

A large tree, 40 to 60 (or 100) feet high, with glabrous young branchlets. Leaf-blades broadly ovate, 4 to 6 inches long, 3 to 4 inches wide, unequally cordate or truncate at the base, dull dark green and glabrous above, lighter green and glabrous beneath except for tufts of hairs in the axils of the principal veins, coarsely glandular-serrate; petioles 1½ to 2 inches long. Flowers 5 to 12, on slender pu-

Fig. 328. Tilia tomentosa Moench.
Leaf, × ½.

bescent pedicels, in cymes, the peduncle 3½ to 4 inches long, the bract 4 to 5 inches long. Fruit ovoid to short-ellipsoidal, ⅜- to ½-inch long, tomentose, without ribs, thick-shelled.

Native in the eastern United States and southeastern Canada. Cultivated as a shade and street tree.

5. **Tilia vulgàris** Hayne. COMMON LINDEN. Fig. 331.

　*T. europaea* L.

A medium-sized to large deciduous tree, 30 to 60 (or 80) feet high, with glabrous young branchlets. Leaf-blades broadly ovate, 2½ to 4½ inches long, 2½ to 3½ inches broad, unequally heart-shaped at the

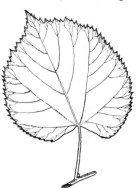

Fig. 329.
Tilia petiolaris Hook. f.
Leaf, × ½.

base, dark green and glabrous above, light green and glabrous beneath except for axillary tufts of hairs, sharply serrate; petioles 1½ to 2 inches long. Flowers 5 to 10, pedicelled, in cymes. Fruit subglobose, about ⅜-inch long, tomentose, thick-shelled.

Supposed to be a hybrid between *T. platy-phyllos* and *T. cordata*. Cultivated in parks, gardens, and along avenues.

6. **Tilia platyphýllos** Scop. BIGLEAF EURO-PEAN LINDEN. Fig. 332.

*T. europaea* L., in part. *T. grandifolia* Ehrh.

A deciduous tree, 30 to 50 (or 100) feet high, with pubescent young branchlets. Leaf-blades broadly ovate to almost round, 3 to 4½ inches long and about as broad, cordate and sometimes unequal at the base, dull dark green and glabrous above or rarely pubescent, light green and pubescent beneath at least on the veins, regularly ser-rate; petioles 1½ to 2 inches long. Flowers 3 to 5 in pendulous cymes. Fruit subglobose to ovoid, about ⅜-inch long, tomentose, 3- to 5-ribbed, hard-shelled.

Fig. 330. Tilia americana L.
Leaf, × ⅓.

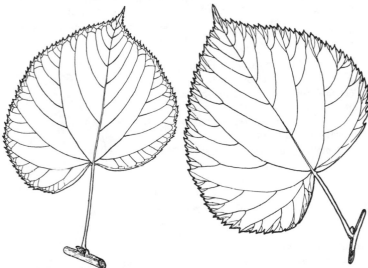

Fig. 331. Tilia vulgaris Hayne.
Leaf, × ½.

Fig. 332. Tilia platyphyllos Scop.
Leaf, × ½.

Native to Europe. A variable species with several varieties grown as park and street trees. Sold usually under the name *Tilia europaea*.

## Malvaceae. Mallow Family
### 1. Lagunària Don
(From its resemblance to *Lagunaea*, an allied genus.)
A genus of a single species.

### 1. Lagunaria pátersoni Don. LAGUNARIA. Fig. 333.
An evergreen tree, 20 to 40 feet high, with ascending branches.

Leaves simple, alternate; the blades oblong-ovate, 2 to 4 inches long, ¾-inch to 1½ inches wide, dark green and roughened above, white-scurfy to finely tomentose beneath, entire; petioles about ½-inch long. Flowers large, about 2 inches broad, hibiscus-like, pink to rose-colored or rarely white, solitary in the leaf axils. Fruit an ovoid rough tomentose capsule about 1 inch long.

Native to Australia and on Norfolk Island. Planted as a specimen tree in parks and gardens for its dense foliage and large pinkish flowers blooming from May to July.

Fig. 333. Lagunaria patersoni Don.
Leaf, flower, × ½.

## Theaceae. Tea Family
### 1. Caméllia L.
(Named for George J. Camellus, a seventeenth-century Moravian Jesuit.)

About 10 species of trees or shrubs, native in tropical and subtropical Asia.

### 1. Camellia japónica L. CAMELLIA. Fig. 334.
*Thea japonica* Nois.

An evergreen shrub or tree, 10 to 40 feet high. Leaves simple, alternate; the blades thick and leathery, elliptic to ovate, 2 to 4 inches long, 1 to 2 inches wide, glabrous, dark green and glossy above, paler beneath, finely serrulate; petioles about ¼-inch long. Flowers large and showy, red in the typical form, 2 to 3½ inches broad. Fruit a woody capsule.

Fig. 334.
Camellia japonica L.
Leaf, × ½.

Native in China and Japan. Many garden forms with red to white and single or double flowers are cultivated. Camellias grow best in semishady and warm moist locations.

## Sterculiaceae. Sterculia Family

### 1. Fremóntia Torr.

(Named in honor of John C. Fremont, a distinguished explorer of western North America.)

One species and one variety, native to California and northern Mexico.

### 1. Fremontia califórnica Torr. FLANNEL BUSH. Fig. 335.

*Fremontodendron californicum* Coville.

Usually a shrub or occasionally a small tree 15 to 25 feet high, tallest in cultivation. The leaves simple, alternate, evergreen; the blades broadly round-ovate, usually 3- or 5-lobed or rarely entire, ½-inch to 3 inches long, ¼-inch to 2 inches wide, or larger on young sterile shoots, usually 3-veined from the base, dark green and soft-pubescent or roughish above, often glabrous in age, covered below with a fine dense whitish or brownish felt; petioles ¼-inch to 1 inch long, very pubescent. Flowers large and showy, yellow or orange-colored, solitary and axillary on the branchlets; the 5 petal-like sepals each with a hairy gland at the base. Fruit an ovoid capsule, ¾-inch to 1 inch long, covered with dense bristly hairs, persistent for several months.

Fig. 335. Fremontia californica Torr. Leaf, flower, × ½.

Native on the lower mountain slopes of the inner Coast Ranges from Shasta County southward to San Diego County, California, and northern Mexico, and on the western side of the Sierra Nevada from Mariposa County to Kern County. Usually rather localized and not abundant. Flannel Bush is one of the most ornamental of the native plants and is becoming more commonly cultivated in gardens and parks. It thrives best in well-watered but well-drained situations. The following variety is more often cultivated than the species.

1*a.* Var. **mexicàna** Jepson.

*Fremontodendron mexicanum* Dav.

Similar to the species but distinguished by the usually larger leaves and sepal-glands without hairs.

## 2. **Stercùlia** L. Bottle Tree

(From the Latin *stercus,* manure, in reference to the odor of the leaves and fruits of some species.)

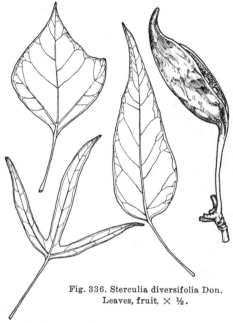

Fig. 336. Sterculia diversifolia Don.
Leaves, fruit, × ½.

Deciduous or evergreen trees. Leaves simple, alternate, petioled; the blades variously lobed or palmately divided. Flowers without petals, the sepals often petal-like, greenish to dull red and scarlet, some very large, in axillary panicles. Fruit composed of 4 or 5 woody or leaf-like carpels becoming follicles.

About 100 species, native in semitropical regions.

KEY TO THE SPECIES

Leaves 2 to 3 inches long, entire or 3- or rarely 5-lobed..1. *S. diversifolia.*
Leaves 4 to 10 inches long, 3- to 7-lobed or rarely entire.
Evergreen trees; leaves deeply 5- or 7-lobed or entire..2. *S. acerifolia*
Deciduous trees; leaves 3- or 5-lobed..........................3. *S. platanifolia.*

**1. Sterculia diversifòlia** Don. BLACK KURRAJONG. Fig. 336.

*Brachychiton populneum* R. Br. *B. diversifolium* R. Br.

A medium-sized evergreen tree, 25 to 60 feet high. Leaf-blades ovate to ovate-lanceolate, 1½ to 3 inches long, 1 to 1½ inches wide, glabrous and glossy, entire or irregularly 3- or 5-lobed; petioles ¾-inch to 2 inches long. Flowers bell-shaped, yellowish white, often reddish-tinged inside, borne in panicles. Follicles ovoid, 1½ to 3 inches long.

Native to Australia. Planted as a street and park tree. Most frequently used in southern California.

**2. Sterculia acerifòlia** A. Cunn. FLAME TREE. VICTORIAN BOTTLE TREE. Fig. 337.

*Brachychiton acerifolium* F. Muell.

A medium-sized evergreen tree, 25 to 60 feet high. Leaf-blades variable, usually 3- to 7-lobed and maple-like, 4 to 8 inches long and as broad, or oblong-lanceolate and entire to coarsely 2- or 3-lobed, glabrous, dark green and glossy; petioles 2 to 6 inches long. Flowers bright scarlet, in large showy clusters. Follicles ovoid, 2 to 4 inches long.

Native to Australia. Planted as a street and garden tree, mostly in southern California.

**3. Sterculia platanifòlia** L. CHINESE PARASOL TREE. Fig. 338.

*Firmiana platanifolia* Schott & Endl. *F. simplex* W. F. Wight.

A deciduous tree, 20 to 50 feet high.

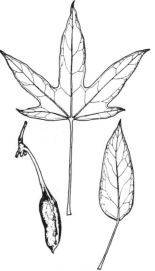

Fig. 337.
Sterculia acerifolia A. Cunn.
Fruit, leaves, × ⅙.

Leaf-blades sycamore-like, 6 to 12 inches long and as broad, palmately 3- or 5-lobed, glabrous or tomentulose; petioles 2 to 6 inches long, pubescent or glabrous. Flowers small, greenish, in terminal panicles. Fruit composed of 4 or 5 follicles, 2 to 4 inches long, separating into expanded leaf-like structures bearing pea-like seeds along the margins.

Native probably to China or Japan. Occasionally cultivated as a lawn or avenue tree in southern California.

## Tamaricaceae. Tamarisk Family

### 1. Támarix L. TAMARISK

(From Tamaris, now Tambro, the name of a river on the borders of the Pyrenees, in the vicinity of which one species abounds.)

Shrubs or small trees. Leaves simple, alternate, usually minute and scale-like, appressed to the slender branchlets. Flowers small, regular, perfect, in racemes or panicled spikes. Fruit a capsule.

Fig. 338. Sterculia platanifolia L.
Leaves, × ¼.

About 75 species, native from western and southern Europe to Asia and Japan. Several species are cultivated for their feathery foliage and clusters of pink flowers. Aside from the following tree-species, the more shrub-like species, **T. gállica** L., FRENCH TAMARISK, is occasionally grown on the Pacific Coast.

KEY TO THE SPECIES

Branches jointed (somewhat resembling *Casuarina*); flowers in panicled
    spikes, blooming in summer........................................1. *T. articulata.*
Branches not jointed; flowers in racemes, blooming in spring.
    Petals and sepals 5................................................................2. *T. juniperina.*
    Petals and sepals 4................................................................3. *T. parviflora.*

### 1. **Tamarix articulàta** Vahl. Athel. Fig. 339.

A small evergreen tree, 20 to 35 feet high, with numerous slender spreading branches and grayish jointed branchlets forming a bushy crown. Leaves very minute, glaucous, sheathing the branchlets, the free portion appearing as a small cusp. Flowers pink, numerous, sessile, borne in panicled spikes; sepals, petals, and stamens 5.

Native in western Asia. Planted as a windbreak in the Sacramento and San Joaquin valleys and in the warm coastal valleys and desert regions of southern California. The numerous grayish, slender, and jointed branchlets and the panicled spikes of flowers appearing in the summer distinguish this species from the others grown on the Pacific Coast.

Fig. 339.
Tamarix articulata Vahl.
Flower-cluster, × 1.
Portion of branchlet
with leaves,
× 3.

### 2. **Tamarix juniperìna** Bunge. Juniper Tamarisk. Fig. 340*d*.

*T. plumosa* Hort.

A deciduous shrub or small tree, 15 to 25 feet high, with numerous slender spreading branches and green feathery branchlets forming a bushy crown.

Fig. 340.
Tamarix parviflora DC.
*a.* Portion of branchlet with leaves, × 9. *b.* Branchlet with flowering spike, × 1½. *c.* Flower, × 4½.
Tamarix juniperina Bunge.
*d.* Flower, × 4½.

Leaves scale-like, ⅛-inch or less long, green, sessile. Flowers pinkish, numerous, almost sessile, appearing in the spring on the branches of the previous year, in spike-like racemes 1 to 2 inches long; sepals, petals, and stamens 5.

Native in northern China and Japan. Planted as a windbreak and as an ornamental in the warmer parts of the Pacific Coast.

3. **Tamarix parviflòra** DC. SMALLFLOWER TAMARISK. Fig. 340*a, b, c*.
An erect-spreading deciduous shrub or small tree, 10 to 18 feet high, with numerous slender branches, gray- or red-brown bark, and green feathery branchlets. Leaves scale-like, ⅛-inch or less long, green, sessile. Flowers numerous, pink, very short-pedicelled, borne

Fig. 341. Lagerstroemia indica L.
Leaf, flower, fruit, × 1.

in the spring in slender racemes 1 to 1½ inches long on branches of the previous season; sepals, petals, and stamens 4.

Native to southern Europe. Planted in the warmer parts of the Pacific Coast as an ornamental for its feathery foliage and masses of pink flowers appearing early in spring.

## Lythraceae. Loosestrife Family

### 1. Lagerstroèmia L.

(Named for Magnus V. Lagerstroem, a Swedish friend of Linnaeus.)
Evergreen or deciduous trees or shrubs. Leaves simple, opposite or the upper ones alternate, entire. Flowers showy, in terminal or axillary clusters; petals 5 to 8, clawed, the limb fringed. Fruit a capsule.

About 30 species, native to southern and eastern Asia, Australia, and the adjacent islands.

KEY TO THE SPECIES

Leaves 1 to 2 inches long; claw of petals long and slender......1. *L. indica.*
Leaves 4 to 8 inches long; claw of petals short.....................2. *L. speciosa.*

1. **Lagerstroemia índica** L. CRAPE-MYRTLE. Fig. 341.

A deciduous shrub or small tree, 10 to 20 feet high. Leaf-blades elliptic or oblong to obovate, 1 to 2 inches long, ¾-inch to 1¼ inches wide, glabrous or pubescent on the veins beneath, sessile or very short-petioled. Flowers white, pink, or purple, about 1¼ inches wide, in panicles 2½ to 8 inches long. Capsules subglobose, about ⅜-inch long.

Native to Asia and Australia. Cultivated for its mass of flowers blooming in summer.

2. **Lagerstroemia speciòsa** Pers. QUEEN CRAPE-MYRTLE. Fig. 342.

A medium-sized to large evergreen tree, 40 to 80 feet high. Leaf-blades leathery, oblong to ovate, 4 to 8 inches long, 2 to 4 inches wide, glabrous, short-petioled. Flowers large, 2 to 3 inches broad, varying from pink to purple, in large terminal panicles. Capsules subglobose, about 1 inch long.

Native to Australia, India, and southern China. Occasionally cultivated as a specimen tree in southern California.

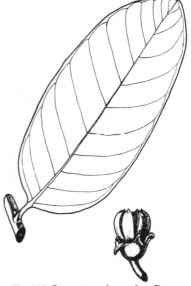

Fig. 342. Lagerstroemia speciosa Pers.
Leaf, fruit, × ½.

## Punicaceae. Pomegranate Family

### 1. Pùnica L.

(From the Latin *Punicus,* Carthaginian; *Malum punicum,* Apple of Carthage, was an early name for the pomegranate.)

Two species, native from the Mediterranean region to the Himalayas. Widely distributed in subtropical and tropical regions by long cultivation.

1. **Punica granàtum** L. POMEGRANATE. Fig. 343.

A small deciduous tree or commonly a large shrub. Leaves simple, opposite or some slightly alternate and fascicled on short lateral branchlets; the blades oblanceolate or obovate, 1 to 3 inches long, ½-inch to 1 inch wide, glabrous, green and glossy above, paler beneath, entire; petioles about ⅛-inch long. Flowers large and showy, 1 to 1½ inches broad, orange-red, with numerous stamens, solitary or in small axillary clusters. Fruit a spheroidal thick-skinned several-celled berry, 2 to 3 inches in diameter, brownish red.

Native from Persia to northwestern India. Cultivated as a garden

Fig. 343. Punica granatum L.
Leaf, × 1. Fruit, × ½.

and park plant because of its ornamental flowers and fruit. Grown in orchards around Lindsay, California, for its edible fruits, which yield a wine-red juice.

## Myrtaceae. Myrtle Family

### 1. Callistèmon R. Br. BOTTLEBRUSH

(From the Greek *kallistos*, most beautiful, and *stemon*, a stamen, in reference to the long graceful stamens.)

Evergreen shrubs or small trees. Leaves simple, alternate, linear or lanceolate, entire, subsessile. Flowers showy, with long protruding distinct stamens, in dense spikes, resembling a bottle brush. Fruit persistent woody capsules opening at the summit, arranged in a cylindrical spike-like cluster the axis of which continues beyond the cluster as a foliage shoot.

About 25 species, native to Australia. The callistemons are frequently planted in gardens, parks, and school grounds. They endure considerable drought and thrive best in well-drained soils in warm climates.

Leaves with a prominent midrib and 2 lateral veins; flower-clusters bright red; capsules contracted at summit...............1. *C. lanceolatus.*
Leaves with a prominent midrib but the lateral veins somewhat obscure.
Flower-clusters bright red; capsules not contracted at summit
2. *C. speciosus.*
Flower-clusters pale or greenish yellow, or rarely light pink
3. *C. salignus.*

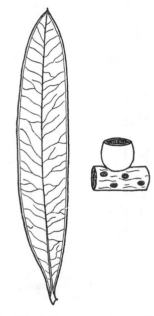

Fig. 344. Callistemon lanceolatus DC. Portion of branch with leaves and fruits, × ½. Single leaf, × 1. Single fruit, × 1.

Fig. 345. Callistemon speciosus DC. Leaf, × 1½. Single capsule on branchlet, × 2.

## 1. Callistemon lanceolàtus DC. LEMON BOTTLEBRUSH. Fig. 344.

*C. citrinus* Stapf. *Metrosideros semperflorens* Lodd. *Metrosideros floribunda* (trade name).

A tall shrub or small tree, 10 to 30 feet high. Leaves lanceolate, 1 to 2½ inches long, about ¼-inch wide, sharp-pointed at the apex, dark green and glabrous on both surfaces, the midrib and 2 lateral veins prominent. Flower-clusters 2 to 4 inches long, bright red, the stamens about 1 inch long. Capsules ovoid, contracted at the summit.

Native to Australia. Frequently cultivated in gardens and parks. The spikes are less dense than those of *C. speciosus*.

**2. Callistemon speciòsus** DC. SHOWY BOTTLEBRUSH. Fig. 345.

A large shrub or small tree, 10 to 40 feet high. Leaves lanceolate, 1½ to 4 inches long, about ¼-inch wide, obtuse or acute at the apex, pale green and glabrous, the midrib prominent but the lateral veins obscure. Flower-clusters very dense, 2 to 4 inches long, bright red, the stamens about 1 inch long. Capsules nearly globose, scarcely contracted at the summit.

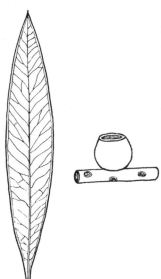

Native to western Australia. Cultivated in gardens and parks for its highly colored flowering spikes, which bloom in December and January and again in May and June. According to Miss Alice Eastwood (*Leaflets of Western Botany*, Vol. I, No. 3, p. 18, 1932), our plants referred to this species are *Callistemon viminalis* Cheel.

**3. Callistemon salígnus** DC. WILLOW BOTTLEBRUSH. Fig. 346.

A large shrub or small tree, 10 to 20 feet high. Leaves lanceolate, 1½ to 3 inches long, ¼- to ¾-inch wide, long-acute at the apex, glabrous, the midrib prominent. Flower-clusters pale yellow or pale pink, 1 to 2½ inches long, the stamens ½-inch or less long. Capsules globose.

Fig. 346. Callistemon salignus DC. Leaf, × 1. Single capsule, × 2.

Native to Australia. The variety **viridiflòrus** F. Muell., fig. 347, has sharp-pointed leaves 1 to 2 inches long, thick, obscurely veined, and greenish yellow flower-clusters. It is usually more tree-like than the species.

**2. Eucalýptus** L'Her. GUM TREE. EUCALYPTUS

(From the Greek *eu*, good, and *kalypto*, to cover as with a lid, in reference to the sepals and petals fused into a cap which falls off as the flower opens.)

Evergreen trees. Leaves simple, alternate, entire, or those on the young shoots opposite. Flowers conspicuous, with numerous protruding stamens protected in the bud by a lid formed of the united sepals and petals, borne singly or in stalked umbels, panicles, or corymbs. Fruit a woody capsule opening at the summit by 3 to 6 valves.

About 300 species, native to Australia and the Malayan region. About 76 species and several varieties are grown in California, the 23 species named below being the most commonly cultivated. It is very difficult to identify the species of *Eucalyptus* unless bark, leaves, buds, flowers, and fruits are available. They are cultivated as shade, ornamental, and avenue trees, as windbreaks, and as a source of timber, fuel, oil, and honey. The climatic conditions of Oregon, Washington, and British Columbia are not suitable for the successful growing of eucalypti.

Fig. 347.
Callistemon salignus var.
viridiflorus F. Muell.
Leaf, single capsule, × 1.

### KEY TO THE SPECIES

I. Leaves with 3 to 5 almost parallel veins from the base
1. *E. coriacea.*

II. Leaves with 1 main vein from the base
A. *Leaves decidedly paler beneath*
1. Peduncles distinctly flattened
Valves of capsules exserted beyond the rim; leaves 1½ inches or less wide.
Capsules ¼-inch wide................................................................2. *E. resinifera.*
Capsules ⅓- to ½-inch wide...............2a. *E. resinifera* var. *grandiflora.*
Valves of capsules not exserted; leaves 1½ to 3½ inches wide
3. *E. robusta.*
2. Peduncles rounded or slightly angular
Capsules usually in panicles composed of umbellate clusters.
Leaves with the lateral veins nearly at right angles to the midrib; flowers red or pink; capsules 1 to 1½ inches long....4. *E. ficifolia.*
Leaves with the lateral veins oblique; flowers whitish; capsules about ¼-inch long................................................................5. *E. paniculata.*
Capsules in simple peduncled umbels.
Lateral veins almost at right angles to the midrib; capsules ovoid; pedicels about ¼-inch long....................................6. *E. diversicolor.*
Lateral veins more oblique; capsules short-oblong, commonly longitudinally streaked; pedicels ⅛-inch or less long
7. *E. corynocalyx.*
B. *Leaves alike on both surfaces*
Leaves less than 2 times longer than broad; capsules in panicles
8. *E. polyanthemos.*

Leaves more than 2 times longer than broad.
 Leaves less than ½-inch wide............9. *E. amygdalina* var. *angustifolia*.
 Leaves ½-inch or more wide.
  Capsules solitary (rarely 2 or 3) and usually sessile in the leaf-axils, about ¾-inch broad..................................10. *E. globulus*.
  Capsules in umbels, heads, or panicles.
   Leaves when crushed emitting odor of lemon; capsules in panicles
            11. *E. maculata* var. *citriodora*.
   Leaves without lemon odor; capsules in simple umbels or heads.
    1. Capsules usually 3 in each umbel (rarely to 5)
  Capsules about ¼-inch long, short-stalked or sessile, the valves protruding beyond the rim..................................12. *E. viminalis*.
  Capsules about ½-inch long, long-pedicelled.
   Rim of capsule ascending; lateral veins widely spreading from the midrib....................................................13. *E. longifolia*.
   Rim of capsule thick or descending; lateral veins more oblique.
    Bark rough, furrowed, and persistent....................14. *E. sideroxylon*.
    Bark not rough and furrowed, light colored............15. *E. leucoxylon*.
    2. Capsules usually more than 3 in each umbel or head
  Valves not exserted beyond the rim of the capsule.
   Peduncles decidedly flattened......................................16. *E. goniocalyx*.
   Peduncles round or angled..........17. *E. eugenioides*. Also 23. *E. gunni*.
  Valves exserted beyond the rim of the capsule or at least level with it.
   Teeth of valves about ½-inch long, united into a cone at summit
                18. *E. cornuta*.
   Teeth of valves ⅛-inch long or less, not united at summit.
    Lid of bud 2 to 4 times as long as the body..........19. *E. tereticornis*.
    Lid of bud not more than 2 times as long as the body.
     Capsules distinctly pedicelled, the pedicels ⅛- to ⅜-inch long.
      Lid cone-shaped, bluntly pointed at apex................20. *E. rudis*.
      Lid hemispheric, abruptly and acutely pointed at apex
              21. *E. rostrata*.
     Capsules sessile or subsessile, the pedicels when present less than
      ⅛-inch long....................22. *E. stuartiana*. Also 23. *E. gunni*.

1. **Eucalyptus coriàcea** A. Cunn. Leatherleaf Gum. Tumbledown Gum. Fig. 348.

A medium-sized tree, 30 to 50 feet high, with spreading branches, deciduous outer bark, and smooth gray inner bark. Leaf-blades thick, narrowly lanceolate to ovate-lanceolate, 4 to 8 inches long, ½-inch to 1¼ inches wide, glabrous and equally green on both surfaces, with 3 to 5 nearly parallel veins from the base; petioles about ½-inch long. Flowers whitish, in peduncled umbels. Peduncles terete, ⅜- to ⅝-inch long. Capsules 5 to 10 in the umbels, sessile, pear-shaped, about ⅜-inch long, with the valve-teeth included below the depressed rim.

Cultivated as a specimen tree.

2. **Eucalyptus resinífera** Sm. Mahogany Gum. Red-Mahogany. Fig. 349.

A medium-sized tree, 30 to 50 feet high, with rough dark reddish persistent bark. Leaf-blades thick and leathery, lanceolate, somewhat curved, 3½ to 6 inches long, ¾-inch to 1¼ inches wide, distinctly paler beneath, the margins slightly revolute, the lateral veins widely spreading from the midrib; petioles ½-inch to 1¼ inches long. Flowers whitish, in peduncled umbels. Peduncles distinctly flattened. Capsules 4 to 10 in the umbels, short-pedicelled, hemispheric, about ¼-inch wide, the valves exserted beyond the broad rim.

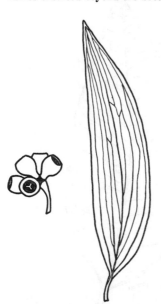 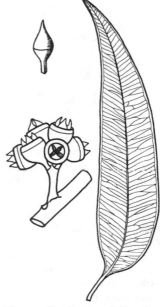

Fig. 348.
Eucalyptus coriacea A. Cunn.
Fruit-cluster, leaf, × ½.

Fig. 349. Eucalyptus resinifera Sm.
Fruit-cluster, × 1. Bud with
operculum, × 1. Leaf, × ½.

Cultivated in the coastal regions of California. Will not endure drought or severe frost conditions.

2a. Var. **grandiflòra** Benth. Fig. 350.

Capsules ⅓- to ½-inch wide, with a raised rim.

3. **Eucalyptus robústa** Sm. SWAMP-MAHOGANY. BROWN GUM. Fig. 351.

A medium-sized to tall tree, 40 to 70 feet high, with rough dark brown persistent bark, spreading habit, and heavy foliage. Leaf-blades leathery, ovate-lanceolate, 3 to 6 inches long, 1½ to 2½ inches wide, dark green and glabrous above, paler beneath, the lateral veins spreading almost at right angles to the midrib; petioles ½-inch to 1 inch long.

Flowers creamy white, in peduncled umbels. Peduncles stout and distinctly flattened. Capsules 4 to 10 in the umbels, pedicelled, goblet-shaped, about ½-inch long, the valves included below the thin rim.

Fig. 350.
Eucalyptus resinifera var.
grandiflora Benth.
Fruit, × ½.

Cultivated as a street tree in the coastal regions of California, where it thrives best in moist habitats. Needs much water if planted in the interior.

4. **Eucalyptus ficifòlia** F.v.M. CRIMSON-FLOWERED EUCALYPTUS. SCARLET GUM. Fig. 352.

A small tree, 15 to 35 feet high, with rough furrowed dark bark and handsome flowers and foliage. Leaf-blades thick, ovate to ovate-lanceolate, 3 to 5 inches long, 1 to 2½ inches wide, dark green above, distinctly paler beneath, the lateral veins spreading almost at right angles from the midrib; petioles ½-inch to 1¼ inches long. Flowers pink to crimson, in panicles. Pedun-

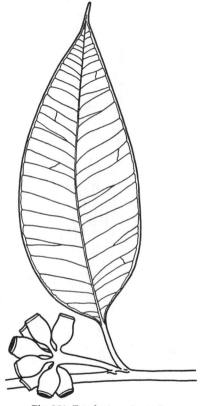

Fig. 351. Eucalyptus robusta Sm.
Portion of branchlet with fruit and leaf,
× ½.

cles and pedicels terete or slightly angled. Capsules pedicelled, broadly urn-shaped, ¾-inch to 1½ inches long and almost as broad, the valves included below the compressed rim.

Commonly cultivated as a street tree. Can withstand drought but not severe frost conditions.

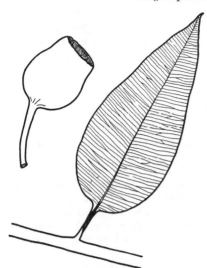

Fig. 352. Eucalyptus ficifolia F. v. M.
Capsule, leaf, $\times$ ½.

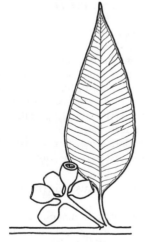

Fig. 354.
Eucalyptus diversicolor
F. v. M.
Portion of branchlet with
fruit-cluster and leaf,
$\times$ ½.

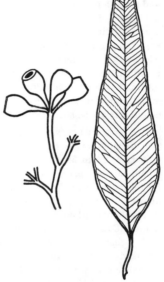

Fig. 353. Eucalyptus paniculata Sm.
Fruit, $\times$ 1. Leaf, $\times$ ½.

**5. Eucalyptus paniculàta** Sm. WHITE IRONBARK. Fig. 353.

A medium-sized tree, 25 to 75 feet high, with rough furrowed dark grayish persistent bark. Leaf-blades lanceolate, often curved, 3 to 6 inches long, ½-inch to 1 inch wide, glabrous, paler beneath; petioles ½-inch to 1 inch long. Flowers small, numerous, in panicles or axillary umbels. Capsules pedicelled, ovoid, truncate at the apex, about ¼-inch broad, the valves enclosed below the thin rim.

Occasionally planted in groves or as a street tree. Does not endure severe drought conditions.

**6. Eucalyptus diversícolor** F. v. M. KARRI. KARRI GUM. Fig. 354.

A medium-sized tree, 30 to 80 feet high, with deciduous outer bark and smooth gray or whitish inner bark. Leaf-blades ovate or lanceolate, 2½ to 4½ inches long, 1 to 1½ inches wide, dark green and shining above, distinctly paler beneath, the lateral veins spreading almost at right angles from the midrib; petioles ½-inch to 1 inch long. Flowers whitish, in peduncled umbels. Peduncles slightly angular. Capsules 5 to 10 in the umbels, pedicelled, ovoid, nearly ½-inch long, the valves included below the flat rim.

Cultivated in the coastal regions of California. Does not endure the conditions of the hot interior valleys. Said to be a good bee tree.

**7. Eucalyptus corynocàlyx** F. v. M. SUGAR GUM. Fig. 355.

*Eucalyptus cladocalyx* F. v. M.

A tree, 25 to 50 feet high, with deciduous outer bark and smooth cream-colored inner bark. Leaf-blades ovate-lanceolate to lanceolate, 3 to 6 inches long, ¾-inch to 1¼ inches wide, dark green and shining above, dull and paler beneath, the lateral veins oblique; petioles ½-inch to 1 inch long. Flowers whitish, in peduncled umbels. Peduncles terete or angled. Capsules 5 to 10 and short-stalked in the umbels, short-ovoid or urn-shaped to ellipsoidal, nearly ½-inch long, some longitudinally streaked, the valves included below the thin rim.

Planted as a roadside tree in southern California. Can withstand drought conditions but not severe frosts. Flowers in August and September.

**8. Eucalyptus polyánthemos** Schau. REDBOX. AUSTRALIAN-BEECH. Fig. 356.

A much branched tree, 30 to 60 feet high, often with more than one trunk from the base, and with somewhat furrowed grayish persistent bark. Leaf-blades broadly oval to ovate-lanceolate, 2 to 4 inches long, 1½ to 3 inches wide, gray-green and dull on both surfaces. Flowers small, whitish, in panicled umbels. Capsules truncate-ovoid, ¼-inch or less long, the valves included below the thin rim.

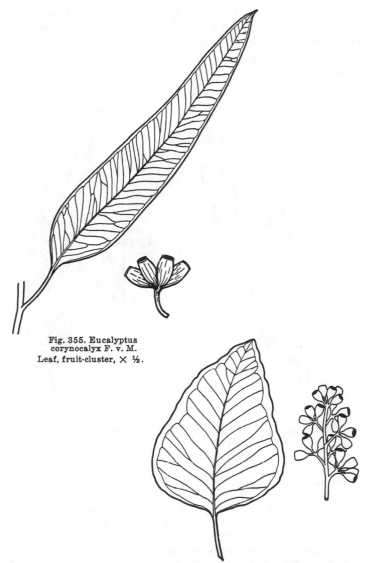

Fig. 355. Eucalyptus
corynocalyx F. v. M.
Leaf, fruit-cluster, × ½.

Fig. 356. Eucalyptus polyanthemos Schau.
Leaf, fruit-cluster, × ½.

Frequently cultivated for its silvery drooping foliage and abundant bloom. Can withstand drought and frost conditions. A good shade tree, suitable for windbreaks, and a good source of honey.

9. **Eucalyptus amygdálina** Labill. PEPPERMINT GUM. ALMOND EUCALYPTUS.

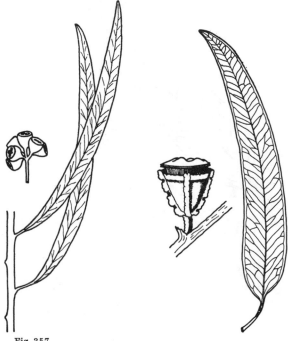

Fig. 357
Eucalyptus amygdalina
var. angustifolia F. v. M.
Leaves, fruit-cluster, × ¾.

Fig. 358.
Eucalyptus globulus Labill.
Fruit, × ⅔. Leaf, × ⅓.

A tall tree, 70 to 120 feet high, with rough persistent bark on old trunks and drooping branchlets. Leaf-blades narrowly lanceolate, 2 to 4 inches long, about ¼-inch wide, dark green and glabrous on both surfaces, with peppermint odor; petioles ⅛- to ¼-inch long. Flowers small, whitish, in peduncled umbels. Peduncles terete. Capsules 4 to many in the umbels (the umbels often appearing in paniculate clusters owing to the fall of the leaves), short-pedicelled, truncate-ovoid, about ¼-inch long, the valves included below, or about even with the flat or slightly concave rim.

The variety **angustifòlia** F. v. M., fig. 357, is more commonly cultivated than the species. It has deciduous outer bark and smooth
whitish inner bark.

10. **Eucalyptus glóbulus** Labill. BLUE GUM. Fig. 358.

A tall tree, 70 to 140 feet high, with deciduous outer bark and smooth
gray or whitish inner bark. Leaf-blades lanceolate, often curved, 6 to 13
inches long, 1 to 1¾ inches wide, equally green above and beneath;
petioles ¾-inch to 1½ inches long. Leaf-blades of young growth sessile,
opposite, white-mealy. Flowers large,
white, mostly solitary. Capsules solitary or 2 or 3 together, sessile, hemispheric, ¾-inch to 1 inch broad, angular, warty-rough, the valves included
within the broad rim.

The most commonly planted eucalyptus in the world. The constant shedding
of the outer bark and the numerous
falling fruit-caps make this plant an
undesirable species.

11. **Eucalyptus maculàta** var. **citrio-
dòra** Bailey. LEMON-SCENTED GUM. Fig.
359.

   *E. citriodora* Hook.

A medium-sized tree, 25 to 60 feet
high, with whitish or reddish gray outer
bark, flaking off in patches and exposing the smooth inner bark, thus giving
the trunk a spotted appearance. Leaf-
blades narrowly lanceolate, 4 to 6 inches
long, ½- to ¾-inch wide, lemon-scented
when crushed, equally green on both
surfaces, the lateral veins nearly parallel but oblique to the midrib; petioles

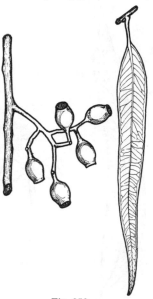

Fig. 359.
Eucalyptus maculata var.
citriodora Bailey.
Fruit-cluster, leaf, × ½.

½-inch to 1 inch long. Flowers creamy white, numerous, in panicled
umbels. Capsules pedicelled, urn-shaped, about ½-inch long, contracted
at the summit, the valves included below the thin flat rim.

Thrives best in moist coastal regions free from extreme frost conditions. A profuse bloomer and an excellent source of honey.

12. **Eucalyptus viminàlis** Labill. MANNA GUM. Fig. 360.

A tall graceful tree, 50 to 80 feet high, with variable bark, that of
the trunk and main branches usually persistent, rough, furrowed, and

dark, but sometimes deciduous in long strips exposing the greenish or cream color of the inner bark. Branchlets usually long-drooping.

Fig. 360.
Eucalyptus viminalis Labill.
Leaf, × ½. Fruit-cluster, × 1.

Leaf-blades lanceolate, somewhat curved, 4 to 7 inches long, ½-inch to 1 inch wide, equally green on both surfaces; petioles ½-inch to 1 inch long. Flowers whitish, in peduncled umbels. Capsules usually 3 in the umbels, sessile or very short-pedicelled, globular or top-shaped, about ¼-inch long, the valves protruding beyond the flat or rounded rim.

Commonly cultivated. Grows rapidly and endures a variety of climatic conditions.

13. **Eucalyptus longifòlia** Link & Otto. WOOLLYBUTT. Fig. 361.

A medium-sized to tall tree, 30 to 70 feet high, with rough grayish or tan-colored persistent bark. Leaf-blades long-lanceolate, usually curved, 4 to 8 inches long, ½-inch to 1 inch wide, equally green on both surfaces, the lateral veins widely spreading from the midrib; petioles ½-inch to 1 inch long. Flowers whitish, in peduncled umbels. Capsules usually 3 in the umbels, long-pedicelled, bell-shaped or semiovoid, about ½-inch long, the valves narrow and included within the ascending rim.

Thrives best along the coast. The flowers, which bloom for many months, are a source of honey.

14. **Eucalyptus sideróxylon** Cunn. RED IRONBARK. Fig. 362.

*E. leucoxylon* var. *sideroxylon* Auct.

A medium-sized tree, 30 to 80 feet high, usually not branched for some distance, with rough furrowed dark persistent bark. Leaf-blades narrowly lanceolate, often curved, 4 to 6 inches long, ½-inch to 1 inch wide, equally green on both surfaces, often glaucous; petioles

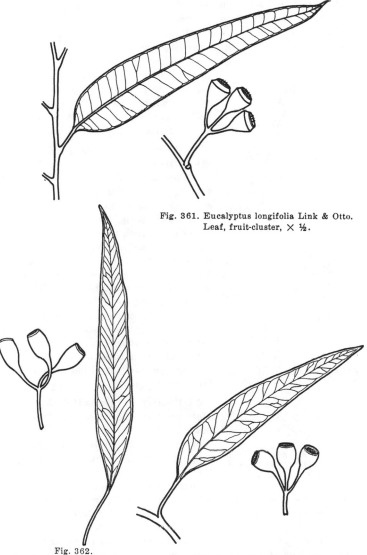

Fig. 361. Eucalyptus longifolia Link & Otto.
Leaf, fruit-cluster, × ½.

Fig. 362.
Eucalyptus sideroxylon Cunn.
Fruit-cluster, leaf, × ½.

Fig. 363. Eucalyptus leucoxylon F. v. M.
Leaf, fruit-cluster, × ½.

½-inch to 1 inch long. Flowers white or yellowish, or pink in some varieties, in umbels. Capsules usually 3 (2 to 5) in the umbels, long-pedicelled, semiovoid to obovoid, about ½-inch long, the valves included below the thick or descending rim.

Cultivated near the coast. Will thrive in dry soils. Variety **ròsea** Hort. (*E. leucoxylon* var. *rosea* Hort.) has pink flowers. Variety **pállens** Auct. (*E. leucoxylon* var. *pallens* Benth.) has reddish flowers and silvery foliage.

### 15. Eucalyptus leucóxylon F. v. M.
WHITE IRONBARK. Fig. 363.

A medium-sized to tall tree, 25 to 60 feet high, often with a crooked trunk and smooth light-colored deciduous bark. Leaf-blades lanceolate, 3 to 5½ inches long, ½-inch to 1 inch wide, equally green on both surfaces, the lateral veins spreading obliquely from the midrib; petioles ¼-inch to 1 inch long. Flowers white or pink, in umbels. Capsules 2 to 7 (usually 3) in the umbels, long-pedicelled, obovoid, about ½-inch long, scarcely contracted at the summit, the valves included below the thick or descending rim.

One of the hardiest of the eucalypti, this species can withstand drought and frost conditions.

### 16. Eucalyptus goniocàlyx F. v. M.
MOUNTAIN GUM. BASTARD BOX TREE. Fig. 364.

A tall tree, 30 to 70 feet high, with rough dark usually persistent bark. Leaf-blades lanceolate, usually curved, 4 to 6 inches long, ½-inch to 1 inch wide, equally pale green on both surfaces; petioles ½-inch to 1 inch long. Flowers whitish, in peduncled umbels. Peduncles flattened, about 1 inch long. Capsules 5 to 14 in the umbels, nearly sessile, ovoid, truncate at the summit, ¼- to ⅓-inch broad, valves included or about level with the narrow depressed rim.

Grows well in the coastal regions of California.

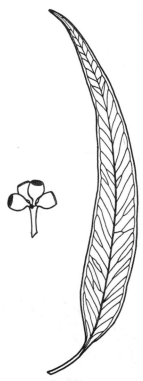

Fig. 364.
Eucalyptus goniocalyx F. v. M.
Fruit-cluster, leaf, × ½.

**17. Eucalyptus eugenioìdes** Sieb. WHITE STRINGYBARK. Fig. 365.

A tree, 40 to 60 feet high, with rough fibrous gray persistent trunk-bark and dense foliage. Leaf-blades ovate to ovate-lanceolate, 2 to 6 inches long, ½-inch to 1 inch wide, usually oblique at the base, equally green on both surfaces; petioles ¼- to ¾-inch long. Flowers whitish, in peduncled umbels. Peduncles terete. Capsules more than 3 in the umbels, short-pedicelled, nearly globular, truncate at the summit, about ¼-inch broad, the valves slightly enclosed below the flat or raised rim.

Thrives best in the coastal regions of California.

Fig. 365. Eucalyptus eugenioides Sieb. Fruit-cluster, leaf, × ½.

**18. Eucalyptus cornùta** Labill. YATE TREE. Fig. 366.

A profusely branched medium-sized tree, 20 to 50 feet high, with a crooked trunk and rough deciduous or persistent bark. Leaf-blades broadly lanceolate, 2 to 5 inches long, ½- to ¾-inch wide, equally green on both surfaces; petioles ½-inch to 1 inch long. Flowers greenish yellow, in peduncled head-like clusters. Capsules numerous, sessile in the clusters, short-cylindric, ⅓- to ½-inch long, the valves much exserted beyond the depressed rim and joined into a beak-like

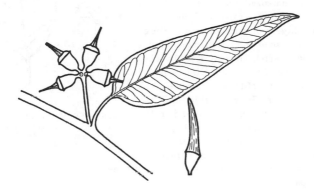

Fig. 366. Eucalyptus cornuta Labill. Portion of branchlet with fruit-cluster and leaf, bud with operculum, × ½.

projection about ½-inch long; lid cylindrical or horn-like, 1 to 1½ inches long.

Thrives in the coastal regions and in the dry interior valleys where frost conditions are not too severe or prolonged.

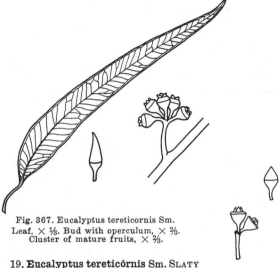

Fig. 367. Eucalyptus tereticornis Sm.
Leaf, × ⅓. Bud with operculum, × ⅔.
Cluster of mature fruits, × ⅔.

### 19. Eucalyptus tereticórnis Sm. SLATY GUM. FOREST GRAY GUM. Fig. 367.

A medium-sized tree, 30 to 70 feet high, with smooth gray bark deciduous in thin layers or persistent and rough near the base of the trunk. Leaf-blades lanceolate, often curved, 4 to 6 inches long, ½-inch to 1 inch wide, equally green on both surfaces; petioles ½-inch to 1 inch long. Flowers whitish, in peduncled umbels. Capsules 4 to 8 in the umbels, short-stalked, top-shaped or nearly globular, about ½-inch broad, the valves exserted about ⅛-inch beyond the broad brim; lid slender, 2 to 4 times as long as the body of the capsule.

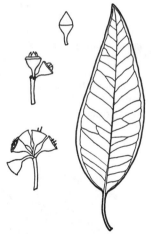

Fig. 368. Eucalyptus rudis Endl.
Cluster of young fruit, mature fruit, bud with operculum, leaf, × ½.

A hardy species, thriving in the hot interior valleys and coastal regions.

20. **Eucalyptus rùdis** Endl. DESERT GUM. Fig. 368.

A medium-sized to tall tree, 50 to 80 feet high, with rough gray usually persistent bark. Leaf-blades thin, broadly or narrowly lanceolate, often curved, 3 to 6 inches long, ¾-inch to 2¼ inches wide, equally dull green on both surfaces; petioles ½-inch to 1 inch long. Flowers whitish, in peduncled umbels. Capsules 4 to 8 in the umbels,

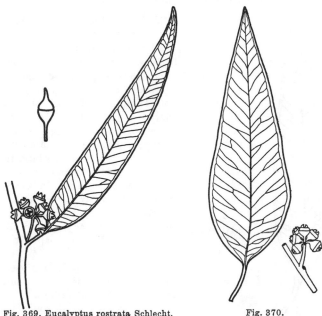

Fig. 369. Eucalyptus rostrata Schlecht.
Portion of branchlet with fruit-cluster
and leaf, × ½. Bud with operculum,
× 1.

Fig. 370.
Eucalyptus stuartiana
F. v. M.
Leaf, fruit-cluster, × ½.

short-pedicelled (the pedicels ⅛- to ⅜-inch long), top-shaped or broadly turbinate, ⅓- to ½-inch wide, the valves exserted about ⅛-inch beyond the slightly ascending rim; lid conical, blunt, about as long as the capsule.

A hardy species if given plenty of water.

21. **Eucalyptus rostràta** Schlecht. RED GUM. CREEK GUM. Fig. 369.

A large tree, 50 to 100 feet high, with usually persistent more or less rough and furrowed bark near the base, occasionally sloughing off in patches, or smooth and deciduous above. Leaf-blades narrowly lanceolate, 4 to 6 inches long, ⅝-inch to 1¼ inches wide, equally green on both surfaces; petioles ½-inch to 1 inch long. Flowers whitish, in stalked umbels. Peduncles slender, terete, about 1 inch long. Capsules 5 to 13 in the umbels, long-pedicelled, nearly globular, about ¼-inch in diameter, the valves protruding beyond the broad ascending rim; lid hemispheric, terminated by a beak or rarely conical.

Thrives under a variety of climatic conditions.

Fig. 371. Eucalyptus gunni Hook.
Leaf, fruit-cluster, × ¾.

**22. Eucalyptus stuartiàna** F'. v. M. APPLE-SCENTED GUM. Fig. 370.

A medium-sized tree, 30 to 60 feet high, with more or less fibrous and rough bark and dense drooping foliage emitting an odor of apples when crushed. Leaf-blades lanceolate, often curved, 3 to 6 inches long, ½-inch to 1 inch broad, equally green on both surfaces; petioles ½-inch to 1 inch long. Flowers greenish white, in peduncled umbels. Capsules 4 to 8, sessile or nearly so in the umbels, semi-ovoid, ¼- to nearly ½-inch wide, truncate at the summit, the valves exserted about ⅛-inch beyond the flat or slightly rounded rim; lid cone-shaped, bluntly pointed.

Thrives in the coastal regions. Can withstand frost conditions but not the heat of the interior valleys.

**23. Eucalyptus gúnni** Hook. CIDER GUM. Fig. 371.

A medium-sized tree, 40 to 75 feet high, with rough brown bark which continually flakes off, and dense dark green foliage. Leaf-blades thick and rather stiff, ovate, 2 to 3½ inches long, ½-inch to 1 inch wide, dark green and shining on both surfaces; petioles ½- to ¾-inch long. Flowers whitish, in peduncled umbels. Capsules 5 to 12 in the umbels, top-shaped, about ¼-inch broad; the valves usually included but sometimes level with or slightly exserted beyond the thin rim.

The variety **undulàta** Auct. has leaves 2 to 4½ inches long, 2 inches wide, often wavy-margined. The variety **acérvula** Deane & Maiden has yellowish green foliage and capsules ¼- to ½-inch broad.

### 3. Eugènia L.

(Named in honor of Prince Eugene of Savoy, a patron of botany.)

Evergreen trees or shrubs. Leaves simple, opposite, entire. Flowers usually white or cream-colored, solitary or in racemes or panicles, with numerous conspicuous stamens. Fruit a berry, adnate to the calyx-tube.

More than 800 species, native in the tropics and subtropics.

KEY TO THE SPECIES

Leaves oblong-lanceolate, 2½ to 4 inches long, acuminate; branchlets
    without wart-like growths near base of petioles......1. *E. paniculata.*
Leaves short-elliptic, 1 to 2½ inches long, short-acute; branchlets with
    wart-like growths near base of the petioles..............2. *E. myrtifolia.*

1. **Eugenia paniculàta** Jacq. Fig. 372.

*E. hookeri* Hort. *E. hookeriana* Hort.

A small evergreen tree or large shrub, 10 to 20 feet high. Branchlets glabrous, without wart-like growths. Leaf-blades leathery, oblong-lanceolate, 2½ to 4 inches long, ¾-inch to 1½ inches wide, dark green and often tinged with red especially when young, glabrous on both surfaces; petioles ⅛- to ¼-inch long. Flowers about ½-inch broad. Fruit ovoid, about ¾-inch long, rose-purple.

Native to Australia. Commonly cultivated in gardens and parks. Cannot stand freezing weather.

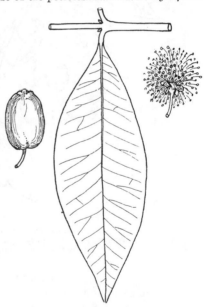

Fig. 372. Eugenia paniculata Jacq.
Fruit, leaf, flower, × ¾.

2. **Eugenia myrtifòlia** Sims. AUSTRALIAN BRUSH-CHERRY. Fig. 373.

*E. australis* Wendl.

An evergreen shrub or small tree, 8 to 25 (or 40) feet high. Branchlets glabrous, with wart-like growths near the bases of the petioles. Leaf-blades leathery, short-elliptic, 1 to 2½ inches long, ½- to ¾-

inch wide, dark green, tinged with red when young, glabrous on both surfaces; petioles ⅛- to ¼-inch long. Flowers about ½-inch wide. Fruit ovoid, ½- to ¾-inch long, rose-violet or reddish.

Native to Australia. Commonly cultivated in gardens and parks. Can be trimmed to form hedges or encouraged by cultivation to grow into a medium-sized tree.

Fig. 373. Eugenia myrtifolia Sims.
Portion of branch with fruit, leaves, and wart-like
growths on the stem, × ½.

#### 4. **Feijòa** Berg.

(Named in honor of J. da Silva Feijo, director of the Natural History Museum at San Sebastián, Spain.)

Two species of shrubs or small trees, native to South America.

#### 1. **Feijoa sellowiàna** Berg. FEIJOA. PINEAPPLE-GUAVA. Fig. 374.

A small evergreen tree or large shrub, 10 to 20 feet high. Leaves simple, opposite; the blades oval-oblong to elliptic, 1¾ to 3 inches long, about 1 inch wide, dark green and glossy above, white-tomentose and prominently veined beneath, entire; petioles about ¼-inch long. Flowers solitary, showy, 1 to 1½ inches broad, the 4 fleshy petals white-tomentose outside and purplish within, the numerous stamens and single style red. Fruit an oblong or spheroidal berry, 1½ to 2½ inches long, terminated by the persistent calyx-lobes, dull green and sometimes tinged with red.

Native to southern Brazil, Paraguay, Uruguay, and Argentina. Cultivated in the warmer parts of California for its edible fruits and ornamental flowers and foliage.

Fig. 374. Feijoa sellowiana Berg.
Leaves, fruit, × ½.

### 5. Leptospérmum Forst.

(From the Greek *leptos,* slender, and *sperma,* a seed, in reference to the small slender seeds.)

About 25 species of shrubs or small trees, native to Australia, New Zealand, and the Malay Archipelago.

1. **Leptospermum laevigàtum** Muell. AUSTRALIAN TEA TREE. Fig. 375.

A large shrub or small tree, 10 to 25 feet high. Leaves simple, alternate, almost sessile, oblanceolate or oblong-lanceolate to elliptic, ¾-inch to 1½ inches long, ¼- to ½-inch wide, 3-veined from the base, gray-green and glabrous on both surfaces, entire. Flowers white, ½- to ¾-inch broad, solitary or in groups of 2 or 3. Fruit a leathery or woody capsule, about ¼-inch long, opening at the summit.

Native to Australia. Extensively cultivated in gardens and parks.

### 6. Melaleúca L. MELALEUCA

(From the Greek *melas*, black, and *leukos*, white, in reference to the black trunk and white branches of one species.)

Evergreen shrubs or small trees. Leaves simple, alternate or opposite, entire. Flowers showy, with long protruding stamens united at their bases into 5 groups opposite the petals, in heads or spikes resembling a bottle brush. Fruit a woody capsule opening at the summit, arranged in persistent clusters beyond which the stem continues as a foliage shoot.

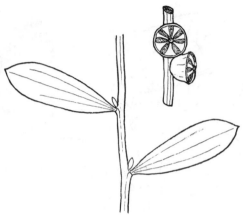

Fig. 375. Leptospermum laevigatum Muell.
Leaves, fruit, × 1.

About 100 species, native to Australia. The melaleucas belong to the bottlebrush group of ornamentals and are cultivated because of their showy flower-clusters and their adaptability to drought conditions.

KEY TO THE SPECIES

Leaves 2 to 4 inches long..................................................1. *M. leucadendron.*
Leaves less than 2 inches long.
    1. Leaves ovate or lance-ovate, sharp-pointed, sessile by a broad base, nearly ¼-inch wide; flower-clusters creamy white
                        2. *M. styphelioides.*
    2. Leaves obovate-oblong, obtuse or abruptly pointed; flower-clusters pink....................................................................................3. *M. nesophila.*
    3. Leaves linear or elliptic-oblong to lanceolate.
        Leaves typically ⅝-inch or less long; spikes ½-inch to 1 inch long.
            Leaves alternate, about ½-inch long; flowers yellowish white
                          4. *M. ericifolia.*
        Leaves opposite, ¼- to ½-inch long; flowers lilac..5. *M. decussata.*
        Leaves ¾-inch to 1½ inches long; spikes 1 to 2 inches long
                          6. *M. armillaris.*

1. **Melaleuca leucadéndron** L. CAJEPUT TREE. Fig. 376.

A small or medium-sized tree, 20 to 40 feet high, with spongy bark flaking off in long strips, and pendulous branchlets. Leaf-blades elliptic or oblong-lanceolate, 2 to 4 inches long, ⅜- to ¾-inch wide, with 3 to 7 nearly parallel veins from the base, pale green and glabrous on both surfaces, the young leaves usually silky; petioles about ¼-inch

Fig. 376. Melaleuca leucadendron L.
Portion of branchlet with leaves
and fruits, × ½.

long, gradually broadening upward to the blade. Flower-clusters 1¼ to 4 inches long, yellowish white. Capsules very short-cylindric, about ³⁄₁₆-inch high.

Occasionally cultivated in parks and gardens, especially in southern California.

2. **Melaleuca styphelioìdes** Sm. RIGIDLEAF MELALEUCA. Fig. 377.

A large shrub or small tree, 12 to 30 feet high, with thick spongy bark flaking off in layers. Leaves rigid, ovate, about ½- to ¾-inch long, nearly ¼-inch wide, sharp-pointed, sessile by a broad usually

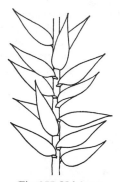

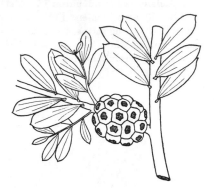

Fig. 377. Melaleuca
styphelioides Sm.
Portion of branchlet
with leaves, × 1.

Fig. 378. Melaleuca nesophila F. Muell.
Portion of branchlet with leaves and
fruit-cluster, × 1.

twisted base. Flower-clusters 1 to 2 inches long, yellowish white, the stem-axils growing out before the flowering is completed. Capsules

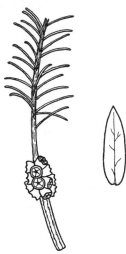

nearly globular, about ⅛-inch in diameter, crowned by the persistent calyx-teeth.

Occasionally cultivated in parks and gardens.

### 3. Melaleuca nesóphila F. Muell. PINK MELALEUCA. Fig. 378.

A shrub or small tree, 6 to 30 feet high, with a slender trunk and thick spongy bark. Leaves obovate-oblong, ½-inch to 1 inch long, nearly ¼-inch wide, obtuse or abruptly pointed at the apex, tapering to a subsessile base, pale green and glabrous on both surfaces. Flower-clusters head-like, about 1 inch in diameter, pink or rose-colored. Capsules globose, about ⅛-inch in diameter, in ovoid or globular compact clusters.

Commonly cultivated in gardens and parks for its dense foliage and pink flowers.

### 4. Melaleuca ericifòlia Sm. HEATH MELALEUCA. Fig. 379.

Fig. 379.
Melaleuca ericifolia Sm.
Portion of branchlet with
mature leaves and fruit-
cluster, × ¾.
Single juvenile leaf, × 1½.

A large shrub or small tree, 10 to 25 feet high, with soft somewhat fibrous bark, spreading branches, and pale green heather-like foliage. Leaves usually alternate, linear, typically ⅝-inch or less long, less than 1/16-inch wide

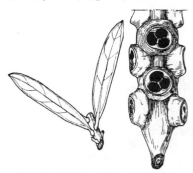

Fig. 380. Melaleuca decussata R. Br.
Leaves, fruit, × 2.

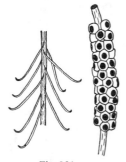

Fig. 381.
Melaleuca armillaris Sm.
Leaves, fruit-cluster, × ½.

(to 3/16-inch on young shoots), usually recurved from the middle, sessile. Flower-clusters ½-inch to 1 inch long, yellowish white. Capsules very short-cylindric, less than 3/16-inch high, clustered in a dense spike.

Occasionally cultivated in gardens and parks. This species suckers abundantly.

**5. Melaleuca decussàta** R. Br. LILAC MEL-ALEUCA. Fig. 380.

A large spreading shrub or small tree, 8 to 20 feet high, with shreddy bark, usually pendulous branches, and glabrous bright green foliage. Leaves opposite, elliptic-oblong or lanceolate, ¼- to ½-inch long, ⅛-inch or less wide, tapering to a narrow sessile base. Flowers lilac, in cylindrical clusters ½-inch to 1 inch long, the axis continuing as a leafy shoot. Capsules about ¼-inch broad, approximately ⅛-inch high, and partly embedded in the stem.

Commonly cultivated in California gardens and parks.

**6. Melaleuca armillàris** Sm. DROOPING MELALEUCA. Fig. 381.

Fig. 382. Metrosideros robusta A. Cunn.
Leaf, fruit-cluster, × 1.

A tall shrub or small tree, 8 to 25 feet high, usually with drooping branchlets. Leaves narrowly linear, ½-inch to 1 inch long, less than 1/16-inch wide, erect or recurved at the apex, sessile. Flower-clusters

1 to 2 inches long, white. Capsules densely clustered, partly sunken into the stem.

Occasionally cultivated in gardens and parks. This is considered the best of the white-flowering melaleucas because of its drooping branchlets and less conspicuous fruit-clusters.

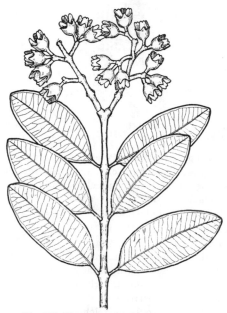

Fig. 383. Metrosideros tomentosa A. Rich.
Portion of branchlet with leaves and
fruits, × ½.

### 7. **Metrosidèros** Banks

(From the Greek *metra*, heart of a tree, and *sideros*, iron, in reference to the hardness of the wood.)

Evergreen trees or shrubs. Leaves simple, opposite, entire, thick, leathery. Flowers showy, with numerous long-exserted white, red, or crimson stamens, in terminal clusters. Fruit a leathery capsule.

About 20 species, native to New Zealand, Australia, and the Pacific Islands.

KEY TO THE SPECIES

Leaves 1 to 1½ inches long, glabrous beneath........................1. *M. robusta.*
Leaves 2 to 4 inches long, white-tomentose beneath.........2. *M. tomentosa.*

**1. Metrosideros robústa** A. Cunn. RATA. Fig. 382.

An evergreen tree, 30 to 60 feet high. Leaf-blades elliptic-ovate to elliptic-lanceolate, 1 to 1½ inches long, ½-inch to 1 inch wide, obtuse at the apex, very leathery, glabrous on both surfaces; petioles ¼-inch or less long. Flowers dark red, in dense terminal clusters. Capsules about ½-inch long, surrounded at the base by the remains of the top-shaped calyx.

Native to New Zealand. Occasionally cultivated in California parks.

**2. Metrosideros tomentòsa** A. Rich. NEW ZEALAND CHRISTMAS TREE. Fig. 383.

A much branched tree, 30 to 60 feet high. Leaf-blades variable, usually oblong-lanceolate or elliptic to elliptic-ovate, 1 to 4 inches long,

Fig. 384. Psidium guajava L. Portion of branchlet with leaf (lower surface) and fruit, × ½.

½-inch to 1½ inches wide, obtuse or acute at the apex, usually tomentose beneath. Flowers dark red, in dense terminal clusters. Capsules about ⅜-inch long, surrounded at the base by the tomentose calyx, opening from the summit.

Native to New Zealand. Occasionally cultivated in California parks and gardens.

**8. Psídium** L. GUAVA

(From the Greek *psidion*, the name of the pomegranate, in reference to the similar fruits.)

Evergreen shrubs or small trees. Leaves simple, opposite, entire. Flowers large, white, with numerous stamens, solitary or 2 or 3 together on slender peduncles. Fruit a berry, crowned by the persistent calyx.

About 150 species, native in tropical and subtropical America.

KEY TO THE SPECIES

Leaves pubescent beneath, the veins prominent; branchlets 4-angled
1. *P. guajava.*

Leaves glabrous beneath, the veins not prominent; branchlets terete
2. *P. cattleianum.*

**1. Psidium guajàva** L. GUAVA. Fig. 384.

An evergreen shrub or small tree, 10 to 25 feet high, with 4 angled young branchlets. Leaf-blades oblong or oblong-elliptical to oval, 2½ to 5 inches long, 1½ to 2 inches wide, rounded or abruptly acute at the apex, dark green and glabrous above, white-pubescent or tomentose and

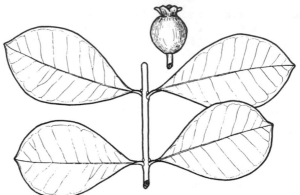

Fig. 385. Psidium cattleianum Sabine.
Leaves, fruit, × ½.

with raised almost straight veins beneath,
entire; petioles about ¼-inch long.
Flowers white, about 1 inch wide, solitary or 2 or 3 in the leaf-axils. Berry globose to pear-shaped, 1 to 3 inches long,
yellow.

Native in tropical America. Grown for
the edible fruit and as an ornamental.

2. **Psidium cattleiànum** Sabine. STRAW-
BERRY GUAVA. Fig. 385.

An evergreen shrub or small tree, 10
to 25 feet high, with terete young branchlets. Leaf-blades leathery, obovate to
broadly elliptic, 1½ to 3½ inches long,
¾-inch to 1½ inches wide, glabrous on
both surfaces, entire; petioles ¼-inch or
less long. Flowers white, solitary, ½-inch
to 1 inch broad. Berry globose or obovoid, ½-inch to 1½ inches long, and
purplish red (yellow in variety **lucidum**
Hort.). Native to Brazil. Cultivated in
southern California for the edible fruits
and as an ornamental.

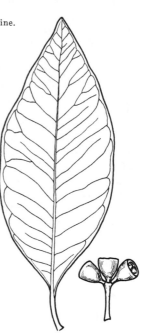

Fig. 386.
Tristania conferta R. Br.
Leaf, fruit-cluster, × ½.

### 9. **Tristània** R. Br.
(Named in honor of Jules M. C. Tristan, a French botanist.)

About 25 species of trees or shrubs, native to Australia, New Caledonia, and the Malay Archipelago.

1. **Tristania conférta** R. Br. BRISBANE-BOX. Fig. 386.

A medium-sized to large evergreen tree, 40 to 75 (or 100) feet high. Leaves simple, alternate, often crowded at the ends of the branches; the blades ovate-lanceolate or elliptical, 3½ to 6 inches long, 1½ to 2½ inches wide, acuminate, glabrous, entire; petioles 1 to 1½ inches long. Flowers white, about ½-inch wide, with numerous stamens, in axillary peduncled few-flowered clusters. Fruit a hemispheric capsule, about ½-inch wide, opening at the summit.

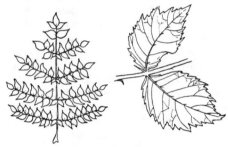

Fig. 387. Aralia spinosa L.
Leaf, × ¼₄. Pair of leaflets, × ½.

Native to Australia. Planted in parks and gardens. The woody capsules resemble those of the genus *Eucalyptus*.

## Araliaceae. Ginseng Family

### 1. Aràlia L.

(Name of uncertain origin.)

About 20 species of herbs, shrubs, or trees, native to North America, Asia, and Australia.

1. **Aralia spinòsa** L. HERCULES' CLUB. DEVIL'S WALKING STICK. Fig. 387.

A small deciduous tree, 15 to 30 feet high, or often shrub-like, with very prickly and pithy stems, branches, and branchlets. Leaves twice-pinnately compound, alternate, 1½ to 2½ feet long, the petioles stout, prickly, about 1½ feet long; leaflets ovate, 2 to 3 inches long, 1 to 1½ inches wide, dark green above, glaucous beneath, serrate, sometimes slightly prickly on the midvein. Flowers small, whitish, in umbels forming large panicles 8 to 16 inches long. Fruit a globose berry-like drupe, about ¼-inch in diameter, black.

Native in the midwestern and southern United States. Occasionally cultivated as an ornamental plant in parks and gardens for its large flower-clusters and large compound leaves.

## 2. Tetrápanax Koch

(From the Greek *tetra*, four, in reference to the four petals, and *Panax*, an herb with all-healing qualities.)

A single species.

### 1. Tetrapanax papyríferum Koch. RICEPAPER PLANT.

*Fatsia papyrifera* Benth. & Hook. *Aralia papyrifera* Hook.

A small straggly evergreen tree or large shrub, 10 to 20 feet high, with dense more or less deciduous tomentum on the young growth. Leaves simple, alternate; the blades broadly ovate to heart-shaped in outline, 8 to 16 inches broad, more or less deeply and palmately 5- or 7-lobed, the lobes toothed, pale green and tomentose above, becoming almost glabrous, densely whitish or brownish tomentose beneath; petioles 6 to 12 inches long. Flowers small, greenish, in umbels forming a large tomentose panicle. Fruit a globose berry-like drupe, about ¼-inch in diameter, black.

Native to Formosa. Used in landscaping patios and about structures of the Spanish type.

## Cornaceae. Dogwood Family

### 1. Córnus L. DOGWOOD

(From the Latin *cornus*, horn, in reference to the toughness of the wood of some species.)

Mostly deciduous trees or shrubs, or rarely herbs. Leaves simple, opposite or rarely alternate, entire or finely serrulate. Flowers small, in terminal clusters. Fruit a small drupe with a 2-celled stone.

About 50 species, native in the temperate region of the northern hemisphere, one in Peru.

#### KEY TO THE SPECIES

Leaves evergreen, 1½ to 4 inches long, narrowed at both ends
1. *C. capitata.*
Leaves deciduous, 3 to 6 inches long.
Involucral bracts 4, 1½ to 2 inches long, notched or truncate at apex, enclosing the flower-buds in winter..............................2. *C. florida.*
Involucral bracts 6 (rarely 4 or 5), 2 to 3 inches long, usually abruptly acute at apex, not enclosing the flower-buds in winter
3. *C. nuttalli.*

### 1. Cornus capitàta Wall. EVERGREEN DOGWOOD. Fig. 388.

A small tree, 15 to 30 feet high. Leaves opposite, evergreen; the blades thick and leathery, elliptic to oblong, 2 to 4 inches long, 1 to 2 inches wide, narrowed at both ends, finely close-pubescent above, densely whitish pubescent beneath; petioles ¼- to ½-inch long. Involucral bracts ovate, acute at the apex, creamy white. Fruits joined into a globular fleshy head about 1 inch in diameter.

Native in the Himalayas. Occasionally planted as a garden or park tree.

2. **Cornus flórida** L. FLOWERING DOGWOOD. Fig. 389.

A bushy tree, 15 to 30 feet high, with green branchlets. Leaves opposite, deciduous; the blades ovate to elliptic, 3 to 6 inches long, 1½ to 2 inches wide, bright green and slightly appressed-pubescent above,

Fig. 388. Cornus capitata Wall.
Leaf, fruit-cluster, × ½.

Fig. 389. Cornus florida L.
Leaf, fruit, × ½.

pale and glaucous beneath, pubescent on the veins; petioles ½- to ¾-inch long. Flowers greenish white; involucral bracts 4, white or pinkish, obovate, 1½ to 2 inches long, notched or truncate and mucronate at the apex, enclosing the flower-buds in winter. Fruit ovoid, about ½-inch long, in dense heads.

Native in southeastern Canada, the eastern and midwestern United States, and in the mountains of northern Mexico. Cultivated in parks and gardens. The variety **rùbra** André has red or pink involucral bracts.

3. **Cornus núttalli** Aud. PACIFIC DOGWOOD. MOUNTAIN DOGWOOD. Fig. 390.

A tree, 10 to 40 feet high, with slender light green branchlets becoming dark reddish purple or brown. Leaves opposite, deciduous; the blades thin, elliptic-obovate or ovate to almost round, 3 to 5 inches long, 1½ to 3 inches wide, bright green and slightly appressed-pubescent above, whitish and glaucous or pubescent beneath, entire or slightly serrulate; petioles ¼- to ½-inch long. Flowers small, yellowish green, crowded in a head-like cluster; involucral bracts 6 (rarely 4 or 5), white, sometimes tinged with green or pink, obovate to oblong, 2 to 3 inches long, usually abruptly acute, not enclosing the winter-buds. Fruit scarlet, ellipsoidal, about ½-inch long, in dense head-like clusters.

Native in the coastal mountains from British Columbia to southern California and on the western slope of the Sierra Nevada from 2000 to 6500 feet elevation. Its distribution in the South Coast Ranges and in southern California is very scattered. Cultivated as an ornamental and

Fig. 390. Cornus nuttalli Aud.
Leaf, fruit, × ¾.

as a street tree in Oregon and Washington. Not hardy in cultivation in the lowlands of California except in the redwood belt. Flowers in May.

## Nyssaceae. Tupelo Family

### 1. Nýssa L. TUPELO

(Nyssa, the Greek name of a nymph, in reference to the swampy habitat of one species.)

Six species of trees, 4 native in North America and 2 in Asia.

1. **Nyssa sylvática** Marsh. TUPELO. SOUR-GUM. Fig. 391.

*N. multiflora* Wang.

A deciduous tree, 30 to 50 (or to 120) feet high, with slender more or less spreading and pendulous branches. Leaves simple, alternate, usually crowded at the ends of the branchlets; the blades lanceolate, elliptic, or obovate, 2 to 5 inches long, ¾-inch to 2½ inches wide, dark green and glossy above, pale beneath, entire or seldom coarsely dentate; petioles slender or stout, ¼-inch to 1¼ inches long. Flowers small, greenish white, unisexual; the staminate in many-flowered peduncled clusters; the pistillate in 2- or few-flowered peduncled axillary clusters. Fruit an ovoid drupe, ⅜- to ⅝-inch long, bluish black, the pulp thin and acrid.

Fig. 391. Nyssa sylvatica Marsh.
Leaf, × ¾. Fruit, × 1½.

Native in swampy regions or wooded mountain slopes from southern Ontario south to Michigan, Missouri, Oklahoma, Texas, and Florida. A very ornamental tree, occasionally cultivated in parks.

## Ericaceae. Heath Family

### 1. Arbùtus L.

(From *arboise*, a Celtic word for rough fruit, in reference to the rough fruit of the Arbute Tree, under which, according to Horace, idle men delight to lie.)

Evergreen trees or shrubs. Leaves simple, alternate. Flowers globular or urn-shaped, in terminal panicles. Fruit a globose many-seeded berry with mealy flesh and granular surface.

About 20 species, native in the Orient, North America, the Mediterranean region, and the Canary Islands.

332     *Pacific Coast Trees*

Leaves 2 to 3 inches long, sharply serrate......................1. *A. unedo.*
Leaves 3 to 6 inches long, entire or finely serrate on young shoots, glaucous beneath......................2. *A. menziesi.*

**1. Arbutus unèdo** L. STRAWBERRY TREE. Fig. 392.

A small tree, 8 to 12 feet high. Leaf-blades coriaceous, oblong-ovate, 2 to 3½ inches long, 1 to 2 inches wide, glabrous, dark green and

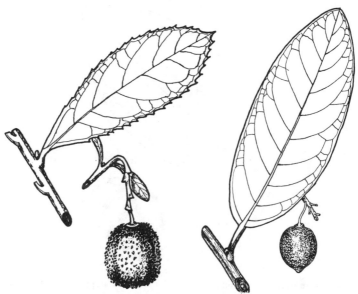

Fig. 392. Arbutus unedo L.
Leaf, fruit, × ¾.

Fig. 393. Arbutus menziesi Pursh.
Leaf, × ½. Fruit, × 1.

glossy above, paler beneath, sharply serrate; petioles ¼- to ½-inch long. Flowers white or tinged with red, globular, about ¼-inch long. Fruit about ¾-inch in diameter, bright red, resembling a strawberry.

Native to Ireland and southern Europe. Cultivated in parks and gardens for the dark glossy foliage, clusters of whitish flowers, and bright red fruits.

**2. Arbutus ménziesi** Pursh. MADRONE. MADROÑO. Fig. 393.

A medium-sized to large tree, 20 to 100 feet high, with smooth terra cotta bark, or the bark rough and dark brown on old trunks. Leaf-blades very thick and leathery, elliptic or oblong to oblong-ovate, 3 to 6 inches long, 1¾ to 2¾ inches wide, dark green and shining above, glaucous beneath, entire or finely serrate especially on young shoots; petioles ½-inch to 1 inch long. Flowers white, widely urn-shaped, about ¼-inch long, in large terminal clusters. Fruit ⅓- to ½-inch in diameter, semifleshy, orange-colored to red.

Native in the foothills and on lower mountain slopes from British Columbia southward in the Coast Ranges to southern California, in

Fig. 394. Diospyros kaki L.
Leaf, fruit, × ½.

the Sierra Nevada to the south fork of the Tuolumne River, and on Santa Cruz Island. A very handsome tree occasionally cultivated in parks and gardens.

## Ebenaceae. Ebony Family

### 1. Diospyros L. PERSIMMON

(From the Greek *Dios*, Jove, and *pyros*, grain, in reference to the edible fruit.)

Deciduous or evergreen trees or shrubs. Leaves simple, alternate, entire, usually leathery. Flowers in axillary clusters or the pistillate solitary. Fruit a large juicy edible berry.

About 200 species, chiefly native in the subtropics and tropics.

334      *Pacific Coast Trees*

Leaves decurrent on the petiole, pubescent beneath, with several distinct
veins from near the base..................................................................1. *D. kaki.*
Leaves not decurrent on the petiole, glabrous beneath when mature,
with 1 distinct vein from the base..............................2. *D. virginiana.*

**1. Diospyros kàki** L. KAKI PERSIMMON. Fig. 394.

A medium-sized deciduous tree, 20 to 40 feet high, with brownish
pubescent branches. Leaf-blades ovate-elliptic or oblong-ovate to
obovate, 3 to 7 inches long, 2 to 3½ inches wide, dark green and
glabrous above, pubescent beneath, decurrent on the petiole, with
few to several distinct upward-curving veins from near the base;
petioles about ⅜-inch long. Flowers yellowish white, ½- to ¾-inch
long. Fruit ovoid, 2 to 3 inches long, orange-yellow or reddish, with
orange-colored pulp and flattened seeds.

Native to Japan and China. Cultivated
in gardens for its foliage and ornamental
edible fruits. Flowers in May and June.

**2. Diospyros virginiàna** L. COMMON
PERSIMMON. Fig. 395.

A medium-sized deciduous tree, 25 to
50 feet high, with spreading or pendulous
branches. Leaf-blades ovate-oblong to
oval or elliptic, 2½ to 6 inches long, 1½
to 3 inches wide, dark green and shining
above, paler and glabrous beneath; petioles ½-inch to 1 inch long. Flowers greenish yellow, ½- to ¾-inch long. Fruit
globose or obovoid, ¾-inch to 2 inches
long, yellow or pale orange, with yellowish
brown pulp and flattened oblong seeds.

Fig. 395.
Diospyros virginiana L.
Leaf, × ½.

Native in the eastern United States from Rhode Island south to
Florida, west to Kansas and Texas. Several varieties with fruit varying in shape and size are in cultivation. Flowers in May and June.

## Oleaceae. Olive Family

### 1. Fráxinus L. ASH

(The ancient Latin name of the ash.)

Deciduous trees or shrubs. Leaves odd-pinnately compound or rarely
reduced to a single leaflet, opposite. Flowers bisexual, dioecious, or
polygamous, small, greenish or greenish white, in panicles. Fruit
usually a 1-seeded samara with the wing at one end or rarely all around.

About 40 species, native in the temperate regions of the northern hemisphere.

KEY TO THE SPECIES

Leaves simple or rarely with 2 or 3 leaflets........................1. *F. anomala.*
Leaves odd-pinnately compound.
    Leaflets sessile or subsessile................................................2. *F. oregona.*
    Leaflets stalked.
        Branchlets quadrangular; leaflets usually 5, serrate, glabrous;
            petals 2.......................................................................3. *F. dipetala.*
        Branchlets not quadrangular.
            Leaflets usually 7 (5 to 9).
                Leaflets 2½ to 6 inches long; petals absent.
                    Leaflets pubescent beneath; branchlets and petioles densely
                        pubescent; petiolules ⅛- to ¼-inch long
                                      4. *F. pennsylvanica.*
                    Leaflets glaucous and either glabrous or pubescent beneath;
                      branchlets glabrous and glossy; petiolules 3⁄16- to 5⁄8-
                      inch long.
                      Leaflets irregularly serrate...........................5. *F. lanceolata.*
                      Leaflets entire or slightly toothed...............6. *F. americana.*
                Leaflets 2 to 3¼ inches long; petals 4.......................7. *F. ornus.*
             Leaflets usually 5 (or 3 to 7), lanceolate, 2 to 4 inches long
                                        8. *F. velutina.*

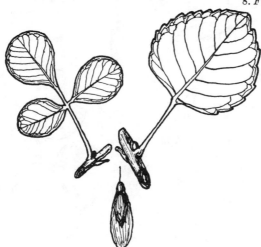

Fig. 396. Fraxinus anomala Wats.<br>Compound leaf, × ½. Single leaf, × 1.<br>Fruit, × 1.

**1. Fraxinus anómala** Wats. DWARF ASH. SINGLELEAF ASH. Fig. 396.
A shrub or small tree, 15 to 20 feet high, with 4-angled branchlets.
Leaves usually reduced to a single leaflet or sometimes with 2 or 3.

Leaflets broadly ovate to almost round, 1¼ (or ½-inch) to 2 inches long, 1 (or ½-inch) to 2 inches in width, dark green and glabrous above, paler beneath, entire or sparingly crenate-serrate; petioles ⅝-inch to 1½ inches long. Flowers polygamous, greenish, without petals, in short compact pubescent panicles. Samaras ½- to ¾-inch long, with a rounded wing surrounding the body.

Native on the Panamint and Providence mountains in Inyo and San Bernardino counties of California, east to Colorado and Texas.

2. **Fraxinus oregòna** Nutt. OREGON ASH. Fig. 397.

A tree, 30 to 70 feet high, with velvety or pilose to nearly glabrous

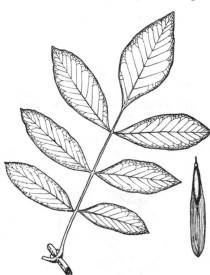

Fig. 397. Fraxinus oregona Nutt. Leaf, × ¼. Fruit, × ⅔.

terete branchlets. Leaves 5 to 12 inches long; leaflets 5 or 7 (rarely 9), ovate to elliptic or rarely obovate, 3 to 6 inches long, 1 to 1½ inches wide, abruptly pointed, light green and nearly glabrous above, paler and usually pubescent beneath, entire or obscurely serrate above the middle, lateral leaflets usually sessile, the terminal petiolulate. The flowers dioecious and without petals, in compact glabrous panicles. Samaras

Fig. 398. Fraxinus dipetala H. & A. Leaf, fruit, × ½.

oblanceolate to oblong-lanceolate, 1 to 2 inches long, borne in crowded clusters, the wing terminal.

Native along stream banks and moist valley flats in the Sierra

Nevada foothills, the Coast Ranges, and higher mountains of southern California, north to British Columbia. Occasionally cultivated as a park and street tree.

3. **Fraxinus dipétala** H. & A. FOOTHILL ASH. MOUNTAIN ASH. CALIFORNIA SHRUB ASH. Fig. 398.

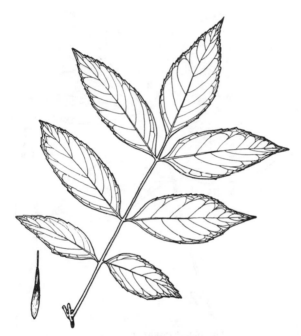

Fig. 399. Fraxinus pennsylvanica Marsh.
Leaf, fruit, × ⅓.

A shrub or small tree, 6 to 18 feet high, with quadrangular branchlets. Leaves 2 to 5½ inches long; leaflets 3 to 9 (rarely 1), oblong-ovate or ovate, ½-inch to 1½ inches long, ⅜- to ¾-inch wide, glabrous above and beneath, serrate or entire below the middle, short-stalked. Flowers bisexual, with 2 white petals, in compound axillary clusters. Samaras oblanceolate or elliptical, ¾-inch to 1 inch long, the wing terminal and often notched at the apex.

Native to California, in canyons and on the foothills and lower
mountains of the middle and inner North Coast Ranges, the Sierra
Nevada, the inner South Coast Range, and the San Bernardino Mountains. Flowers from March to June.

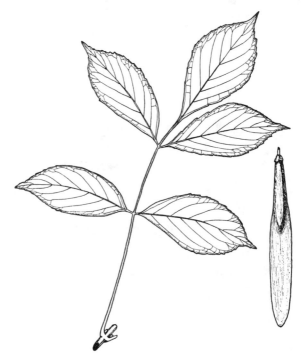

Fig. 400. Fraxinus americana L.
Leaf, × ⅛. Fruit, × 1.

### 4. **Fraxinus pennsylvánica** Marsh. RED ASH. Fig. 399.

*F. pubescens* Lam.

A tree, 30 to 50 feet high, with densely pubescent branchlets and
petioles. Leaves 10 to 12 inches long; leaflets usually 7 (5 to 9),
oblong-lanceolate, ovate, or obovate, 4 to 6 inches long, 1 to 1½
inches wide, narrowed at the apex to a long slender point, light yellow-green and glabrous above, pale and pubescent beneath, crenate-
serrate, or entire below the middle; petiolules ⅛- to ¼-inch long or

the terminal leaflet with a petiolule up to 1 inch long. Flowers dioecious, without petals, in rather compact tomentose panicles. Samaras oblong-obovate or lanceolate to elliptic, 1 to 2½ inches long, the wings terminal and usually rounded at the apex, sometimes emarginate or tipped with a slender bristle.

Native in the eastern United States and southeastern Canada. Occasionally cultivated in parks.

### 5. **Fraxinus lanceolàta** Borkh. GREEN ASH.

*F. pennsylvanica* var. *lanceolata* Sarg.

Branchlets and petioles glabrous. Leaflets lanceolate to ellipticoblong, bright green and glabrous above, paler and glabrous beneath, or pubescent along the midrib.

Native with the preceding species, but more common in the western part of its range.

### 6. **Fraxinus americàna** L. WHITE ASH. Fig. 400.

*F. alba* Marsh.

A tree, 60 to 80 feet high, with round glabrous glossy branchlets. Leaves 8 to 12 inches long; leaflets usually 7 (5 to 9), ovate to oblanceolate, 3 to 5 inches long, 1½ to 3 inches wide, dark green and glabrous above, pale and either glabrous or pubescent beneath, nearly entire or finely and obscurely serrulate; petiolules ¼- to ½-inch long, or the terminal one up to 1 inch long. Flowers dioecious, without petals, in glabrous branched clusters.

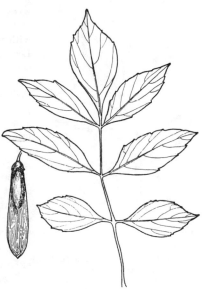

Fig. 401. Fraxinus ornus L.
Fruit, × 1. Leaf, × ⅓.

Samaras lanceolate or oblanceolate, 1 to 2½ inches long, the wing terminal and pointed or slightly notched at the apex.

Native in southeastern Canada and the midwestern United States, from Michigan and Minnesota southward to Texas and Florida. Often cultivated in parks and gardens.

**7. Fraxinus órnus** L. FLOWERING ASH. Fig. 401.

A small or medium-sized tree, 15 to 35 feet high. Leaves 6 to 10 inches long; leaflets usually 7 (5 to 9), elliptic to ovate, the terminal one obovate, 1½ to 3 inches long, ½-inch to 1 inch wide, abruptly pointed, glabrous above, glabrous beneath except for a rusty pubescence at the base of the midrib. Flowers perfect, fragrant, with 4 white petals, in dense terminal panicles 3 to 5 inches long. Samaras lanceolate or narrow-oblong, ¾-inch to 1 inch long, the wing terminal, truncate or notched at the apex.

Native in southeastern Europe and western Asia. Cultivated for the large clusters of fragrant white flowers, which appear in May and June.

Fig. 402. Fraxinus velutina Torr.
Leaf, fruit, × ½.

**8. Fraxinus velùtina** Torr. ARIZONA ASH. Fig. 402.

A small tree, 15 to 30 feet high, with velvety pilose terete branchlets. Leaves 4 to 8 inches long, with pubescent petioles; leaflets usually 5 (3 to 7), narrowly ovate to elliptic, 1 to 1½ inches long, ¾-inch to 1 inch wide, pale green and glabrous above, paler and tomentose beneath, finely crenulate-serrate above the middle. Flowers dioecious, without petals, greenish, in elongated pubescent panicles. Samaras elliptical to oblong-ovate, about 1 inch long, the wing terminal.

Native on the desert mountain slopes of southeastern California and eastward to Texas.

*8a*. Var. **coriàcea** Jepson. LEATHER-LEAF ASH.

*F. coriacea* Wats.

Similar to the species, but may be distinguished by its thicker, leathery, and more coarsely serrate leaflets and less pubescent branchlets. It has the same range as the species but is not common.

## 2. Ligùstrum L. PRIVET

(From the Latin *ligulare,* to tie, in reference to the use made of the flexible shoots.)

About 50 species of shrubs or small trees, native in eastern Asia, the Himalayas, Australia, Europe, and northern Africa.

1. **Ligustrum lùcidum** Ait. GLOSSY PRIVET. Fig. 403.

*L. japonicum* var. *macrophyllum* L.

An evergreen shrub or small tree, 10 to 25 feet high. Leaves simple, opposite; the blades thick and leathery, ovate-lanceolate, 2½ to 5 inches long, 1¼ to 2 inches wide, glabrous, acuminate, rounded at the base, entire, somewhat trough-shaped and dark green and shining above, paler beneath; petioles ¼- to ½-inch long. Flowers small, white, in large terminal panicles 4 to 8 inches long. Fruit a black berry-like drupe about ¼-inch in diameter.

Native to China and Japan. Cultivated in gardens and parks. Some varieties have variegated leaves.

Fig. 403.
Ligustrum lucidum Ait.
Fruit-cluster, leaf, × ½.

### 3. **Òlea** L. OLIVE

(The classical Latin name of the olive.)

About 30 species of evergreen shrubs and small trees, native in warm temperate regions of the Old World.

1. **Olea europaèa** L. COMMON OLIVE. Fig. 404.

A small or medium-sized tree, 15 to 30 feet high, with smooth gray bark and spreading branches forming a broad compact rounded crown.

Fig. 404.
Olea europaea L.
Fruit, leaf, × ½.

Leaves simple, opposite, evergreen; the blades leathery, elliptical to oblong-lanceolate, 1½ to 2¾ inches long, ⅜- to ⅝-inch wide, tapering at the base, dull green and glabrous above, white- or silver-tomentose beneath, entire; petioles about ⅛-inch long. Flowers small, yellowish white, in axillary branching clusters. Fruit an ovoid or subglobose drupe, ½-inch to 1½ inches long, black when mature.

Native in the eastern Mediterranean region and in western Asia. Cultivated in groves for its edible fruits and in gardens for its gray-green foliage. Flowers in April and May.

### 4. **Syrínga** L. LILAC

(From the Greek *syrinx*, meaning a pipe or tube. Pipes are made from stems of the genus *Philadelphus* to which the name *Syringa* is often but incorrectly applied.)

About 30 species, native in Europe and Asia. The lilacs are mostly deciduous shrubs but sometimes they become tree-like. They are among the most popular ornamental woody plants because of their large clusters of fragrant purple, white, yellowish, or lilac flowers.

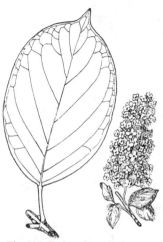

Fig. 405. Syringa japonica Decne.
Leaf, × ½. Flower-cluster, × ⅒.

### 1. Syringa japónica Decne. Fig. 405.

A small tree, 15 to 30 feet high, with upright branches forming a pyramidal crown. Leaves simple, opposite, deciduous; the blades broadly elliptic or ovate, 3 to 6 inches long, 1½ to 2½ inches wide, rounded to subcordate at the base, short-acuminate, dark green and glabrous above, paler and pubescent beneath, at least when young, and entire; petioles ½- to ¾-inch long. Flowers small, yellowish white, borne in large terminal panicles 8 to 12 inches long. Fruit a small leathery ovoid capsule.

Native to Japan. Occasionally cultivated as a park, garden, or street tree, especially in Washington and Oregon. This lilac is the only real tree-species of the genus *Syringa*.

## Apocynaceae. Dogbane Family

### 1. Nèrium L. OLEANDER

(From *Nerion*, the Greek name of the oleander.)

Three species and several varieties, native from the Mediterranean region eastward to Japan. The oleanders are usually shrubs, but on the Pacific Coast, especially in the warmer regions of California, they are often tree-like.

### 1. Nerium oleánder L. OLEANDER. Fig. 406.

An erect evergreen shrub or small tree, 8 to 20 feet high. Leaves simple, usually in whorls of 3; the blades thick and leathery, narrowly oblong-lanceolate, 3 to 5½ inches long, ¾-inch to 1 inch wide, acute or acuminate, tapering at the base, transversely feather-veined, glabrous, dark green above, paler beneath, entire; petioles ¼- to ½-inch long. Flowers large, 1½ to 3 inches broad, white, pink, red, or purple, often double, in terminal branching clusters. Fruit composed of 2 elongated follicles.

Native in the Mediterranean region. Extensively cultivated and well suited for street and garden planting if kept free of scale. Thrives best in the warmer parts of the Pacific Coast.

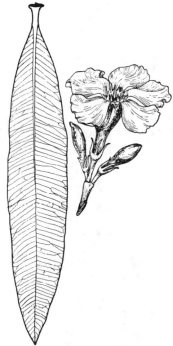

Fig. 406. Nerium oleander L.
Leaf, flower, × ½.

## Scrophulariaceae. Figwort Family

### 1. **Paulòwnia** Sieb. & Zucc.

(Named after Anna Paulowna, princess of the Netherlands.)
About 8 species of trees, native in China.

1. **Paulownia tomentòsa** Steud. ROYAL PAULOWNIA. Fig. 407.
 *P. imperialis* Sieb. & Zucc.

A medium-sized deciduous tree, 25 to 40 feet high. Leaves simple, opposite; the blades broadly heart-shaped, 5 to 12 inches long, 4 to 8 inches broad, light green and pubescent above, paler and tomentose beneath, entire or sometimes 3-lobed; petioles 4 to 6 inches long. Flowers showy, purplish, 1½ to 2 inches long, fragrant, in terminal panicles 6 to 10 inches long. Fruit an ovoid woody capsule, 1 to 1½ inches long, 2-celled, with numerous small winged seeds.

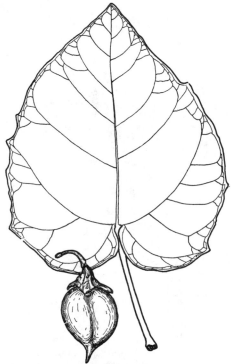

**Fig. 407. Paulownia tomentosa Steud.**
Fruit, × ⅔. Leaf, × ⅛.

Native to China. Cultivated in parks and gardens. The leaves, shape of flowers, and habit of growth remind one of the genus *Catalpa*. Flowers in April and May before the leaves appear.

## Bignoniaceae. Bignonia Family
### 1. Catálpa Scop.

(The American Indian name of one species, probably a corruption of the Indian word, *Catawba*, the name of a tribe in Georgia and the Carolinas.)

Deciduous or rarely evergreen trees or shrubs. Leaves simple, opposite or sometimes whorled. Flowers large, showy, white or mottled with pink, purple, or yellow, in terminal panicles or racemes. Fruit an elongated terete pendent capsule with many seeds, each seed bearing a white hairy tuft at each end.

About 10 species, native to eastern North America, the West Indies, and eastern Asia.

KEY TO THE SPECIES

Leaves abruptly short-acuminate; the flowers spotted with yellow and purple; wings of seeds narrowed..............................1. *C. bignonioides.*
Leaves long-acuminate; flowers nearly white, with only faint brownish spots; wings of seeds broad............................................2. *C. speciosa.*

1. **Catalpa bignonioides** Walt. COMMON CATALPA. Fig. 408.

A small or medium-sized tree, 20 to 50 feet high. Leaf-blades broadly ovate, 6 to 12 inches long, 4 to 7 inches wide, rounded or heart-shaped at the base, abruptly short-acuminate, dark green and almost glabrous above, paler and pubescent beneath, strong-scented, entire or rarely laterally lobed; petioles nearly as long as the blades. Flowers white, with purple streaks and spots in the throat, 1 to 1¾ inches long, in large erect terminal panicles. Capsules linear-cylindric, 6 to 12 inches long. Seeds about 1 inch long, with pointed wings terminated by tufts of white hairs.

Native to Georgia, Florida, Alabama, and Mississippi. Naturalized in many parts of the eastern United States. Commonly cultivated on the Pacific Coast as a garden, park, or street tree. Flowers in May and June.

Fig. 408. Catalpa bignonioides Walt.
Flower, leaf, × ⅓.

2. **Catalpa speciòsa** Engelm. WESTERN CATALPA. Fig. 409.

A large tree, 40 to 70 feet high. Leaf-blades broadly ovate, 8 to 12 inches long, 6 to 8 inches wide, heart-shaped at the base, long-pointed at the apex, dark green and glabrous above, soft-pubescent beneath, entire or with 1 or 2 lateral teeth; petioles 4 to 6 inches long. Flowers nearly white, usually spotted externally with purple and internally with yellow streaks and purple spots, about 2 inches long, in open few-

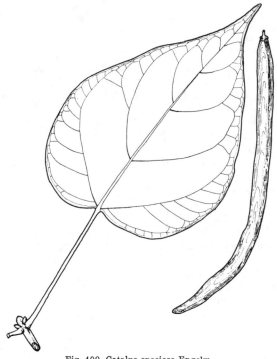

Fig. 409. Catalpa speciosa Engelm.
Leaf, fruit, × ⅓.

flowered panicles 4 to 6 inches long. Capsules long-cylindric, 8 to 18 inches long. Seeds about 1 inch long, with wings rounded at the ends and terminated by fringes of white hairs.

Native in southern Indiana, Illinois, Kentucky, Tennessee, Missouri, and Arkansas. Cultivated in parks and gardens for the broad leaves and large showy flowers which blossom in May and June.

## 2. Chilópsis Don

(From the Greek *cheilos*, lip, and *opsis*, resemblance, in reference to the 2-lipped corolla.)

A single species, allied to *Catalpa*.

### 1. Chilopsis lineàris DC. DESERT-WILLOW. Fig. 410.

*C. saligna* Don.

A slender-branched deciduous shrub or small tree, 10 to 20 feet high.

Fig. 410. Chilopsis linearis DC.
Portion of branchlet with
leaves and fruit, × ⅓.

Fig. 411. Jacaranda ovalifolia R. Br.
Fruit, leaf, flower, × ⅓.

Leaves simple, opposite or whorled or some alternate, sessile, linear or linear-lanceolate, 2 to 5 inches long, ⅛- to ¼-inch wide, often sickle-shaped, pale gray- or yellow-green, glabrous, entire. Flowers showy, pink or whitish, with purplish markings, 1 to 2 inches long, borne in short terminal clusters. Fruit a pendent linear capsule, 4 to 10 inches long, with many hairy-winged seeds.

Native in desert washes and stream beds of the deserts of southern California, eastward to Nevada, Arizona, New Mexico, and western Texas, south to Lower California. Rarely cultivated.

### 3. Jacaránda Juss.

(The Brazilian name.)

About 50 species of trees and shrubs, native in the tropics of America.

**1. Jacaranda ovalifòlia** R. Br. GREEN-EBONY. Fig. 411.

*J. mimosaefolia* D. Don.

A medium-sized to large tree, 30 to 60 feet high, deciduous only in early spring. The leaves twice- (or rarely once-) pinnately compound, opposite, and with 20 to 40 pinnae; the leaflets 18 to 48 on each pinna,

oblong, ¼- to ⅓-inch long, short-pointed, pubescent, the terminal one long-acuminate and about ½-inch long. Flowers large, showy, tubular, blue, about 2 inches long, numerous, in loose terminal panicles 6 to 8 inches long. Fruit an ovoid capsule, 1 to 2 inches long.

Native to Brazil. Cultivated in southern California for its feathery fern-like foliage and clusters of large blue flowers appearing in June and July.

## Myoporaceae. Myoporum Family

### 1. Myóporum Banks

(From the Greek *muein*, to close, and *poros*, pore, in reference to the translucent dots on the leaves.)

About 30 species of shrubs or trees, native from Japan to Australia.

Fig. 412.
Myoporum laetum
Forst.
Leaf, × ½.
Fruit, × 1.

**1. Myoporum laètum** Forst. Fig. 412.

An evergreen shrub or small tree. Leaves simple, alternate; the blades lanceolate to elliptic-oblong, 2 to 4 inches long, ½-inch to 1½ inches wide, tapering at both ends, glabrous, with translucent dots, dark green and shining above, slightly paler beneath, entire or serrate above the middle; petioles ½-inch or less long. Flowers white, spotted with purple, about ½-inch wide, 2 to 6 in axillary clusters. Fruit a small ovoid drupe, about ⅜-inch long, black or reddish purple.

Native to New Zealand. Cultivated in California parks.

## Caprifoliaceae. Honeysuckle Family

### 1. Sambùcus L. ELDER

(From the Greek *sambuke*, a musical instrument, made of elder wood.)

Deciduous shrubs, small trees, or rarely perennial herbs, with stout pithy branches. Leaves odd-pinnately compound, opposite. Flowers small, in terminal compound clusters. Fruit berry-like, with 3 to 5 1-seeded nutlets.

About 20 species of trees and shrubs, native in temperate and subtropical regions of both hemispheres.

KEY TO THE SPECIES

Flowers in a flat-topped cluster; fruit blue or black, covered with a whitish bloom; leaflets entire near apex.
    Leaves glabrous................................................................1. *S. caerulea.*
    Leaves pubescent................................................................2. *S. velutina.*
Flowers in a dome-shaped cluster; fruit red; leaflets pubescent beneath, serrate to the very apex........................3. *S. racemosa* var. *callicarpa.*

Fig. 413. Sambucus caerulea Raf.<br/>Fruit-cluster, leaf, × ½.

**1. Sambucus caerùlea** Raf. BLUEBERRY ELDER. Fig. 413.

*S. glauca* Nutt.

A large deciduous shrub or a small tree, 15 to 30 feet high. Leaves 4 to 7 inches long; leaflets 5 or 7, somewhat leathery, oblong-lanceolate or ovate, 1 to 4 inches long, glabrous, sharply serrate except at the very apex, sessile or short-stalked, usually unequal at the base, the lower leaflets sometimes pinnately divided. Flowers white, in flat-topped clusters 2 to 6 inches wide. Fruit globose, about ¼-inch in diameter, blue, covered with a whitish bloom.

Native on canyon slopes and in open woods of the lower mountain slopes, foothills, and valleys from British Columbia southward to south-

Fig. 414. Sambucus velutina D. & H.
Leaf, × ⅓.

ern California, and eastward to Utah. Rarely cultivated. Flowers from May to August.

## 2. **Sambucus velùtina** D. & H. Fig. 414.

*S. mexicana* Presl. *S. canadensis* var. *mexicana* Sarg.

A deciduous shrub or small tree, 10 to 20 feet high, with whitish pubescent young branchlets. Leaves 6 to 9 inches long; leaflets 5 to 9, usually thick and leathery, lanceolate, acute at the apex, 2 to 6 inches long, 1 to 2 inches wide, soft-pubescent above and below, finely and regularly serrate except at the apex, unequal at the base, short-petiolulate, the lower leaflets sometimes divided. Flowers pale yellow, fragrant, in flat-topped clusters 3 to 12 inches wide. Fruit globose, about ⅛-inch in diameter, black, with a pale bloom.

Native in the Sierra Nevada and the mountains of southern California from 3000 to 8000 feet elevation, and probably extending eastward in Arizona and New Mexico to Texas, and southward to Mexico. Nowhere abundant. Flowers from June to August.

3. **Sambucus racemòsa** var. **callicárpa** Jepson. Redberry Elder. Fig. 415.

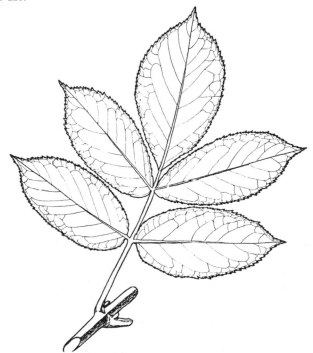

Fig. 415. Sambucus racemosa var. callicarpa Jepson.
Leaf, × ½.

A deciduous shrub or small tree, 10 to 25 feet high. Leaves 5 to 8 inches long; leaflets 5 or 7, rather thinnish, oblong-ovate or obovate, 2 to 6 inches long, dark green and glabrous or slightly pubescent above, paler and pubescent beneath, sharply serrate to the very apex, sessile or short-stalked. Flowers cream-colored, in terminal dome-shaped clusters 2 to 3 inches long. Fruit about ¼-inch in diameter, red.

Native along the coast from San Mateo County, California, northward to Washington. Rarely cultivated. Flowers from late April to late June. Fruit ripens from July to September

# GLOSSARY OF BOTANICAL TERMS

*Abruptly pinnate.* A pinnate leaf ending in a pair of leaflets.

*Achene.* A one-seeded dry fruit from a simple or compound ovary which does not open.

*Acorn.* A nut partly surrounded by a fibrous or woody cup; the fruit of an oak.

*Acuminate.* Gradually diminishing to the apex.

*Acute.* Terminating in an angle, usually less than a right angle, but not prolonged.

*Adherent.* The growing together of different organs, as the stamens to the petals.

*Aggregate fruit.* A collection of ovaries from a single flower ripening into a single fruit, as in the blackberry.

*Angiospermae.* Plants having their seeds enclosed in an ovary.

*Angiospermous.* Having the seeds borne in an ovary, as in Angiosperms.

*Angiosperms.* Angiospermae.

*Anther.* The pollen-bearing pouch at the top of the stamen

*Apetalous.* Without petals.

*Apex.* The end or summit, as the end of a leaf.

*Appressed.* Pressed close to the stem or other organ.

*Arborescent.* Tree-like.

*Armed.* Bearing spines, thorns, or similar structures.

*Aromatic.* With a pleasant odor.

*Auriculate.* With small ear-shaped lobes or appendages.

*Awl-shaped.* Narrow and tapering to a point like a shoemaker's awl; subulate.

*Awn-tipped.* Tipped with a bristle-like appendage or awn.

*Axillary.* In an axil or angle, as a bud in the axil formed by the leaf and stem.

*Berry.* A fleshy fruit from a compound ovary with a fleshy pulp throughout.

*Bifid.* Divided halfway into two.

*Bipinnate.* Twice pinnate.

*Bract.* A modified or undeveloped leaf associated with the flower or flower-cluster; a narrow structure subtending the cone-scales in cone-bearing trees.

*Bur.* A prickly fruit like that of a chestnut or beech.

*Buttressed.* Trunks swollen at the base.

*Callous-tipped.* Tipped with hardened swellings

*Calyces.* Plural of calyx.

*Calyx.* The sepals collectively; the outermost series of floral envelopes.

*Campanulate.* Bell-shaped.

*Capsule.* A dry fruit from a compound ovary, splitting along two or more lines.

*Carpel.* A simple pistil or an element of a compound pistil; a modified leaf constituting a pistil.

*Catkin.* A slender compact cluster of unisexual flowers, as in the willow.

*Chaparral.* The thorny or rigid shrubs usually growing on dry mountain slopes.

*Ciliate.* Having the margin fringed with hairs

*Claw.* The petiole-like base of petals and sepals.

*Coalesced.* United organs of the same kind, as the union of petals.

*Compound.* Composed of more than one similar part united into a single organ.

*Cone.* A collection of ovule-bearing scales, bracts, or carpels, usually applied to the fruits of pines, firs, and other Coniferae, but also to those of alders, birches, etc.

*Coniferae.* Cone-bearing trees; gymnosperms.

*Cordate.* Heart-shaped, with a broad and notched base.

*Coriaceous.* Leathery.

*Corymb.* A type of inflorescence in which the lower divisions or pedicels are elongated so as to form a flat-topped cluster.

*Cotyledon.* A leaf of the embryo within the seed-coat.

*Crenate.* Notched with rounded teeth.

*Crown.* The upper part of the tree composed of the branches and foliage; the junction of stem and root in a seed plant.

*Cuneate.* Wedge-shaped.

*Cusp.* A sharp, rigid point.

*Cyclic.* Arranged in whorls; relating to a cycle.

*Deciduous.* Said of leaves that drop off before winter.

*Decurrent.* Running down, as when leaves extend down the stem beyond the point of attachment.

*Decussate.* In opposite pairs alternately arranged at right angles.

*Dehiscent.* Splitting open naturally at maturity.

*Dentate.* Toothed with the teeth directed outward.

*Denticulate.* Minutely dentate.

*Dioecious.* Staminate and pistillate flowers borne upon separate plants, as in the willow.

*Drupe.* A fleshy fruit from a simple ovary, with a fleshy outside (exocarp) surrounding a hard stone or pit (endocarp), as in the peach.

*Drupelets.* Small drupes, as the sections of a blackberry.

*Eccentric.* Said of midribs when not in the middle; out of the center.

*Ellipsoidal.* Shaped like an ellipsoid, that is, a solid bounded by a surface all plane sections of which are elipses or circles.

*Elliptic.* Like an ellipse; longer than broad and with regularly curving margins and rounded ends.

*Elliptical.* Elliptic.

*Emarginate.* With a notch cut out, usually at the tip.

*Entire.* With an even margin, not toothed.

*Erose.* Having the margin irregularly notched as if gnawed out.

*Even-pinnate.* A pinnate leaf ending in a pair of leaflets.

*Evergreen.* With leaves throughout the year.

*Exfoliating.* To peel off in scales or flakes.

*Exocarp.* The outer fleshy part of a drupe or stone-fruit.

*Exotic.* Not native; introduced from abroad.

*Facial leaves.* The small scale-like leaves of certain members of the Cypress Family, borne on the flat sides of the branches, not on the edges.

*Falcate.* Sickle-shaped.

*Fascicled.* In clusters, as the needles of pine.

*Feather-veined.* With the veins of the leaf arising from the sides of the midrib.

*Filament.* A thread-like structure; the stalk of the stamen.

*Flexuous.* Bent alternately in opposite directions; more or less zigzag.

*Foliate.* Having leaves.

*Foliolate.* Having leaflets, as 3-foliolate.

*Follicle.* A dry fruit from a simple pistil, splitting along one line.

*Four-ranked.* In four rows.

*Funicle.* The stalk that attaches the ovule or seed to the ovary wall.

*Glabrate.* Nearly glabrous or becoming glabrous in age.

*Glabrous.* Not hairy, bald.

*Glaucescent.* Nearly glaucous or becoming so.

*Glaucous.* Covered with a whitish or grayish bloom which easily rubs off.

*Globose.* Spherical in form.

*Globular.* Globose.

*Gymnospermae.* Plants with their seeds not enclosed in an ovary.

*Gymnospermous.* Having seeds not enclosed in an ovary.

*Gymnosperms.* Gymnospermae.

*Head.* A type of inflorescence, as in the sunflower.

*Hesperidium.* The fruit of the Citrus Family, the orange, etc.

*Hispid.* With stiff hairs.

*Hybrid.* A cross between two nearly related species.

*Imbricated.* Overlapping like shingles on a roof.

*Indehiscent.* Not breaking open at maturity.

*Indigenous.* Native to.

*Inferior.* Said of ovaries when grown to the calyx or receptacle; below the other floral parts.

*Inflorescence.* The arrangement of flowers on the stem.

*Internode.* Portion of stem or branch between two nodes.

*Involucrate.* With an involucre.

*Involucre.* A circle of bracts subtending a flower-cluster or some fruits.

*Involute.* Having the edges of the leaves rolled inward.

*Keeled.* With a central ridge or keel on the back of an organ.

*Lanceolate.* Lance-shaped, about 4 to 6 times as long as wide and broader below the middle.

*Leaf-blade.* The expanded part of the leaf.

*Leaf-rachis.* The central axis of a pinnately compound leaf; the continuation of the petiole.

*Leaflet.* A division of a compound leaf.

*Legume.* A dry fruit from a simple pistil, splitting along two lines.

*Life Zones.* Regions determined by temperature, water, altitude, latitude, and other factors, in which plants and animals live. See Introduction, pp. 16–20.

*Linear.* Long and narrow with parallel sides.

*Lobe.* Any division of an organ, often rounded.

*Membranaceous.* Membranous.

*Membranous.* Thin and rather soft, like an animal membrane.

*Midrib.* The central vein or rib of a leaf.

*Monocotyledons.* Plants that have seeds with one seed-leaf; the foliage leaves are usually parallel-veined.

*Monoecious.* Staminate and pistillate flowers borne on the same plant.

*Montane.* Growing in the mountains.

*Mucronate.* Tipped with a short abrupt point.

*Netted-veined.* Reticulated; with any system of closely interwoven cross veins.

*Node.* The place of attachment of leaves on the stem.

*Nut.* A dry fruit from a compound ovary, with a hard wall not splitting open.

*Nutlet.* Small nut.

*Ob-.* A Latin prefix, signifying an inversion.

*Oblanceolate.* Inversely lanceolate, the broadest part beyond the middle.

*Oblique.* Unequal-sided.

*Oblong.* About 3 or 4 times as long as wide and with nearly parallel sides.

*Obovate.* Inversely ovate.

*Obtuse.* Blunt or rounded.

*Odd-pinnate.* A pinnate leaf ending in a single leaflet.

*Orbicular.* Circular in outline.

*Oval.* Broad-elliptic, with rounded ends.

*Ovary.* Part of the pistil bearing the ovules.

*Ovate.* Of the shape of a longitudinal section of a hen's egg, with the broad end basal.

*Ovoid.* Egg-shaped; solid ovate, or less correctly solid oval.

*Ovulate.* Bearing ovules; said of the female flowers of Gymnosperms because they have no pistils.

*Ovule.* The structure in the ovary which when matured becomes the seed.

*Palmate.* Radiately lobed or divided, like the fingers from the palm of the hand.

*Palmately compound.* The leaflets all borne at the summit of the petiole.

*Palmately lobed.* Radiately lobed.

*Palmately veined.* The primary veins radiating from the summit of the petiole.

*Panicle.* A compound flower-cluster.

*Papilionaceous.* Flowers with the shape of sweet pea flowers, that is, having a large upper petal (banner), two smaller lateral petals (wings), and two lower slightly joined petals forming a keel.

*Papillae.* Small nipple-shaped protuberances.

*Parallel-veined.* Veins as in a grass leaf.

*Pedicel.* Stalk of the individual flowers in a cluster.

*Peltate.* Umbrella- or toadstool-shaped, as a leaf attached by its lower surface to a stalk instead of by its base.

*Perfect.* Flowers with both stamens and pistils.

*Perianth.* Collectively the corolla and calyx; floral envelopes.

*Petals.* The inner cycle of floral envelopes, usually colored.

*Petiolate.* With a petiole.

*Petiole.* The stalk of a leaf.

*Petiolule.* The stalk of a leaflet.

*Phyllodia.* Flattened petioles or leaf-rachises performing the function of leaf-blades, as in many species of *Acacia.*

*Pinna.* A division of a compound leaf.

*Pinnate.* With the divisions of a compound leaf arranged along the sides of the central axis or rachis.

*Pinnately compound.* Pinnate.

*Pinnately veined.* With the veins arising from the sides of the midrib.

*Pistillate.* Flowers with pistils but without stamens.

*Plane.* Flat and even.

*Plumose.* With long branched hairs; feathery.

*Pod.* A dry fruit that splits open.

*Polygamo-dioecious.* With perfect, staminate, and pistillate flowers on different plants.

*Polygamous.* With perfect, staminate, and pistillate flowers on the same plant (polygamo-monoecious).

*Pome.* A fleshy fruit from a compound ovary, with papery or cartilaginous carpels surrounded by a fleshy adherent calyx and receptacle, as in the apple.

*Puberulent.* Minutely pubescent.

*Pubescent.* With soft or downy hairs.

*Pungent.* Terminating in a rigid sharp point.

*Raceme.* A type of inflorescence with a central elongating axis giving off pedicels of about equal length.

*Rachis.* The central axis of a compound leaf; the continuation of the petiole; the axis of a raceme or panicle; the continuation of the peduncle.

*Receptacle.* The summit of the stalk that bears the parts of a flower; the enlarged summit of the stem bearing flowers, as in the Sunflower Family.

*Reflexed.* Abruptly bent or turned downward or backward.

*Remote.* Scattered; not close together on the axis.

*Resin.* Hardened turpentine.

*Reticulate-veined.* Netted-veined; with a network of veins.

*Revolute.* Rolled back from the margins.

*Rhomboid.* Approaching a quadrangular outline, with the lateral angles obtuse.

*Rhomboidal.* Rhomboid.

*Ringed.* Said of palms when the scars of the old leaf-bases form distinct rings around the trunk.

*Samara.* A winged nut, achene, or any other indehiscent fruit, as in the ash and maple.

*Scabrous.* Rough to the touch.

*Scale.* A small thin structure, usually a degenerate leaf.

*Scurfy.* With small bran-like scales on the epidermis.

*Seed.* The structure developed from the ovule.

*Serrate.* Saw-toothed, the teeth directed forward or upward.

*Serrulate.* Minutely serrate.

*Sessile.* Without a stalk.

*Sheath.* An enrolled part of an organ, as the lower part of a leaf-blade or expanded petiole enclosing the stem for some distance; the tissue surrounding the cluster of pine needles.

*Simple.* Of one part; not branched.

*Sinuate.* With a wavy or recessed margin.

*Sinuses.* Plural of sinus, a recess or indentation, as between the lobes of some oak leaves.

*Spike.* A cluster of flowers without pedicels and set along a central axis.

*Spine.* A sharp-pointed and hardened or woody structure.

*Spinose.* Having spines.

*Stamen.* One of the essential organs of a flower, bearing pollen in the anther.

*Staminate.* Flowers with stamens but without pistils.

*Stipules.* Small appendages borne in pairs at the base of the petiole of a leaf.

*Stoloniferous.* With basal shoots disposed to send out roots; with stolons or suckers.

*Stomata.* Plural of stoma, an aperture in the epidermis for gas exchange.

*Stomatiferous.* Bearing stomata.

*Straight-veined.* With the lateral veins from the midrib running nearly parallel to one another.

*Style.* The usually narrowed part of the pistil between the ovary and stigma.

*Sub-.* Latin prefix, meaning somewhat, nearly, under, or below.

*Subtending.* To extend under.

*Subulate.* Awl-shaped.

*Terete.* Round in cross section.

*Thorn.* A sharp and hardened or woody structure, usually a modified branch.

*Timber line.* The upper limit of tree growth on mountains.

*Tomentose.* With dense, short, soft hairs or matted wool.

*Tomentulose.* Slightly tomentose.

*Trifoliolate.* Having 3-leaflets.

*Truncate.* Cut off squarely at the end.

*Twice pinnate.* Two times pinnately compound.

*Two-ranked.* Disposed in two vertical rows.

*Umbel.* A flower-cluster with the divisions arising from the summit of a common peduncle.

*Umbellate.* Having the flowers in umbels.

*Umbo.* A protuberance or point at the center of the enlarged terminal portions of cone-scales.

*Unarmed.* Without spines, prickles, or thorns.

*Undulate.* Wavy-margined.

*Valvate.* Opening by valves; said of parts when they meet without overlapping.

*Vein.* A division or branch of the midrib; a strand of vascular tissue.

*Villous.* With long weak hairs.

*Viscid.* Sticky with a tenacious secretion.

*Whorled.* With the organs borne in a circle.

# ADDENDA

## ADDITIONAL NATIVE SPECIES WHICH OCCASIONALLY ARE TREE-LIKE

Àcer glàbrum Torr. Sierra Maple. Mountains of California, Oregon, Washington, British Columbia, and east to the Rocky Mountains.

Amelánchier alnifòlia Nutt. Western Service Berry. Mountains of California, Oregon, Washington, British Columbia, and east to the Rocky Mountains.

Arctostáphylos gláuca Lindl. Greatberried Manzanita. South Coast Ranges and the lower mountain slopes of southern California.

Arctostáphylos manzaníta Parry. Common Manzanita. North Coast Ranges and the Sierra Nevada foothills of California.

Arctostáphylos víscida Parry. Whiteleaf Manzanita. Inner North Coast Ranges and foothills of the Sierra Nevada, north to southern Oregon.

Canòtia holacántha Torr. Canotia. Providence Mountains of Inyo County, California, and east to Arizona.

Ceanòthus arbòreus Greene. Tree Ceanothus. Santa Catalina, Santa Cruz, and Santa Rosa islands.

Ceanòthus párryi Trel. Parry Blueblossom. North Coast Ranges of California.

Ceanòthus spinòsus Nutt. Greenbark Ceanothus. Coastal mountains of southern California.

Ceanòthus thyrsiflòrus Esch. Blueblossom. Coast Ranges of southern Oregon and south to Monterey County, California.

Ceanòthus velùtinus var. laevigàtus T. & G. Varnishleaf Ceanothus. Southern Oregon and North Coast Ranges of California.

Cephalánthus occidentàlis L. Button Bush. Inner valleys and the Sierra Nevada foothills of northern California, north to Canada, and east to New Brunswick.

Gárrya ellíptica Dougl. Silktassel Bush. Coast Ranges of California from San Luis Obispo County north to Lincoln County, Oregon.

Nicotiàna gláuca Graham. Tree Tobacco. Naturalized in California from Paraguay, Argentina, and Bolivia.

Quércus dumòsa Nutt. Scrub Oak. California, in the chaparral regions.

Rhámnus anonaefòlia Greene. Sierra Buckthorn. Northern Sierra Nevada of California.

Rhámnus cròcea var. ilicifòlia Greene. Hollyleaf Buckthorn. Inner North Coast Range, the Sierra Nevada foothills, and coastal southern California.

Rhámnus cròcea var. insulàris Sarg. Island Buckthorn. Santa Catalina and Santa Cruz islands.

Rhododéndron califórnicum Hook. California Rose-Bay. Coast Rhododendron. Along the coast from northern California to Washington.

Rhús integrifòlia Benth. & Hook. Lemonade Berry. Mahogany Sumac. Coastal southern California, northern Lower California, Santa Catalina and Santa Barbara islands.

## INTERESTING AND RARE INTRODUCED SPECIES

The plants listed here deserve a much greater use on the Pacific Coast than they now receive. Those marked with a star (*) should be planted in situations in which the climatic conditions are similar to those of Santa Barbara, California. However, some of these species do well as far north as San Francisco Bay.

### TREE FERNS

*Alsóphila austràlis R.Br. Australian Tree Fern. Tasmania and Australia.

*Dicksònia antárctica Labill. Tree Fern. Tasmania and Australia.

### BAMBOO AND PALMS

*Erythèa brándegei Purpus. Brandegee Palm. Lower California.

*Hówea belmoreàna Becc. Belmore Palm. Lord Howe Island.

*Phyllóstachys bambusoìdes Sieb. & Zucc. Japanese Timber Bamboo. China and Japan.

### GYMNOSPERMS

*Ágathis austrális Salisb. Kawri-Pine. Tall tree. New Zealand.

*Ágathis orientàlis Lam. Large tree. East Indies.

*Ágathis robústa Hook. Australian Dammar-Pine. Tall tree. Australia.

*Araucària cóoki R.Br. Large tree. New Caledonia.

Juníperus oxycèdrus L. Prickly Juniper. Large bushy shrub or small tree. Mediterranean region.

Làrix larícina Koch. American Larch. Medium-sized tree. Canada south to Pennsylvania, west to Illinois.

Làrix leptólepis Murr. (*L. kaempferi* Sarg.) Japanese Larch. Medium-sized tree. Japan.

Pìnus montàna Mill. Swiss Mountain Pine. Small tree. Central Europe.

### ANGIOSPERMS (Broad-leaved Trees)

*Abrophýllum órnans Hook. f. Shrub or small tree. New South Wales.

*Acàcia péndula Cunn. Weeping Acacia. Small tree. Queensland and New South Wales.

Acer oblóngum Wall. Himalayan Maple. Medium-sized tree. Himalayas.

Aésculus pàvia L. Red Buckeye. Shrub or small tree. Virginia to Florida and Louisiana.

*Aléctryon excélsum Gaertn. Titoki. Medium-sized tree. New Zealand.

*Aleurìtes cordàta R.Br. Small or medium-sized tree. Southeastern Asia and adjacent islands.

*Aleurìtes moluccàna Willd. Candlenut Tree. Medium-sized tree, with long spreading branches. Probably Malay region.

*Alstònia scholàris R.Br. Medium-sized to large tree. India.

*Annòna cherimòla Mill. Cherimoya. Small tree. Andes of Peru and Ecuador.

*Bauhínia grandiflòra Juss. Largeflower Bauhinia. Small tree. South America.

*Bauhínia variegàta L. Buddhist Bauhinia. Small tree. India and China.

Bétula nìgra L. River Birch. Medium-sized tree. Massachusetts to Florida and Kansas.

*Caesalpìnia echinàta L. Brazilwood. Small tree. Brazil.

*Calodéndrum capénsis Thunb. Medium-sized tree. South Africa.

Carpìnus caroliniàna Walt. American Hornbeam. Small to medium-sized tree. Eastern North America to Texas.

*Casimiròa edùlis Llav. & Lex. White Sapote. Medium-sized to large tree. Mexico.

Castanospérmum austràle Cunn. Moreton Bay-Chestnut. Medium-sized tree. Australia.

*Daphniphýllum macropòdum Miq. A shrub or small tree. Japan and China.

Dombèya natalénsis Sond. Cape Weddingflower. Shrub or small tree. Natal.

Eucalýptus algieriénsis Trabut. Algerian Gum. Small tree, of hybrid origin. Algeria.

Eucalýptus botryoídes Smith. Bangalay. Medium-sized to large tree. Australia.

*Eucalýptus calophýlla R.Br. Port Gregory Gum. Medium-sized tree. Australia.

*Eucalýptus tetráptera Turcz. Shrub or small tree. Australia.

*Eucalýptus torquàta Luehm. Coolgardie White Gum. Western Australia.

*Fìcus panduràta Hort. Fiddleleaf Fig. Small tree. Tropical Africa.

Fráxinus excélsior L. European Ash. Medium-sized to large tree. Europe and western Asia.

*Harpephýllum cáffrum Bernh. Small tree. South Africa.

Hibíscus arnottiànus Gray. Shrub or small tree. Hawaii.

364 *Pacific Coast Trees*

Hicòria laciniòsa Sarg. (*Carya laciniosa* Engler & Graebn.) Shellbark Hickory. Large tree. Eastern United States to Iowa and Oklahoma.

Hippóphae rhamnoìdes L. Common Sea-Buckthorn. Shrub or small tree. Europe and Asia.

*Malvaviscus arbòreus Cav. Waxmallow. Tall shrub or small tree. South America.

*Melaleùca genistifòlia Smith. Small tree. Australia.

Mèlia azédarach L. Chinaberry. Small to medium-sized tree. Himalayan region.

Myóporum insulàre R.Br. (*M. serratum* R.Br.) Myoporum. Shrub or rarely a small tree. Australia.

Pistàcia chinénsis Bunge. Chinese Pistache. Medium-sized tree. China.

Prùnus spinòsa L. Blackthorn. Spreading shrub or small tree. Central and southern Europe and northern Africa to Persia and Siberia.

Prùnus subhirtélla var. pendula Tanaka. Weeping Higan Cherry. Small tree. Japan.

Quércus bícolor Willd. Swamp White Oak. Medium-sized to large tree. Quebec to Georgia, west to Michigan and Arkansas.

Quércus phéllos L. Willow Oak. Medium-sized tree. New York to Florida, west to Missouri and Texas.

Quércus serràta Thunb. Bristletooth Oak. Small tree. Japan, Korea, and China.

Quércus velùtina Lam. Black Oak. Large tree. Maine to Florida, west to Minnesota and Texas.

Quillàja saponària Molina. Soapbark Tree. Large tree. Chile.

Rhús cótinus L. (*Cotinus coggygria* Scop.) Common Smoke Tree. Dense shrub or small tree. Japan and China.

Sálix álba L. White Willow. Medium-sized to large tree. Europe, Asia, and northern Africa.

*Schòtia brachypétala Sond. Large shrub or small tree. South Africa.

*Schòtia latifòlia Jacq. Large shrub or small tree. South Africa.

*Sóphora tetráptera Ait. Fourwing Sophora. Shrub or small tree. New Zealand.

Stercùlia bídwilli Hook. (*Brachychiton bidwilli* Hook.) Shrub or tree. Australia.

Tricuspidària depéndens Ruiz. & Pav. (*Crinodendron dependens* Schneid.) Small to medium-sized tree. Chile.

Xanthóceras sorbifòlia Bunge. Yellowhorn. Shrub or small tree. North China.

# REFERENCES

ABRAMS, LEROY

1910. "A Phytogeographic and Taxonomic Study of Southern California Trees and Shrubs," *Bulletin of the New York Botanical Garden*, Vol. VI, No. 21.

1923. *Illustrated Flora of the Pacific States*, Vol. I. Stanford University Press.

AMERICAN ASSOCIATION FOR THE ADVANCEMENT OF SCIENCE

1915. *Nature and Science on the Pacific Coast*, ed. under the auspices of the Pacific Coast Committee. Paul Elder and Co., San Francisco.

AMERICAN JOINT COMMITTEE ON HORTICULTURE NOMENCLATURE

1924. *Standardized Plant Names*. The Committee, Salem, Mass.

APGAR, A. C.

1892. *Trees of the Northern United States*. American Book Co.

BAILEY, L. H.

1917. (ed.) *The Standard Cyclopedia of Horticulture*. 6 vols. The Macmillan Co., 2d ed.

1924. *Manual of Cultivated Plants*. The Macmillan Co.

1928. *Cultivated Evergreens*. The Macmillan Co., new revised ed.

1930. (ed.) *The Standard Cyclopedia of Horticulture*. 3 vols. The Macmillan Co.

1933. *How Plants Get Their Names*. The Macmillan Co.

BAILEY, L. H., and ZOE, ETHEL

1930. *Hortus*. The Macmillan Co.

BENSON, G. T.

1930. "The Trees and Shrubs of Western Oregon," *Contributions from the Dudley Herbarium*, Vol. II. Stanford University Press.

BERRY, E. W.

1923. *Tree Ancestors*. Williams and Wilkins Co.

BRITTON, N. L., and BROWN, A.

1913. *Illustrated Flora of the Northern United States and Canada.* 3 vols. Charles Scribner's Sons, 2d ed.

CAMPBELL, D. H.

1926. *Outline of Plant Geography*. The Macmillan Co.

ENGLER, A., and PRANTL, K.

1926—to date. *Die Natürlichen Pflanzenfamilien*. Verlag von Wilhelm Engelmann, Leipzig, 2d ed.

GRIMWADE, R.

   1920. *An Anthography of the Eucalyptus.* Angus and Robertson, Ltd., Sydney.

HALL, H. M.

   1910. "Studies in Ornamental Trees and Shrubs," *University of California Publications in Botany.* Vol. IV, No. 1.

HALL, H. M., and GRINNELL, J.

   1919. "Life Zone Indicators in California," *Proceedings of the California Academy of Sciences,* Ser. 4, Vol. IX, No. 2.

HALL, H. M., and HALL, C. C.

   1912. *A Yosemite Flora.* Paul Elder and Co., San Francisco.

INGHAM, N. D.

   1908. "Eucalyptus in California," *University of California Publications in the College of Agriculture,* Bulletin No. 196.

JEPSON, W. L.

   1909. *Trees of California.* Cunningham, Curtis, and Welch, San Francisco.

   1910. *The Silva of California.* (*Memoirs of the University of California,* Vol. II.) The University Press, Berkeley.

   1914–1934. (ed.) Madroño, *Journal of the California Botanical Society,* Vols. I and II.

   1925(*c*). *Manual of the Flowering Plants of California.* Associated Students' Store, University of California, Berkeley.

MCCLATCHIE, A. J.

   1902. "Eucalyptus Cultivated in the United States," *United States Department of Agriculture, Forest Service,* Bulletin No. 35. Government Printing Office, Washington, D. C.

MAIDEN, J. H.

   1903—to date. *A Critical Revision of the Genus Eucalyptus.* 8 vols. so far published. Government Printer, Sydney, New South Wales.

MOWRY, H.

   1926. "Palms of Florida," *University of Florida Experiment Station,* Bulletin No. 184. Gainsville.

NEHRLING, H.

   1933. *The Plant World in Florida.* The Macmillan Co.

PARISH, S. B.

   1907. "A Contribution Toward the Knowledge of the Genus Washingtonia," *Botanical Gazette,* Vol. 44, pp. 408–434.

PIPER, C. V.

   1906. "Flora of the State of Washington," *Contributions from the United States National Herbarium,* Vol. XI. Government Printing Office, Washington, D. C.

PIPER, C. V., and BEATTIE, R. K.

  1915. *Flora of the Northwest Coast.* State College of Washington, Pullman.

PRATT, M. B.

  1921. *Shade and Ornamental Trees of California.* California State Board of Forestry, Sacramento.

REHDER, A.

  1927. *Manual of Cultivated Trees and Shrubs.* The Macmillan Co.

SARGENT, C. S.

  1922. *Manual of Trees of North America.* Houghton Mifflin Co., 2d ed.

SAUNDERS, C. F.

  1926. *Trees and Shrubs in California Gardens.* Robert M. McBride and Co.

SHAW, G. R.

  1914. "The Genus Pinus," *Publications of the Arnold Arboretum,* No. 5. Riverside Press, Cambridge.

STANDLEY, P. C.

  1920–1926. "Trees and Shrubs of Mexico," *Contributions from the United States National Herbarium,* Vol. XXIII. Government Printing Office, Washington, D. C.

SUDWORTH, G. B.

  1908. *Forest Trees of the Pacific Slope.* United States Department of Agriculture, Forest Service. Government Printing Office, Washington, D. C.

TRELEASE, WM.

  1924. "The American Oaks," *Memoirs of the National Academy of Sciences,* Vol. XX.

WALTHER, E.

  1928. "A Key to the Species of Eucalyptus Grown in California," *Proceedings of the California Academy of Sciences,* Ser. 4, Vol. XVII, No. 3.

# LISTS OF TREES

## RECOMMENDED FOR VARIOUS USES
### ON THE PACIFIC COAST

BY

## H. W. SHEPHERD
Associate Professor of Landscape Design
in the University of California

SPECIES marked with an asterisk (*) should be planted in the milder regions of the Pacific Coast. They thrive best where the climatic conditions are comparable to those of Santa Barbara, California, and other restricted frost-free districts such as the milder sections of San Francisco Bay region.

Species marked with a dagger (†) are listed in the Addenda but are not "keyed out" or included in the descriptive account of the Pacific Coast trees. They are rarely seen in cultivation, but deserve more use on the Pacific Coast.

The species followed by the letters O. W. BC. thrive in cultivation only in the cooler and moister regions of Oregon, Washington, and British Columbia.

### 1. ACCENT TREES

(NOTE: An accent tree is one noticeably different from those which form its setting. Change of scale, form, texture, and color are factors involved.)

#### CONIFER

| Botanical Name | Common Name |
|---|---|
| *Cephalotaxus drupacea | Japanese Plum-Yew |
| *Cephalotaxus fortunei | Chinese Plum-Yew |
| Chamaecyparis lawsoniana | Lawson-Cypress |
| Chamaecyparis lawsoniana var. alumi | Scarab-Cypress |
| Chamaecyparis lawsoniana var. erecta | Erect Lawson-Cypress |
| Cupressus sempervirens var. stricta | Columnar Italian Cypress |

## CONIFER

| Botanical Name | Common Name |
|---|---|
| Libocedrus chilensis | Chilean Incense-Cedar |
| Libocedrus decurrens | California Incense-Cedar |
| *Podocarpus elongata | Podocarpus |
| *Podocarpus macrophylla | Yew Podocarpus |
| *Sciadopitys verticillata | Umbrella-Pine |
| Taxus baccata | English Yew |
| Taxus brevifolia | Pacific Yew |
| Taxus cuspidata | Japanese Yew |
| Thuja occidentalis | American Arborvitae |
| Thuja orientalis | Oriental Arborvitae |
| Thuja plicata | Giant Arborvitae |

## DECIDUOUS

| | |
|---|---|
| Populus nigra var. italica | Lombardy Poplar |

## 2. ALKALI-TOLERANT TREES
### DECIDUOUS

| | |
|---|---|
| Acer platanoides | Norway Maple |
| Catalpa speciosa | Western Catalpa |
| Fraxinus velutina | Arizona Ash |
| Gleditsia triacanthos | Honey Locust |
| Koelreuteria paniculata | Goldenrain Tree |
| Lagerstroemia indica | Crape-Myrtle |
| Liquidambar styraciflua | Sweet Gum |
| †Melia azedarach | Chinaberry |
| Melia azedarach var. umbraculiformis | Texas Umbrella Tree |
| Morus, species | Mulberry |
| Platanus acerifolia | London Plane Tree |
| Populus fremonti | Fremont Cottonwood |
| Robinia pseudoacacia | Common Locust |

### EVERGREEN

| | |
|---|---|
| *Acacia longifolia | Sydney Golden Wattle |
| *Acacia melanoxylon | Blackwood Acacia |
| *Acacia pycnantha | Golden Wattle |
| *Acacia saligna | Willow Acacia |
| *Callistemon, species | Bottlebrush |
| *Casuarina cunninghamiana | Cunningham Beefwood |
| *Casuarina equisetifolia | Horsetail Tree |

## EVERGREEN

| *Botanical Name* | *Common Name* |
|---|---|
| *Casuarina stricta | Beefwood, She-oak |
| *Ceratonia siliqua | Carob |
| *Cinnamomum camphora | Camphor Tree |
| *Eucalyptus amygdalina | Almond Eucalyptus |
| *Eucalyptus amygdalina var. angustifolia | Narrow-leaf Almond Eucalyptus |
| †*Eucalyptus botryoides | Bangalay |
| *Eucalyptus cornuta | Yate Tree |
| *Eucalyptus corynocalyx | Sugar Gum |
| *Eucalyptus goniocalyx | Mountain Gum |
| *Eucalyptus gunni | Cider Gum |
| *Eucalyptus resinifera | Mahogany Gum |
| *Eucalyptus robusta | Brown Gum |
| *Eucalyptus rostrata | Creek Gum |
| *Eucalyptus rudis | Desert Gum |
| *Eucalyptus tereticornis | Slaty Gum |
| *Eucalyptus viminalis | Manna Gum |
| *Ficus macrophylla | Moreton Bay Fig |
| *Leptospermum laevigatum | Australian Tea Tree |
| *Melaleuca ericifolia | Heath Melaleuca |
| *Melaleuca leucadendron | Cajeput Tree |
| *Melaleuca nesophila | Pink Melaleuca |
| *†Myoporum insulare | Myoporum |
| *Myoporum laetum | Myoporum |
| *Nerium oleander | Oleander |
| *Olea europaea | Olive |
| *Parkinsonia aculeata | Parkinsonia |
| *Pittosporum crassifolium | Karo |
| *Prosopis juliflora var. glandulosa | Honey Mesquite |
| *Prunus ilicifolia | Hollyleaf Cherry |
| *Quercus suber | Cork Oak |
| *Tamarix articulata | Athel |

## 3. AUTUMN-FOLIAGE-PRODUCING TREES

### RED

| | |
|---|---|
| Acer buergerianum | Trident Maple |
| Acer rubrum | Red Maple |
| Cornus florida | Flowering Dogwood |
| Crataegus crusgalli | Cockspur Thorn |
| Liquidambar styraciflua | Sweet Gum |

### RED

| *Botanical Name* | *Common Name* |
|---|---|
| †Pistacia chinensis | Chinese Pistache |
| Quercus coccinea | Scarlet Oak |
| Quercus rubra | Red Oak |
| Quercus kelloggi | California Black Oak |

### YELLOW

| | |
|---|---|
| Acer dasycarpum | Silver Maple |
| Acer saccharum | Sugar Maple |
| Ailanthus glandulosa | Tree of Heaven |
| Betula alba | European White Birch |
| Castanea dentata | American Chestnut |
| Cedrela sinensis | Chinese Cedrela |
| Diospyros virginiana | Common Persimmon |
| Fraxinus americana | American Ash |
| Fraxinus velutina | Arizona Ash |
| Ginkgo biloba | Maidenhair Tree |
| Hicoria ovata | Shagbark Hickory |
| Juglans, species | Walnut |
| Larix, species | Larch |
| Liriodendron tulipifera | Tulip Tree |
| Magnolia acuminata | Cucumber Tree |
| Populus, species | Poplar, Cottonwood |
| Quercus macrocarpa | Mossycup Oak |
| Robinia, species | Locust |
| Salix, species | Willow |
| Sassafras variifolium | Sassafras |

## 4. AVENUE TREES
### DECIDUOUS

| | |
|---|---|
| Acer macrophyllum | Oregon Maple |
| Acer platanoides | Norway Maple |
| Acer pseudoplatanus | Sycamore Maple |
| Aesculus carnea | Red Horsechestnut |
| Aesculus hippocastanum | Horsechestnut |
| Fagus sylvatica | European Beech |
| Fraxinus americana | American Ash |
| Fraxinus velutina | Arizona Ash |
| Ginkgo biloba | Maidenhair Tree |
| Liriodendron tulipifera | Tulip Tree |
| Platanus acerifolia | London Plane Tree |
| Tilia americana | American Linden |

## EVERGREEN

| Botanical Name | Common Name |
|---|---|
| *Casuarina cunninghamiana | Cunningham Beefwood |
| *Cinnamomum camphora | Camphor Tree |
| *Eucalyptus polyanthemos | Redbox |
| *Eucalyptus viminalis | Manna Gum |
| Magnolia grandiflora | Southern Magnolia |
| Quercus agrifolia | California Live Oak |
| Quercus chrysolepis | Maul Oak |
| Quercus ilex | Holly Oak |
| *Quercus suber | Cork Oak |
| Quercus wislizeni | Interior Live Oak |
| *Schinus molle | Pepper Tree |

### PALMS

| | |
|---|---|
| *Cocos plumosa | Plume Palm |
| *Livistona chinensis | Chinese Fan Palm |
| *Phoenix canariensis | Canary Date Palm |
| *Phoenix dactylifera | Date Palm |
| *Washingtonia filifera | California Washington Palm |
| *Washingtonia gracilis | Mexican Washington Palm |

## 5. BORDER TREES FOR HOME GROUNDS

(NOTE: All pittosporums do best along the coast in frostless belts. Most acacias do not tolerate frost.)

### BROAD-LEAF EVERGREEN

| | |
|---|---|
| *Acacia baileyana | Cootamundra Wattle |
| *Acacia cultriformis | Knife Acacia |
| *Acacia cyanophylla | Blueleaf Acacia |
| *Acacia podalyriaefolia | Pearl Acacia |
| *Acacia pravissima | Screwpod Acacia |
| *Acacia pruinosa | Bronze Acacia |
| *Acacia pycnantha | Golden Wattle |
| *Acacia verticillata | Star Acacia |
| *Albizzia lophantha | Plume Albizzia |
| *Arbutus unedo | Strawberry Tree |
| *Camellia japonica | Common Camellia |
| *Cassia tomentosa | Woolly Senna (Winter yellow bloom) |
| *Ceratonia siliqua | Carob |
| *Citrus, species | Orange, Lemon, Grapefruit, Kumquat |
| *Cornus capitata | Evergreen Dogwood |

BROAD-LEAF EVERGREEN

| Botanical Name | Common Name |
|---|---|
| *Eugenia myrtifolia | Australian Brush-Cherry |
| *Hakea laurina | Sea-urchin Hakea |
| *Hymenosporum flavum | Hymenosporum |
| *Lagunaria patersoni | Lagunaria |
| *Lyonothamnus floribunda | Catalina Ironwood |
| *Macadamia ternifolia | Queensland Nut |
| *Melaleuca, species | |
| *†Myoporum insulare | Myoporum |
| *Myoporum laetum | Myoporum |
| Myrica californica | California Wax-Myrtle |
| *Olea europaea | Olive |
| *Photinia serrulata | Low Photinia |
| *Pittosporum crassifolium | Karo |
| *Pittosporum eugenioides | Tarata |
| *Pittosporum phillyraeoides | Willow Pittosporum |
| *Pittosporum tenuifolium | Tawhiwhi |
| *Pittosporum undulatum | Orange Pittosporum |
| *Pittosporum viridiflorum | Cape Pittosporum |
| Prunus caroliniana | Carolina Cherry |
| Prunus ilicifolia | Hollyleaf Cherry |
| Prunus lyoni | Catalina Cherry |
| *†Quillaja saponaria | Soapbark Tree |
| Ulmus parvifolia | Chinese Elm |

CONIFER

| | |
|---|---|
| *Callitris robusta | Cypress-Pine |
| Cupressus arizonica | Arizona Cypress |
| Juniperus occidentalis | Western Juniper |
| Libocedrus decurrens | California Incense-Cedar |
| Pinus montana var. mughus | Mugho Pine |
| *Podocarpus elongata | Podocarpus |

DECIDUOUS

| | |
|---|---|
| Acer palmatum, in variety | Japanese Maple |
| *Bauhinia purpurea | Purple Bauhinia |
| Celtis australis | European Hackberry |
| Cercis occidentalis | Western Redbud |
| Cornus florida | Flowering Dogwood |
| Cornus nuttalli | Pacific Dogwood |
| Fraxinus ornus | Flowering Ash |
| Laburnum vulgare | Goldenchain Tree |

## DECIDUOUS

| Botanical Name | Common Name |
|---|---|
| Lagerstroemia indica | Common Crape-Myrtle |
| Lagerstroemia speciosa | Queen Crape-Myrtle |
| Magnolia soulangeana | Saucer Magnolia |
| Magnolia stellata | Star Magnolia |
| Malus fusca | Oregon Crab |
| Prunus demissa | Western Chokecherry |
| Prunus emarginata | Bitter Cherry |
| Prunus subcordata | Sierra Plum |
| Punica granatum | Pomegranate |
| Robinia hispida var. macrophylla | Smooth Rose-Acacia |
| Sambucus caerulea | Blueberry Elder |
| Sambucus racemosa | Redberry Elder |
| Sorbus aucuparia | European Mountain-Ash |
| *Zizyphus jujuba | Common Jujube |

## 6. COLUMNAR AND FASTIGIATE TREES

### CONIFER

| | |
|---|---|
| Cephalotaxus drupacea var. fastigiata | Columnar Plum-Yew |
| Cupressus guadalupensis | Guadalupe Cypress |
| Cupressus sempervirens var. stricta | Columnar Cypress |
| Juniperus communis var. hibernica | Irish Juniper |
| Libocedrus decurrens | California Incense-Cedar |
| Taxus baccata var. erecta | Broom Yew |
| Taxus baccata var. fastigiata | Irish Yew |
| Thuja occidentalis var. pyramidalis | Pyramidal Western Arborvitae |
| Thuja orientalis var. pyramidalis | Pyramidal Oriental Arborvitae |
| Thuja plicata var. fastigiata | Pyramidal Giant Arborvitae |

### DECIDUOUS

| | |
|---|---|
| Populus alba var. pyramidalis (Populus bolleana) | Bolleana Poplar |
| Populus nigra var. italica | Lombardy Poplar |

### EVERGREEN

| | |
|---|---|
| *Eugenia myrtifolia | Australian Brush-Cherry |
| *Eugenia paniculata | |
| Ilex aquifolium | English Holly |
| Laurus nobilis | Grecian Laurel |

## 7. DROUGHT-TOLERANT TREES

### CONIFER

| *Botanical Name* | *Common Name* |
|---|---|
| Cedrus atlantica | Atlas Cedar |
| Cedrus libani | Cedar of Lebanon |
| Cupressus guadalupensis | Guadalupe Cypress |
| Cupressus macnabiana | Macnab Cypress |
| Cupressus macrocarpa | Monterey Cypress |
| Cupressus sempervirens var. stricta | Columnar Italian Cypress |
| Juniperus californica | California Juniper |
| Libocedrus decurrens | California Incense-Cedar |
| Pinus aristata | Bristlecone Pine |
| Pinus contorta | Shore Pine |
| Pinus coulteri | Coulter Pine |
| Pinus monophylla | Singleleaf Pine |
| †Pinus montana | Swiss Mountain Pine |
| Pinus muricata | Bishop Pine |
| Pinus pinaster | Cluster Pine |
| Pinus pinea | Italian Stone Pine |
| Pinus sabiniana | Digger Pine |
| Pinus torreyana | Torrey Pine |
| Thuja plicata | Giant Arborvitae |

### DECIDUOUS

| | |
|---|---|
| *Acacia greggi | Catclaw |
| Ailanthus glandulosa | Tree of Heaven |
| Catalpa speciosa | Western Catalpa |
| Celtis australis | European Hackberry |
| Cydonia oblonga | Common Quince |
| Fraxinus velutina | Arizona Ash |
| Gleditsia triacanthos | Honey-Locust |
| Juglans californica | California Black Walnut |
| Juglans nigra | Black Walnut |
| Koelreuteria paniculata | Goldenrain Tree |
| Maclura pomifera | Osage-Orange |
| Morus nigra | Black Mulberry |
| Punica granatum | Pomegranate |
| Robinia pseudoacacia | Common Locust |
| Sambucus caerulea | Blueberry Elder |
| Sophora japonica | Chinese Scholar Tree |
| Tamarix parviflora | Smallflower Tamarisk |
| Zelkova serrata | Sawleaf Zelkova |
| *Zizyphus jujuba | Common Jujube |

## EVERGREEN

| *Botanical Name* | *Common Name* |
|---|---|
| *Acacia, species | |
| Arbutus menziesi | Madroño |
| *Arbutus unedo | Strawberry Tree |
| *Callistemon, species | Bottlebrush |
| Castanopsis chrysophylla | Giant Chinquapin |
| *Cassia tomentosa | Woolly Senna |
| *Casuarina, species | Beefwood |
| *Ceratonia siliqua | Carob |
| Cercocarpus betuloides | Hard Tack |
| *Eucalyptus, species | |
| *Fremontia californica var. mexicana | Mexican Flannel Bush |
| *Grevillea robusta | Silk-Oak |
| *Leptospermum laevigatum | Australian Tea Tree |
| Ligustrum lucidum | Glossy Privet |
| Magnolia grandiflora | Southern Magnolia |
| *Melaleuca, species | |
| *†Myoporum insulare | Myoporum |
| *Olea europaea | Olive |
| *Parkinsonia aculeata | Parkinsonia |
| Photinia arbutifolia | Christmas Berry |
| *Photinia serrulata | Low Photinia |
| *Pittosporum crassifolium | Karo |
| *Pittosporum eugenioides | Tarata |
| *Pittosporum phillyraeoides | Willow Pittosporum |
| *Pittosporum undulatum | Orange Pittosporum |
| *Pittosporum viridiflorum | Cape Pittosporum |
| Prunus ilicifolia | Hollyleaf Cherry |
| Prunus lyoni | Catalina Cherry |
| †*Quillaja saponaria | Soapbark Tree |
| Sambucus caerulea | Blueberry Elder |
| *Schinus molle | Pepper Tree |

## PALM AND PALM-LIKE

| | |
|---|---|
| *Cocos australis | Pindo Palm |
| *Cordyline australis | Green Dracena |
| *Erythea edulis | Guadalupe Palm |
| *Phoenix canariensis | Canary Date Palm |
| *Trachycarpus exelsa | Windmill Palm |
| *Washingtonia filifera | California Washington Palm |
| *Washingtonia gracilis | Mexican Washington Palm |

## 8. EROSION-CONTROL TREES
### DECIDUOUS

| Botanical Name | Common Name |
| --- | --- |
| *Cercidium torreyanum | Palo Verde |
| Platanus racemosa | California Plane Tree |
| *Prosopis juliflora | Honey Mesquite |
| Salix hindsiana | Valley Willow |
| Salix laevigata | Red Willow |
| Salix lasiandra | Yellow Willow |
| Salix lasiolepis | Arroyo Willow |
| Salix melanopsis | Dusky Willow |
| Salix nigra var. vallicola | Black Willow |
| Salix piperi | Dune Willow |
| Salix scouleriana | Nuttall Willow |
| Salix sessilifolia | Sandbar Willow, Softleaf Willow |
| Salix sitchensis var. coulteri | Velvet Willow |
| Tamarix parviflora | Tamarix |

### EVERGREEN

| | |
| --- | --- |
| Photinia arbutifolia | Christmas Berry |
| Prunus ilicifolia | Hollyleaf Cherry |

## 9. FLOWERING TREES
### SPRING-FLOWERING TREES (March to May)

| | |
| --- | --- |
| *Acacia cultriformis | Knife Acacia |
| *Acacia koa | Koa |
| *Acacia longifolia | Sydney Golden Wattle |
| *Acacia longifolia var. floribunda | Gossamer Wattle |
| *Acacia pravissima | Screwpod Acacia |
| *Acacia pruinosa | Bronze Acacia |
| *Acacia pycnantha | Golden Wattle |
| Aesculus carnea | Pink Horsechestnut |
| Aesculus hippocastanum | Horsechestnut |
| *Bauhinia purpurea | Purple Bauhinia |
| Catalpa speciosa | Western Catalpa |
| Cercis canadensis | American Redbud |
| Cercis occidentalis | Western Redbud |
| Crataegus mollis | Downy Hawthorn |
| Crataegus oxyacantha, and variety | English Hawthorn |
| Cydonia oblonga | Common Quince |
| *Grevillea robusta | Silk-Oak |
| Laburnum vulgare | Goldenchain Tree |

SPRING-FLOWERING TREES (March to May)

| Botanical Name | Common Name |
|---|---|
| Liriodendron tulipifera | Tulip Tree |
| Magnolia soulangeana | Saucer Magnolia |
| Magnolia stellata | Star Magnolia |
| *Photinia serrulata | Low Photinia |
| Robinia hispida | Rose-Acacia |
| Robinia hispida var. macrophylla | Smooth Rose-Acacia |
| Robinia pseudoacacia | Common Locust |
| Robinia pseudoacacia var. bessoniana | Besson Locust |
| Robinia pseudoacacia var. decaisneana | Decaisne Locust |
| Sorbus aucuparia | European Mountain-Ash |
| †Xanthoceras sorbifolia | Yellowhorn |

SUMMER-FLOWERING TREES (June to August)

| | |
|---|---|
| *Acacia elata | Cedar Wattle |
| *Acacia retinodes | Water Wattle |
| Albizzia julibrissin | Silk Tree |
| †Eucalyptus algeriensis | Algerian Gum |
| *Eucalyptus corynocalyx | Sugar Gum |
| *Eucalyptus ficifolia | Scarlet Gum |
| *Eucalyptus rostrata | Creek Gum |
| *Eucalyptus viminalis | Manna Gum |
| *Hymenosporum flavum | Hymenosporum |
| *Jacaranda ovalifolia | Jacaranda |
| Koelreuteria paniculata | Goldenrain Tree |
| Lagerstroemia indica | Crape-Myrtle |
| *Lagunaria patersoni | Lagunaria |
| *Melaleuca styphelioides | Rigidleaf Melaleuca |
| Nerium oleander | Oleander |
| Paulownia tomentosa | Royal Paulownia |
| *Sterculia acerifolia | Victorian Bottle Tree |

FALL-FLOWERING TREES (September to November)

| | |
|---|---|
| *Acacia retinodes | Water Wattle |
| *Cassia tomentosa | Woolly Senna |
| Lagerstroemia indica | Crape-Myrtle |
| *Parkinsonia aculeata | Parkinsonia |
| Sophora japonica | Chinese Scholar Tree |

WINTER-FLOWERING TREES (December to February)

| *Botanical Name* | *Common Name* |
|---|---|
| *Acacia baileyana | Cootamundra Wattle |
| *Acacia decurrens var. dealbata | Silver Wattle |
| *Acacia podalyriaefolia | Pearl Acacia |
| *Albizzia lophantha | Plume Albizzia |
| *Arbutus unedo | Strawberry Tree |
| *Cassia tomentosa | Woolly Senna |
| *Eucalyptus ficifolia | Scarlet Gum |
| *Eucalyptus sideroxylon var. rosea | Pink-flowering Red Ironbark |
| *Hakea laurina | Sea-urchin Hakea |

FRAGRANT-FLOWERING DECIDUOUS TREES

| | |
|---|---|
| Crataegus, species | Hawthorn |
| Magnolia, species | Magnolia |
| Malus, species | Crabapple |
| Melia, species | Texas Umbrella Tree |
| Prunus, species | Apricot, Cherry, Peach, Plum |
| Robinia, species | Locust |
| Tilia, species | Linden |

## 10. FLOWERING OR BERRIED TREES—WHICH MAY BE TRAINED AS VINES AGAINST A WALL

| | |
|---|---|
| Albizzia julibrissin (Pink bloom and fern-like foliage) | Silk Tree |
| *Erythrina cristagalli (Striking crimson bloom) | Coral Tree |
| Laburnum vulgare (Trained to resemble yellow wisteria) | Goldenchain Tree |
| Morus nigra | Black Mulberry |
| Sambucus racemosa | Redberry Elder |

## 11. FRUIT TREES ADAPTED TO ESPALIER

| | |
|---|---|
| *Citrus aurantium | Seville Orange |
| *Citrus limonia | Lemon |
| Cydonia oblonga | Common Quince |
| Diospyros kaki | Kaki Persimmon |
| Malus, species | Apple |
| *Persea americana | Avocado |
| Prunus cerasus, and varieties | Plum |

## 11. FRUIT TREES ADAPTED TO ESPALIER—*Continued*

| Botanical Name | Common Name |
|---|---|
| Pyrus communis, and varieties | Pear |
| *Zizyphus jujuba | Common Jujube |

## 12. HEAT-TOLERANT TREES

### CONIFER

| | |
|---|---|
| Cedrus libani | Cedar of Lebanon |
| Chamaecyparis lawsoniana | Lawson-Cypress |
| Cupressus arizonica | Arizona Cypress |
| Juniperus californica | California Juniper |
| Libocedrus decurrens | California Incense-Cedar |
| Pinus pinea | Italian Stone Pine |
| Pinus sabiniana | Digger Pine |
| Sequoia gigantea | Big Tree |
| Taxodium distichum | Bald-Cypress |

### DECIDUOUS

| | |
|---|---|
| Aesculus carnea | Red Horsechestnut |
| Aesculus hippocastanum | Horsechestnut |
| Ailanthus glandulosa | Tree of Heaven |
| Carpinus betulus | European Hornbeam |
| Castanea sativa | Spanish Chestnut |
| Catalpa speciosa | Western Catalpa |
| Cedrela sinensis | Chinese Cedrela |
| Diospyros kaki | Kaki Persimmon |
| Diospyros virginiana | Common Persimmon |
| *Ficus carica | Common Fig |
| Fraxinus velutina | Arizona Ash |
| Gleditsia triacanthos | Honey-Locust |
| Juglans californica | California Black Walnut |
| Juglans cinerea | Butternut |
| Koelreuteria paniculata | Goldenrain Tree |
| Lagerstroemia indica | Crape-Myrtle |
| Maclura pomifera | Osage-Orange |
| Melia azedarach var. umbraculiformis | Texas Umbrella Tree |
| Morus alba | White Mulberry |
| Morus nigra | Black Mulberry |
| Platanus acerifolia | London Plane Tree |
| Platanus racemosa | California Plane Tree |
| Populus alba | White Poplar |
| Populus balsamifera | Balsam Poplar |

## DECIDUOUS

| *Botanical Name* | *Common Name* |
|---|---|
| Populus candicans | Balm of Gilead Poplar |
| Populus fremonti | Fremont Cottonwood |
| Pterocarya stenoptera | Chinese Wingnut |
| Quercus douglasi | Blue Oak |
| Quercus lobata | Valley Oak |
| Robinia, species | Locust |
| Salix, species | Willow |
| Sophora japonica | Chinese Scholar Tree |
| Tamarix parviflora | Smallflower Tamarisk |
| Tilia americana | American Linden |
| Tilia platyphyllos | Bigleaf European Linden |
| Tilia tomentosa | Silver Linden |
| Ulmus americana | American Elm |
| Ulmus campestris | English Elm |
| Ulmus glabra | Scotch Elm |
| Ulmus hollandica | Holland Elm |
| Ulmus pumila | Dwarf Asiatic Elm |
| *Zizyphus jujuba | Common Jujube |

## EVERGREEN

| | |
|---|---|
| *Acacia retinodes | Water Wattle |
| *Acacia verticillata | Star Acacia |
| *Arbutus unedo | Strawberry Tree |
| *Casuarina, species | Beefwood |
| *Ceratonia siliqua | Carob |
| *Citrus, species | Orange, Lemon, Grapefruit |
| *Eucalyptus rostrata | Creek Gum |
| *Eucalyptus rudis | Desert Gum |
| *Eucalyptus viminalis | Manna Gum |
| Ligustrum lucidum | Glossy Privet |
| Magnolia grandiflora | Southern Magnolia |
| *Nerium oleander | Oleander |
| *Olea europaea | Olive |
| *Parkinsonia aculeata | Parkinsonia |
| *Photinia serrulata | Low Photinia |
| *Pittosporum phillyraeoides | Willow Pittosporum |
| Quercus ilex | Holly Oak |
| *Quercus suber | Cork Oak |
| Quercus wislizeni | Interior Live Oak |
| *Schinus molle | Pepper Tree |
| Tamarix articulata | Athel |

## 13 HEDGE TREES
### DECIDUOUS

| *Botanical Name* | *Common Name* |
|---|---|
| Acer campestre | Hedge Maple |
| Carpinus betulus | European Hornbeam |
| Crataegus cordata | Washington Hawthorn |
| Crataegus oxyacantha | English Hawthorn |
| Fagus americana | American Beech |
| Gleditsia triacanthos | Honey-Locust |
| †Hippophae rhamnoides | Common Sea-Buckthorn |
| Maclura pomifera | Osage-Orange |
| †Prunus spinosa | Blackthorn |

### CONIFER AND BROAD-LEAF EVERGREEN

| | |
|---|---|
| *Acacia armata | Kangaroo Thorn |
| *Acacia longifolia | Sydney Wattle |
| *Acacia saligna | Willow Acacia |
| *Acacia verticillata | Star Acacia |
| *Ceratonia siliqua | Carob |
| *Cupressus macrocarpa | Monterey Cypress |
| †Eucalyptus algeriensis | Algerian Gum |
| *Eucalyptus polyanthemos | Redbox |
| *Eugenia myrtifolia | Australian Brush-Cherry |
| *Feijoa sellowiana | Feijoa |
| Laurus nobilis | Grecian Laurel |
| Libocedrus decurrens | California Incense-Cedar |
| Ligustrum lucidum | Glossy Privet |
| *Metrosideros robusta | Rata |
| *Metrosideros tomentosa | New Zealand Christmas Tree |
| †Myoporum insulare | Myoporum |
| *Myoporum laetum | Myoporum |
| Myrica californica | California Wax-Myrtle |
| *Olea europaea | Olive |
| *Pittosporum crassifolium | Karo |
| *Pittosporum eugenioides | Tarata |
| *Pittosporum tenuifolium | Tawhiwhi |
| *Pittosporum undulatum | Orange Pittosporum |
| *Pittosporum viridiflorum | Cape Pittosporum |
| Prunus caroliniana | Carolina Cherry-Laurel |
| Prunus ilicifolia | Hollyleaf Cherry |
| Prunus laurocerasus | English Cherry-Laurel |
| Prunus lyoni | Catalina Cherry |

| | |
|---|---|
| *Psidium cattleianum | Guava |
| Quercus agrifolia | California Live Oak |
| Quercus chrysolepis | Maul Oak |
| Quercus ilex | Holly Oak |
| Quercus wislizeni | Interior Live Oak |
| Rhamnus purshiana | Cascara |
| *Tamarix articulata | Athel |
| Taxus baccata var. erecta | Broom Yew |
| Taxus cuspidata | Japanese Yew |
| Umbellularia californica | California-Laurel |

## 14. LAWN SPECIMEN TREES FOR PARKS OR LARGE PRIVATE GROUNDS

### CONIFER

| *Botanical Name* | *Common Name* |
|---|---|
| Abies balsamea | Balsam Fir |
| Abies cephalonica | Greek Fir |
| Abies concolor | White Fir |
| Abies grandis | Lowland Fir |
| Abies nordmanniana | Nordmann Fir |
| Abies numidica | Algerian Fir |
| Abies pectinata | Silver Fir |
| Abies pinsapo | Spanish Fir |
| *Araucaria bidwilli | Bunya-Bunya |
| *Araucaria excelsa | Norfolk Island-Pine |
| Cedrus atlantica | Atlas Cedar |
| Cedrus deodara | Deodar |
| Cedrus libani | Cedar of Lebanon |
| Chamaecyparis lawsoniana | Lawson-Cypress |
| Cupressus guadalupensis | Guadalupe Cypress |
| Larix europaea | European Larch |
| Sciadopitys verticillata | Umbrella-Pine |
| Sequoia gigantea | Giant Sequoia, Big Tree |
| Sequoia sempervirens | Redwood |
| Taxodium distichum | Bald-Cypress |
| Taxodium mucronatum | Montezuma Bald-Cypress |
| Tsuga heterophylla (O.W.BC.) | Western Hemlock |

### EVERGREEN

| | |
|---|---|
| *Callistemon speciosus | Showy Bottlebrush |
| *Ficus macrophylla | Moreton Bay Fig |

EVERGREEN

| *Botanical Name* | *Common Name* |
|---|---|
| Ilex aquifolium | English Holly |
| *Jacaranda ovalifolia | Jacaranda, Green-Ebony |
| *Leucadendron argenteum | Silver Tree |
| Magnolia grandiflora | Southern Magnolia |

DECIDUOUS

| | |
|---|---|
| Aesculus carnea | Red Horsechestnut |
| Aesculus hippocastanum | Horsechestnut |
| Betula alba | European White Birch |
| Catalpa speciosa | Western Catalpa |
| Cornus florida | Flowering Dogwood |
| Cornus nuttalli | Pacific Dogwood |
| *Erythrina cristagalli | Coral Tree |
| Hicoria ovata | Shagbark Hickory |
| Laburnum vulgare | Goldenchain |
| Liquidambar styraciflua | Sweet Gum |
| Magnolia conspicua | Yulan |
| Magnolia soulangeana | Saucer Magnolia |
| Magnolia stellata | Star Magnolia |
| Paulownia tomentosa | Royal Paulownia |
| Sorbus aucuparia | European Mountain-Ash |
| Sterculia platanifolia | Chinese Parasol Tree |

## 15. PEST-RESISTANT TREES

CONIFER

| | |
|---|---|
| Cedrus, species | Cedar |
| Ginkgo biloba | Maidenhair Tree |
| Libocedrus decurrens | California Incense-Cedar |
| Pseudotsuga douglasi | Douglas-Fir |
| Taxodium distichum | Bald-Cypress |
| Tsuga heterophylla (O.W.BC.) | Western Hemlock |

DECIDUOUS

| | |
|---|---|
| Aesculus carnea | Red Horsechestnut |
| Aesculus hippocastanum | Horsechestnut |
| Ailanthus glandulosa | Tree of Heaven |
| Albizzia julibrissin | Silk Tree |
| Celtis australis | European Hackberry |
| Diospyros kaki | Kaki Persimmon |
| Diospyros virginiana | Common Persimmon |
| Gleditsia triacanthos | Honey Locust |

DECIDUOUS

| *Botanical Name* | *Common Name* |
|---|---|
| Koelreuteria paniculata | Goldenrain Tree |
| Liquidambar styraciflua | Sweet Gum |
| Liriodendron tulipifera | Tulip Tree |
| Maclura pomifera | Osage-Orange |
| Magnolia soulangeana | Saucer Magnolia |
| Magnolia stellata | Star Magnolia |
| Magnolia tripetala | Umbrella Magnolia |
| Sassafras variifolium | Sassafras |
| *Tamarix, species | Tamarix |

EVERGREEN

| | |
|---|---|
| *Acacia, species | Acacia |
| *Arbutus unedo | Strawberry Tree |
| *Callistemon, species | Callistemon |
| *Cinnamomum camphora | Camphòr Tree |
| *Eucalyptus, species | Eucalyptus |
| *Leptospermum laevigatum | Australian Tea Tree |
| *Melaleuca, species | Melaleuca |

## 16. PICTURESQUE TREES

(NOTE: Picturesqueness may be defined as a quality produced by the action of the more violent forces of nature. It usually characterizes plant-forms that are chiefly to be accounted for by their obvious resistance to unfavorable environmental or climatic conditions. Age of growth also may cause this quality.)

CONIFER

| | |
|---|---|
| Cedrus atlantica | Atlas Cedar |
| Cedrus libani | Cedar of Lebanon |
| Cupressus macrocarpa | Monterey Cypress |
| Cupressus torulosa | Bhutan Cypress |
| Juniperus californica | California Juniper |
| Pinus contorta | Shore Pine |
| Pinus edulis | Nut Pine |
| Pinus muricata | Bishop Pine |
| Pinus pinaster | Cluster Pine |
| Pinus radiata | Monterey Pine |
| Pinus torreyana | Torrey Pine |

DECIDUOUS

| | |
|---|---|
| Acer compestre | Hedge Maple |
| Aesculus californica | California Buckeye |

## DECIDUOUS

| Botanical Name | Common Name |
|---|---|
| Ailanthus glandulosa | Tree of Heaven |
| *Aralia spinosa | Devil's Walking Stick |
| Diospyros kaki | Kaki Persimmon |
| *Ficus carica | Common Fig |
| Gleditsia triacanthos | Honey-Locust |
| Platanus racemosa | California Plane Tree |
| Quercus garryana | Oregon Oak |
| Quercus lobata | Valley Oak |
| Robinia pseudoacacia | Common Locust |
| Sophora japonica | Chinese Scholar Tree |
| *Zizyphus jujuba | Common Jujube |

## EVERGREEN

| | |
|---|---|
| *Casuarina equisetifolia | Horsetail Tree |
| *Leptospermum laevigatum | Australian Tea Tree |
| *Melaleuca ericifolia | Heath Melaleuca |
| *Olea europaea | Olive |
| *Parkinsonia aculeata | Parkinsonia |
| Quercus agrifolia | California Live Oak |
| Quercus wislizeni | Interior Live Oak |

## PALM AND PALM-LIKE

| | |
|---|---|
| *Cordyline australis | Green Dracena |
| *Dracaena draco | Dragon Tree |
| *Phoenix reclinata | Senegal Date Palm |
| *Washington robusta | Mexican Washington Palm |
| *Yucca brevifolia | Joshua Tree |
| *Yucca mohavensis | Spanish Dagger |

## 17. RAPID-GROWING DECIDUOUS TREES

| | |
|---|---|
| Acer dasycarpum | Silver Maple |
| Acer negundo | Box-Elder |
| Acer platanoides | Norway Maple |
| Acer rubrum | Red Maple |
| Ailanthus glandulosa | Tree of Heaven |
| Catalpa speciosa | Western Catalpa |
| Ginkgo biloba | Maidenhair Tree |
| Gleditsia triacanthos | Honey-Locust |
| Magnolia acuminata | Cucumber Tree |
| Magnolia tripetala | Umbrella Magnolia |
| Morus alba | White Mulberry |

## 17. RAPID-GROWING DECIDUOUS TREES—*Continued*

| *Botanical Name* | *Common Name* |
|---|---|
| Paulownia tomentosa | Royal Paulownia |
| Platanus acerifolia | London Plane Tree |
| Platanus orientalis | European Plane Tree |
| Populus alba | White Poplar |
| Populus nigra var. italica | Lombardy Poplar |
| Quercus alba | White Oak |
| †Quercus velutina | Black Oak |
| †Salix alba | White Willow |
| Salix babylonica | Weeping Willow |
| Sorbus americana | American Mountain-Ash |
| Sorbus aucuparia | European Mountain-Ash |
| Tilia americana | American Linden |
| Ulmus americana | American Elm |

## 18. ROUND-HEADED TREES FORMING GOOD GROUPS

### CONIFER

| | |
|---|---|
| Pinus montana var. mughus | Mugho Pine |
| Pinus muricata | Bishop Pine |
| Pinus pinea | Italian Stone Pine |

### EVERGREEN

| | |
|---|---|
| *Acacia decurrens var. dealbata | Silver Wattle |
| *Acacia elata | Cedar Wattle |
| *Acacia longifolia | Sydney Wattle |
| *Acacia melanoxylon | Blackwood Acacia |
| *Albizzia lophantha | Plume Albizzia |
| Arbutus menziesi | Madroño |
| *Arbutus unedo | Strawberry Tree |
| *Ceratonia siliqua | Carob |
| *Cinnamomum camphora | Camphor Tree |
| *Citrus, species | Orange, Lemon, Grapefruit |
| *Cornus capitata | Evergreen Dogwood |
| *Eucalyptus ficifolia | Scarlet Gum |
| *Eucalyptus polyanthemos | Redbox |
| *Jacaranda ovalifolia | Green-Ebony |
| Lithocarpus densiflora | Tan-Oak |
| *Olea europaea | Olive |
| Quercus agrifolia | California Live Oak |
| Quercus chrysolepis | Maul Oak |
| Quercus ilex | Holly Oak |

## EVERGREEN

| *Botanical Name* | *Common Name* |
|---|---|
| *Quercus suber | Cork Oak |
| Quercus wislizeni | Interior Live Oak |

## DECIDUOUS

| | |
|---|---|
| Acer buergerianum | Trident Maple |
| Acer dasycarpum | Silver Maple |
| Acer macrophyllum | Oregon Maple |
| †Acer oblongum | Himalayan Maple |
| Acer palmatum | Japanese Maple |
| Acer platanoides | Norway Maple |
| Acer rubrum | Red Maple |
| Acer saccharum | Sugar Maple |
| Carpinus betulus | European Hornbeam |
| Castanea sativa | Spanish Chestnut |
| Catalpa speciosa | Western Catalpa |
| Celtis australis | European Hackberry |
| Cornus florida | Flowering Dogwood |
| Cornus nuttalli | Pacific Dogwood |
| Diospyros kaki | Kaki Persimmon |
| Diospyros virginiana | Common Persimmon |
| Fraxinus oregona | Oregon Ash |
| Fraxinus ornus | Flowering Ash |
| Koelreuteria paniculata | Goldenrain Tree |
| Morus nigra | Black Mulberry |
| Quercus coccinea | Scarlet Oak |
| Quercus lobata | Valley Oak |
| Quercus macrocarpa | Mossycup Oak |
| Quercus robur | English Oak |
| Sassafras variifolium | Sassafras |
| Sophora japonica | Chinese Scholar Tree |
| Sterculia platanifolia | Chinese Parasol Tree |
| Tilia cordata | Littleleaf European Linden |
| Ulmus campestris | English Elm |
| Zelkova serrata | Sawleaf Zelkova |

## 19. SAND-HOLDING TREES—FOR SAND DUNES OR SANDY SOIL

| | |
|---|---|
| *Acacia longifolia | Sydney Wattle |
| *Leptospermum laevigatum | Australian Tea Tree |
| Salix lasiolepis | Arroyo Willow |

### 19. SAND-HOLDING TREES—FOR SAND DUNES OR SANDY SOIL—*Continued*

| *Botanical Name* | *Common Name* |
|---|---|
| Salix laevigata | Red Willow |
| Salix piperi | Dune Willow |
| Salix sessilifolia | Sandbar Willow, Softleaf Willow |
| *Tamarix articulata | Athel |
| Tamarix juniperina | Juniper Tamarix |
| Tamarix parviflora | Tamarix |

### 20. SEACOAST-TOLERANT TREES

#### DECIDUOUS

| | |
|---|---|
| Acer platanoides | Norway Maple |
| Acer pseudoplatanus | Sycamore Maple |
| Acer rubrum | Red Maple |
| Carpinus betulus | European Hornbeam |
| Fagus sylvatica | European Beech |
| †Fraxinus excelsior | European Ash |
| Fraxinus velutina | Arizona Ash |
| Laburnum vulgare | Goldenchain Tree |
| Populus alba | White Poplar |
| Populus balsamifera | Balsam Poplar |
| Populus nigra var. italica | Lombardy Poplar |
| †Quercus phellos | Willow Oak |
| Quercus rubra | Red Oak |
| †Salix alba | White Willow |
| Ulmus glabra | Scotch Elm |

#### EVERGREEN

| | |
|---|---|
| *Acacia longifolia | Sydney Wattle |
| *Acacia retinodes | Water Wattle |
| *Cassia tomentosa | Woolly Senna |
| *Casuarina stricta | Beefwood |
| *Cupressus macrocarpa | Monterey Cypress |
| *Eucalyptus cornuta | Yate Tree |
| *Eucalyptus ficifolia | Scarlet Gum |
| *Lagunaria patersoni | Lagunaria |
| *Leptospermum laevigatum | Australian Tea Tree |
| *Melaleuca armillaris | Drooping Melaleuca |
| *Melaleuca ericifolia | Heath Melaleuca |
| *Metrosideros robusta | Rata |
| *†Myoporum insulare | Myoporum |
| *Myoporum laetum | Myoporum |

### EVERGREEN

| *Botanical Name* | *Common Name* |
| --- | --- |
| Myrica californica | California Wax-Myrtle |
| Pinus contorta | Shore Pine |
| Pinus muricata | Bishop Pine |
| Pinus pinaster | Cluster Pine |
| Pinus radiata | Monterey Pine |
| Pinus torreyana | Torrey Pine |
| *Pittosporum crassifolium | Karo |
| *†Quillaja saponaria | Soapbark Tree |
| *Tamarix articulata | Athel |

## 21. SILHOUETTE TREES

(NOTE: Silhouette trees are trees producing thin foliage through which fine structural composition is registered against the horizon.)

### CONIFER

| Cedrus libani | Cedar of Lebanon |
| --- | --- |
| Cupressus funebris | Mourning Cypress |
| Pinus torreyana | Torrey Pine |

### DECIDUOUS

| Betula alba | European White Birch |
| --- | --- |
| Betula papyrifera | Canoe Birch |
| Robinia pseudoacacia | Common Locust |
| Robinia pseudoacacia var. decaisneana | Decaisne Locust |

### EVERGREEN

| *Casuarina cunninghamiana | Cunningham Beefwood |
| --- | --- |
| *Eucalyptus amygdalina var. angustifolia | Narrowleaf Almond Eucalyptus |
| *Eucalyptus maculata var. citriodora | Lemon Gum |
| *Eucalyptus viminalis | Manna Gum |
| *Melaleuca ericifolia | Heath Melaleuca |
| *Melaleuca styphelioides | Rigidleaf Melaleuca |

## 22. STREET TREES

### DECIDUOUS

| Acer buergerianum | Trident Maple |
| --- | --- |
| †Acer oblongum | Himalayan Maple |
| Acer platanoides | Norway Maple |
| Aesculus carnea | Red Horsechestnut |
| Celtis australis | European Hackberry |

### DECIDUOUS

| *Botanical Name* | *Common Name* |
|---|---|
| Crataegus cordata | Washington Hawthorn |
| Fraxinus ornus | Flowering Ash |
| Fraxinus velutina | Arizona Ash |
| Koelreuteria paniculata | Goldenrain Tree |
| Platanus acerifolia | London Plane Tree |
| Prunus cerasifera var. pissardi | Purpleleaf Plum |

### EVERGREEN

| | |
|---|---|
| *Causuarina cunninghamiana | Cunningham Beefwood |
| *Ceratonia siliqua | Carob |
| *Cinnamomum camphora | Camphor Tree |
| *Eucalyptus ficifolia | Scarlet Gum |
| *Eucalyptus polyanthemos | Redbox |
| *Metrosideros robusta | Rata |
| *Olea europaea | Olive |
| *Parkinsonia aculeata | Parkinsonia |
| *Photinia serrulata | Low Photinia |
| *Pittosporum crassifolium | Karo |
| Ulmus parvifolia | Chinese Elm |

## 23. TROPICAL TREES FOR MILD REGIONS

### EVERGREEN AND DECIDUOUS

| | |
|---|---|
| *Araucaria bidwilli | Bunya-Bunya |
| *Bauhinia purpurea | Purple Bauhinia |
| *Citrus aurantium | Seville Orange |
| *Citrus limonia | Lemon |
| *Citrus sinensis | Orange |
| *†Daphniphyllum macropodum | |
| *Eucalyptus ficifolia | Scarlet Gum |
| *Ficus elastica | India Rubber Tree |
| *Ficus macrophylla | Moreton Bay Fig |
| *†Ficus pandurata | Fiddleleaf Fig |
| *Grevillea robusta | Silk-Oak |
| *Jacaranda ovalifolia | Green-Ebony |
| *Macadamia ternifolia | Queensland Nut |
| Magnolia grandiflora | Southern Magnolia |
| Magnolia soulangeana | Saucer Magnolia |
| *†Malvaviscus arboreus | Waxmallow |
| *Persea americana | Avocado |
| *Persea indica | Madeira-Bay |

## EVERGREEN AND DECIDUOUS

| *Botanical Name* | *Common Name* |
|---|---|
| *Pittosporum undulatum | Orange Pittosporum |
| *Podocarpus elongata | Small-leaf Podocarpus |
| *Podocarpus macrophylla | Yew Podocarpus |
| Prunus laurocerasus | English Cherry-Laurel |
| *Schinus molle | Pepper Tree |
| *Sterculia acerifolia | Victorian Bottle Tree |
| Sterculia platanifolia | Chinese Parasol Tree |
| *Tetrapanax papyriferum | Ricepaper Plant |
| Tristania conferta | Brisbane Box |

## PALM AND PALM-LIKE PLANTS

| | |
|---|---|
| *†Alsophilia australis | Australian Treefern |
| *Cocos australis | Pindo Palm |
| *Cocos plumosa | Plume Palm |
| *Cordyline australis | Green Dracena |
| *†Dicksonia antarctica | Treefern |
| *Dracaena draco | Dragon Tree |
| *Erythea edulis | Guadalupe Palm |
| *Glaucothea armata | Blue Palm |
| †*Howea belmoreana | Belmore Palm |
| *Jubaea spectabilis | Syrup Palm |
| *Livistona chinensis | Chinese Fan Palm |
| *Musa ensete | Abyssinian Banana |
| *Phoenix canariensis | Canary Date Palm |
| *Phoenix dactylifera | Date Palm |
| *Phoenix reclinata | Senegal Date Palm |
| *†Phyllostachys bambusoides | Japanese Timber Bamboo |
| *Trachycarpus excelsa | Windmill Palm |
| *Washingtonia filifera | California Washington Palm |
| *Washingtonia gracilis | Mexican Washington Palm |
| *Yucca aloifolia | Spanish Dagger |
| *Yucca mohavensis | California Spanish Dagger |

## 24. TUB OR BOX TREES

### CONIFER

| | |
|---|---|
| Chamaecyparis obtusa | Hinoki-Cypress |
| Chamaecyparis pisifera | Sawara Retinospora |
| Cryptomeria japonica var. elegans | Plume Cryptomeria |
| Pinus montana var. mughus | Mugho Pine |

## CONIFER

| Botanical Name | Common Name |
|---|---|
| *Sciadopitys verticillata | Umbrella-Pine |
| Taxus baccata | English Yew |
| Taxus cuspidata | Japanese Yew |
| Thuja occidentalis | Western Arborvitae |
| Thuja plicata | Giant Arborvitae |

## BROAD-LEAF EVERGREEN

| | |
|---|---|
| *Acacia longifolia | Sydney Golden Wattle |
| *Acacia pycnantha | Golden Wattle |
| *Albizzia lophantha | Plume Albizzia |
| *Camellia japonica | Camellia |
| *Citrus aurantium | Seville Orange |
| *Citrus limonia | Lemon |
| *Citrus sinensis | Orange |
| *Eugenia myrtifolia | Australian Brush-Cherry |
| *Eugenia paniculata | |
| *Ficus elastica | India Rubber Tree |
| *Ficus macrophylla | Moreton Bay Fig |
| *Grevillea robusta | Silk-Oak |
| Ilex aquifolium | English Holly |
| Lagerstroemia indica | Crape-Myrtle |
| Laurus nobilis | Grecian Laurel |
| Ligustrum lucidum | Glossy Privet |
| *Metrosideros robusta | Rata |
| Nerium oleander | Oleander |
| *Olea europaea | Olive |
| *Pittosporum undulatum | Orange Pittosporum |
| *Pittosporum viridiflorum | Cape Pittosporum |
| Prunus laurocerasus | English-Laurel |
| Punica granatum | Pomegranate |
| *Tetrapanax papyriferum | Ricepaper Plant |
| Umbellularia californica | California-Laurel |

## PALM AND PALM-LIKE

| | |
|---|---|
| *Cocos australis | Pindo Palm |
| *Cordyline australis | Green Dracena |
| *Dracaena draco | Dragon Tree |
| *Erythea edulis | Guadalupe Palm |
| *Phoenix reclinata | Senegal Date Palm |
| *Trachycarpus excelsa | Windmill Palm |

## 25. WATER-EDGE, STILL-RIVER, OR LAKE-EDGE TREES

### CONIFER

| *Botanical Name* | *Common Name* |
|---|---|
| Larix europaea | European Larch |
| Taxodium distichum | Bald-Cypress |
| Taxodium mucronatum | Montezuma Bald-Cypress |
| Torreya californica | California-Nutmeg |
| Tsuga heterophylla (O.W.BC.) | Western Hemlock |
| Tsuga mertensiana (O.W.BC.) | Mountain Hemlock |

### DECIDUOUS

| | |
|---|---|
| Alnus, species | Alder |
| Betula alba | European White Birch |
| Cornus florida | Flowering Dogwood |
| Cornus nuttalli | Pacific Dogwood |
| Liquidambar styraciflua | Sweet Gum |
| Populus, species | Poplar and Cottonwood |
| Salix, species | Willow |

### BROAD-LEAF EVERGREEN

| | |
|---|---|
| *Maytenus boaria | Mayten |
| Myrica californica | California Wax-Myrtle |
| Ulmus parvifolia (tardily deciduous) | Chinese Elm |
| Umbellularia californica | California-Laurel |

## 26. WEEPING TREES

### CONIFER

| | |
|---|---|
| Chamaecyparis lawsoniana var. pendula | Weeping Lawson-Cypress |
| Cupressus funebris | Mourning Cypress |
| Picea smithiana | Himalayan Spruce |
| Pinus excelsa var. pendula | Weeping Himalayan Pine |
| Pinus longifolia | Longleaf Pine |
| Sequoia gigantea var. pendula | Weeping Big Tree |

### EVERGREEN

| | |
|---|---|
| *Acacia cultriformis | Knife Acacia |
| *Eucalyptus amygdalina var. angustifolia | Narrow-leaf Almond Eucalyptus |
| *Leptospermum laevigatum | Australian Tea Tree |
| *Maytenus boaria | Mayten |

| *Botanical Name* | *Common Name* |
|---|---|
| *Pittosporum phillyraeoides | Willow Pittosporum |
| *Schinus molle | Pepper Tree |

DECIDUOUS

| | |
|---|---|
| Betula alba var. laciniata | Cutleaf Weeping Birch |
| Cercocarpus betuloides | Hard Tack |
| Gleditsia triacanthos var. bujote | Weeping Honey-Locust |
| Juglans regia var. pendula | Weeping English Walnut |
| Laburnum vulgare var. pendula | Weeping Goldenchain |
| Morus alba var. tartarica pendula | Weeping Russian Mulberry |
| Populus grandidentata var. pendula | Weeping Largetooth Aspen |
| †Prunus subhirtella var. pendula | Weeping Higan Cherry |
| Quercus robur var. pendula | Weeping English Oak |
| Salix babylonica | Weeping Willow |
| Salix babylonica var. annularis | Ringleaf Weeping Willow |
| Sophora japonica var. pendula | Weeping Chinese Scholar Tree |
| Sorbus aucuparia var. pendula | Weeping European Mountain-Ash |
| Tilia petiolaris | Weeping Linden |

## 27. WET-SOIL DECIDUOUS TREES

| | |
|---|---|
| Acer dasycarpum | Silver Maple |
| Acer rubrum | Red Maple |
| Alnus, species | Alder |
| †Betula nigra | River Birch |
| Betula populifolia | Gray Birch |
| †Carpinus caroliniana | American Hornbeam |
| Fraxinus nigra | Black Ash |
| †Hicoria laciniosa | Shellbark Hickory |
| †Larix laricina | American Larch |
| Liquidambar styraciflua | Sweet Gum |
| Platanus occidentalis | American Plane Tree |
| Populus, species | Poplar and Cottonwood |
| †Quercus bicolor | Swamp White Oak |
| Quercus palustris | Pin Oak |
| †Quercus phellos | Willow Oak |
| Salix, species | Willow |
| Taxodium distichum | Bald-Cypress |

## 28. WINDBREAK TREES

### DECIDUOUS

| *Botanical Name* | *Common Name* |
|---|---|
| Acer dasycarpum | Silver Maple |
| Acer negundo | Box Elder |
| †Carpinus caroliniana | American Hornbeam |
| Fagus americana | American Beech |
| Fagus sylvatica | European Beech |
| Larix europaea | European Larch |
| †Larix leptolepis | Japanese Larch |
| Maclura pomifera | Osage-Orange |
| Populus, species | Poplar, Cottonwood |
| Salix, species | Willow |

### CONIFER AND BROAD-LEAF EVERGREEN

| | |
|---|---|
| *Acacia longifolia | Golden Wattle |
| *Acacia retinodes | Water Wattle |
| *Casuarina cunninghamiana | Cunningham Beefwood |
| *Casuarina equisetifolia | Horsetail Tree |
| *Cupressus macrocarpa | Monterey Cypress |
| *†Eucalyptus globulus | Blue Gum |
| *Eucalyptus robusta | Brown Gum |
| *Eucalyptus rostrata | Creek Gum |
| *Eucalyptus rudis | Desert Gum |
| *Eucalyptus viminalis | Manna Gum |
| *Leptospermum laevigatum | Australian Tea Tree |
| *Metrosideros robusta | Rata |
| *†Myoporum insulare | Myoporum |
| *Myoporum laetum | Myoporum |
| Pinus muricata | Bishop Pine |
| Pinus pinaster | Cluster Pine |
| Pinus radiata | Monterey Pine |
| *Pittosporum crassifolium | Karo |
| *Tamarix articulata | Athel |

# INDEX

(The *italicized* words are synonyms.)

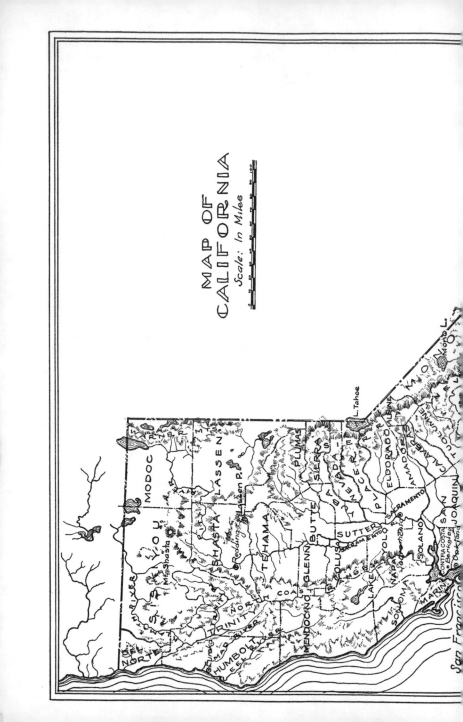

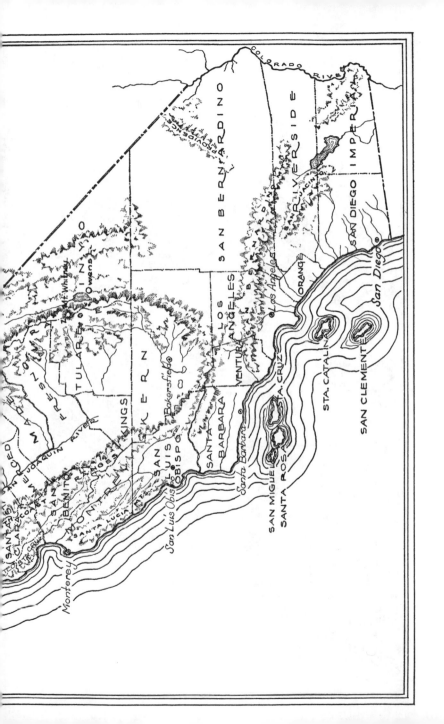

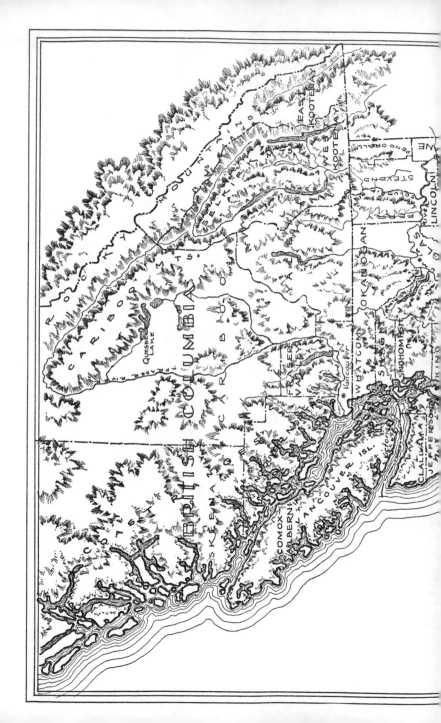

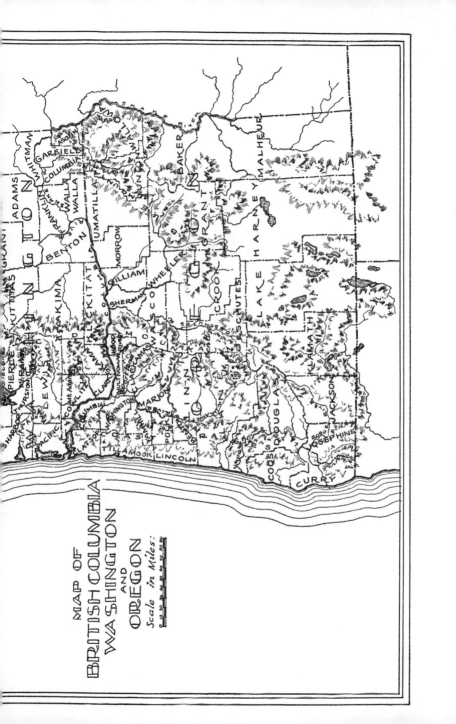

MAP OF
BRITISH COLUMBIA
WASHINGTON
AND
OREGON
Scale in Miles: